Lecture Notes in Physics

Volume 982

The Lecture Notes in Physics

The series Lecture Notes in Physics (LNP), founded in 1969, reports new developments in physics research and teaching-quickly and informally, but with a high quality and the explicit aim to summarize and communicate current knowledge in an accessible way. Books published in this series are conceived as bridging material between advanced graduate textbooks and the forefront of research and to serve three purposes:

- to be a compact and modern up-to-date source of reference on a well-defined topic;
- to serve as an accessible introduction to the field to postgraduate students and nonspecialist researchers from related areas;
- to be a source of advanced teaching material for specialized seminars, courses and schools.

Both monographs and multi-author volumes will be considered for publication. Edited volumes should however consist of a very limited number of contributions only. Proceedings will not be considered for LNP.

Volumes published in LNP are disseminated both in print and in electronic formats, the electronic archive being available at springerlink.com. The series content is indexed, abstracted and referenced by many abstracting and information services, bibliographic networks, subscription agencies, library networks, and consortia.

Proposals should be sent to a member of the Editorial Board, or directly to the responsible editor at Springer:

Dr Lisa Scalone
Springer Nature
Physics
Tiergartenstrasse 17
69121 Heidelberg, Germany
lisa.scalone@springernature.com

More information about this series at http://www.springer.com/series/5304

V. M. (Nitant) Kenkre

Memory Functions, Projection Operators, and the Defect Technique

Some Tools of the Trade for the Condensed Matter Physicist

 Springer

V. M. (Nitant) Kenkre
University of New Mexico
Albuquerque, NM, USA

ISSN 0075-8450   ISSN 1616-6361   (electronic)
Lecture Notes in Physics
ISBN 978-3-030-68666-6   ISBN 978-3-030-68667-3   (eBook)
https://doi.org/10.1007/978-3-030-68667-3

This Springer imprint is published by the registered company Springer Nature Switzerland AG
The registered company address is: Gewerbestrasse 11, 6330 Cham, Switzerland

My learning process, I have noticed, is now stunted without a doubt since my retirement, in the absence of the help of my graduate students who, over a period of almost half a century, allowed me to pretend to teach them, encouraging me by permitting me to guide their dissertations. My sincere thanks to them for educating me through their penetrating questions. I dedicate this book, with my affectionate regards and best wishes, to all of these creative scholars who had the tenacity to complete their dissertations while working with me: V. Seshadri, Yiu-Man Wong, Paul Parris, David Brown, John Andersen, George Tsironis, David Dunlap, Honglu Wu, Ximing Fan, Vassilios Kovanis, Fabio Biscarini*, Srikanth Raghavan, Mark Endicott, Daniel Sheltraw, Joseph Scott, Luca Giuggioli, David Macinnis, Mukesh Tiwari, Ziya Kalay, Luis Felipe Gonzalez, Kathrin Spendier*, Alden Astwood, Satomi Sugaya, Matthew Chase, and Anastasia Ierides. (* marks joint thesis supervision.)

Preface

ग्रंथमभ्यस्य मेधावी ज्ञानविज्ञानतत्परः
पलालमिव धान्यार्थी त्यजेत् ग्रंथमशेषतः

... from the Amritabindu Upanishad, 18th verse...
(One desirous of gaining knowledge should, after studying the book, discard
it just the way one who wishes for the grain discards the husk.)

It is the author's sincere wish that the reader of this book will apply the same critical treatment that the sage who wrote the words quoted above from the *Amritabindu Upanishad* recommended to the spiritual seeker: Use the book to learn but discard it as soon as you have mastered what it teaches you (and proceed then under your own steam).

This book addresses three technical tools that the theoretical condensed matter physicist finds useful in the analysis of various phenomena. The first deals with *Memory Functions*. The tool appears in the formalism of generalized master equations that express the time evolution of probabilities via equations non-local in time. The second consists of *Projection Operators* that allow the extraction of parts of quantities such as the diagonal part of density matrices in statistical mechanics. The third is a specific method, the *Defect Technique*, constructed to solve transport equations in which translational invariance is broken into small regions such as when crystals are doped with impurities. All three form powerful weapons of analysis, immensely useful for the theoretical investigator in condensed matter physics.

The last of the three techniques was originated by the late Elliott Montroll and his co-workers, particularly Renfrey Potts who was one of his early collaborators in this area, and Bruce West who coauthored many noteworthy articles with him. I myself first met the technique in the work of Katja Lindenberg on Frenkel excitons. Projection techniques are typically associated with the name of one of its inventors, the late Robert Zwanzig who, working independently of another inventor of the technique, Nakajima, applied it to the study of a central problem of quantum statistical mechanics: the derivation of a Master equation from the von Neumann evolution of the density matrix. It is difficult to associate one name with the formalism of Memory Functions as several have worked them into their research repertoire, including the present author, who wrote a Springer monograph four decades earlier, on a particular application, to exciton dynamics in molecular crystals, in *Kenkre and Reineker, Springer Tracts in Modern Physics, Vol. 94 (Springer, Berlin, 1982)*.

With his students and colleagues, the author has spent several decades in doing research developing and using these three tools of the trade, separately and in combination. The outcome has included more than 70 research publications, and numerous advances in the theory itself and the theory as applied to experiments in several disparate fields of physics. Without doubt, the tools form a mighty investigative weapon in various areas of the theory of condensed matter as in the physics of excitations in molecular crystals, sensitized luminescence and photosynthesis, charge transport, general aspects of non-equilibrium statistical physics, vibrational relaxation, compaction of granular materials, nuclear magnetic resonance used as a microscopy, and even in theoretical ecology. Arming oneself with this weapon is certain to be profitable for the theoretical investigator.

This present book was written with two purposes in mind. One is to collect and connect the three tools as they turn out to be closely related among themselves. Collecting them into linked structures helps one understand them, and connecting them aids in internalizing the spirit of their application.

The second purpose is to teach them specifically to the budding theoretical physicist in these areas, typified by an advanced graduate student. I have been fortunate to be of some help, in the garb of a research advisor, to several outstanding young physicists. Many have become well-known professors in universities and institutes, one has chaired a department, another has served as a university provost, and all of them have shown creativity and tangible success in their research. Surely, the lion's share of the credit is all their own. It is possible, however, that the operational insights they gained in their research environment have made a small contribution also. What the author has done, conscientiously and intensely over half a century for the training of physics graduate students, since he graduated with his own Ph.D. at the SUNY Stony Brook in 1971, is being attempted in this book again. This time it is hoped that it will benefit students he will meet through the pages of the book rather than in person. The book should thus, in addition to fulfilling the job of acquainting the reader with interesting theoretical research in several fascinating areas of physics, also perform the task of training the future budding investigator. With this intention firmly in mind, calculational exercises have been sprinkled throughout the book at carefully selected places along with information where solutions may be found in the published literature in most of the cases.[1] It

---

[1]There are 27 exercises. They may be found on pages 4, 10, 13, 36, 45, 56, 72, 84 (which contains two), 92, 95, 101, 140, 148, 157, 169, 177, 196, 207, 219, 225, 248, 275, 279, 300, 303, 323 with unobtrusive alerts ⊛ placed in the margins of the book.

should be understood that this is not a textbook that a student might use to learn the introductory elements of condensed matter physics. I think it will help her already at the very beginning of her research career, perhaps as, after passing her preliminary/comprehensive examination, she selects an area and start research on her dissertation.

Outline of the Book

Three tools will be taught in the book. Two of them are so closely allied to each other that presenting them separately would work against the reader. Therefore, the book material is presented in *two* parts, the first of which will be about *Memory Functions* as well as about *Projection Operators* and the ten chapters in this part will deal with now one and now the other of these two tools. The next five chapters will be devoted to the second part, expounding in it the third tool, the *Defect Technique*. Finally, the last, that is, the sixteenth, chapter, will consist of an epilogue. It will contain miscellaneous comments not belonging to the flow of the material in the rest of the book as also a brief restatement of the book's contents. Every chapter will have at its end a brief summary of what has been explained in the chapter. Colleagues who read the manuscript of this specific book, and made helpful suggestions, are acknowledged at the end of that final chapter.

A bibliography of references is provided at the end.

Albuquerque, NM, USA

V. M. (Nitant) Kenkre

Acknowledgments

My understanding of the subjects analyzed in this book comes not only from the inputs of more than two dozen graduate students to whom I have dedicated this work but also from many other individuals. Prominent among them, Max Dresden whose lucid lectures when I was his graduate student were nothing short of thrilling; Nandor Balazs who insisted that one did not understand a subject until one could explain it to someone else in at least three different ways (it is rumored he said six, but I decided long ago to settle for three because of my own limitations); Elliott Montroll who inspired by his casual approach to science and his analytical brilliance matched only by his unique sense of humor; David Dexter, who started bitingly by introducing me to his colleagues as the "young man who makes simple things difficult" but became a dear friend who shared with me his admiration for P. G. Wodehouse and Rex Stout, the "two giants of English literature of the century" as he called them; and Robert Knox, who introduced me to most of what I know about the field of excitons, for his integrity in the practice of science and his unwavering attachment to the importance of experiment in the activities of the theoretical physicist. Only the last of these teachers of mine, Bob Knox, is still living and will remain so, I hope, for many more years.

Contents

About the Author

V. M. (Nitant) Kenkre is Distinguished Professor (Emeritus) of Physics at the University of New Mexico (UNM), USA, retired since 2016. His undergraduate studies were at the IIT, Bombay (India), and his graduate work took place at the SUNY Stony Brook (USA). He was elected Fellow of the American Physical Society in 1998 and Fellow of the American Association for Advancement of Science in 2005 and has won an award from his university for his international work. He was the Director of two centers at UNM: the Center for Advanced Studies for 4 years and then the Founding Director of the Consortium of the Americas for Interdisciplinary Science for 16 years. He was given the highest faculty research award of his university in 2004 and supervised the Ph.D. research of 25 doctoral scientists and numerous postdoctoral researchers.

Through 270 published papers, his research achievements include formalistic contributions to non-equilibrium statistical mechanics, particularly quantum transport theory, observations in sensitized luminescence and exciton/electron dynamics in molecular solids, and solutions to cross-disciplinary puzzles arising in spread of epidemics, energy transfer in photosynthetic systems, statistical mechanics of granular materials, and the theory of microwave sintering of ceramics.

He has interests in comparative religion, literature, and visual art and has often lectured on the first of these. His most recent coauthored book is *Theory of the Spread of Epidemics and Movement Ecology of Animals* (Cambridge University Press, 2020). He has also co-authored a book on exciton dynamics (Springer 1982), coedited another on modern challenges in statistical mechanics (AIP 2003), and published a book on his poetry entitled *Tinnitus*, and two on philosophy: *The Pragmatic Geeta* and *What Is Hinduism*.

The Memory Function Formalism: What and Why

1.1 Introduction to Memory Functions

Complexity in our perceptions of the world around us delights the senses. Here we see a pendulum or a mass attached to a spring oscillates, repeating its motion. There, when we look upwards, we sense a planet circling a star. If we turn inward into the microscopic world, we notice an ammonia molecule shuttling between two states, a Bose-Einstein condensate tunneling between two traps, or perhaps a Bloch electron in a pure crystal moving wave-like among its Wannier states. All these perform *oscillations*. Yet, all these systems, if or when affected by strong damping influences, exhibit decay in their evolution. The arrow of time, ever clear in macroscopic phenomena, forces decays. Reversible equations such as Newton's, Schrödinger's, or von Neumann's, accepted by most to underlie mechanical behavior, generally predict oscillations at the microscopic level. Irreversible equations such as the Boltzmann, Navier-Stokes or the Master equation, lead to approach to equilibrium, i.e., decays. One of the major problems of statistical mechanics, considered by some to be *the central problem* of the field, concerns the perhaps paradoxical coexistence of, and generally the relation between, microscopic short-time oscillations and macroscopic long-time decays. Furthermore, the study of that relation has practical relevance in diverse fields such as photosynthesis, electron conduction, and energy transfer. The memory formalism is a smooth device to view this relation and unify the oscillatory and decay extremes of motions we notice in Nature.

There is no simpler way to understand the power of the memory approach than to notice that the profoundly different time evolutions implied by

$$\frac{d^2 y}{dt^2} + \omega^2 y = 0 \tag{1.1}$$

© The Author(s), under exclusive license to Springer Nature Switzerland AG 2021
V. M. (Nitant) Kenkre, *Memory Functions, Projection Operators, and the Defect Technique*, Lecture Notes in Physics 982,
https://doi.org/10.1007/978-3-030-68667-3_1

on the one hand, with its reversible, oscillatory solutions involving trigonometric functions, and by

$$\frac{dy}{dt} + \Gamma y = 0 \tag{1.2}$$

on the other, which shows irreversible behavior with a clear approach to equilibrium, with an exponential as a solution, are unified in a single swoop by

$$\frac{dy(t)}{dt} + \Gamma \int_0^t ds\, \phi(t - s) y(s) = 0 \tag{1.3}$$

through the existence of a "memory function" $\phi(t)$.

How this can be done should be clear if we take the memory to take, in turn, the extreme limiting forms of a constant and a δ-function. The latter case is trivial since then the non-locality in time goes away and we get (1.2). The former is understood since a time differentiation can produce the oscillatory (1.1). In order to describe the entire intermediate range between the two extremes, one could consider, for instance, that the memory is an exponential in time, $\phi(t) = \alpha e^{-\alpha t}$, with α taking the value ω^2/Γ. The two limits are recovered respectively when $\alpha \to 0$ in one case, and $\alpha \to \infty, \omega \to \infty, \omega^2/\alpha \to \Gamma$ in the other. Also remarkable is the fact that, for fixed values of the parameters, we typically witness oscillations[1] with frequency ω at short times which turn into a decay with rate constant Γ at longer times, a measure of the transition time being $1/\alpha$, the reciprocal of the decay constant of the memory function. The astute reader has surely realized that we have here, in the particular realization of the exponential memory, the system of a damped harmonic oscillator with ω and α describing the frequency and the damping rate respectively, the full evolution equation (for the displacement of the harmonic oscillator) being the standard

$$\frac{d^2y}{dt^2} + \alpha \frac{dy}{dt} + \omega^2 y = 0, \tag{1.4}$$

but with the stipulation that initially dy/dt is vanishing. Whether or not the integro-differential equation (1.3) can be converted, thus, into a differential equation without integral terms (as in Eq. (1.4)), depends on the particular memory $\phi(t)$. Exotic memories such as of algebraic nature, often employed in disordered systems, lead to exotic evolution of $y(t)$ such as a power law form. In such cases it is often not possible to replace Eq. (1.3) by a differential equation including terms only local in time.

---

[1] Surely, for sufficiently large values of α, sustained oscillations are not observed, their signature being left only in the curvature of the time dependence. We refer here to the familiar phenomena of underdamping and overdamping separated by critical damping.

Important to notice is the fact that an exponential memory, with its simple *parameter* limits (infinite and vanishing α leading, respectively, to a δ-function and a constant) describes the unified behavior of oscillations on the one hand, and decay on the other. Also important is the independent observation that, for given fixed values of the parameters, the behavior tends from one kind to the other over the passage of time. The characteristic time of the memory, in this case $1/\alpha$, is the time over which the transition takes place. In other words, the memory formalism has the ability to bring about a unification of different kinds of behavior, both in the space of parameter values, and in time, for a given set of parameter values.

Let us now complicate our system from simple (damped) oscillations to movement on a line. The position of the moving entity is denoted by x and the probability density $P(x, t)$ of being at x at time t obeys the wave equation

$$\frac{\partial^2 P(x, t)}{\partial t^2} = c^2 \frac{\partial^2 P(x, t)}{\partial x^2}. \tag{1.5}$$

Here we encounter waves moving with velocity c. If the string, which may be the medium on the line we are considering, is plucked initially, two pulses may symmetrically travel along the line with speeds $\pm c$, never changing their shape.

Contrast with this the other extreme in which the governing equation for $P(x, t)$ is the diffusion equation

$$\frac{\partial P(x, t)}{\partial t} = D \frac{\partial^2 P(x, t)}{\partial x^2} \tag{1.6}$$

which, if $P(x, t)$ describes a probability density, also typically describes a random walk performed by a particle. Now, an initial plucking will result in a very different kind of motion: the pulse will broaden and become fatter and fatter as the entire shape feels the disturbance. These cases that we are displaying here on a line for simplicity, apply equally well in multi-dimensional space with the second spatial derivative replaced by the Laplacian.

How might we unify these two *quite* different extremes of motion represented by Eqs. (1.5) and (1.6)? Once again, naturally, with the help of memory functions. Consider, for instance, a memory function $\phi(t)$ and with its help a generalization of the diffusion equation (1.6) into

$$\frac{\partial P(x, t)}{\partial t} = D \int_0^t ds\, \phi(t - s) \frac{\partial^2 P(x, s)}{\partial x^2}. \tag{1.7}$$

If ϕ is a constant, we get the wave case, Eq. (1.5). If ϕ is a δ-function, we get the diffusive case, Eq. (1.6). What happens if the memory ϕ is an exponential, i.e., $\phi(t) = \alpha e^{-\alpha t}$, which certainly possesses intermediate properties? We now get the telegrapher's equation

$$\frac{\partial^2 P(x,t)}{\partial t^2} + \alpha \frac{\partial P(x,t)}{\partial t} = c^2 \frac{\partial^2 P(x,t)}{\partial x^2} \tag{1.8}$$

if we take $c^2/\alpha = D$.

An initial plucked string will evolve into one with pulses traveling symmetrically as in the wave case (case of Eq. (1.5)) but with their amplitudes damped as they travel outwards, and, additionally, a part slumping at the initial point of disturbance getting fatter and fatter as in the diffusive case (case of Eq. (1.5)). A straightforward analysis employing Fourier transforms shows that the solution for an initial condition $P(x,0) = \delta(x)$ for a memory equation (1.7) with exponential memory, in other words Eq. (1.8), will be[2]

$$P(x,t) = e^{-\alpha t/2} \left[\frac{\delta(x+ct) + \delta(x-ct)}{2} + T(x,t) \right]. \tag{1.9}$$

In Eq. (1.9), $T(x,t)$ vanishes identically for $ct \le |x|$ and equals, otherwise,

$$T(x,t) = \left(\frac{\alpha}{4c}\right) \left[I_0\left(\frac{\alpha}{2c}\sqrt{c^2t^2 - x^2}\right) + \frac{ct}{\sqrt{c^2t^2 - x^2}} I_1\left(\frac{\alpha}{2c}\sqrt{c^2t^2 - x^2}\right) \right]. \tag{1.10}$$

Here, I_0 and I_1 are modified Bessel functions of order 0 and 1, respectively.

We display this solution in Fig. 1.1, all units used being arbitrary. In both panels we can see the location of the initial δ-function disturbance. Coherent travel in the form of the so-called 'light-cones' which depict the localized disturbance moving in both directions at the medium speed c is clear in front as is the incoherent Gaussian spread in the back after sufficient time has elapsed. The two panels describe two different degrees of coherence decided by different ratios of values of α and c (see caption).

This unification of coherent (wave-like) and incoherent (diffusive) motion provided by memory functions can also be understood or described quite simply by examining the mean square displacement $\langle x^2 \rangle$ defined as $\int_{-\infty}^{+\infty} x^2 P(x,t)\,dx$. Multiplying Eq. (1.7) by x^2 and doing the integration, under the reasonable assumption that the probability density and its spatial derivative vanish as $x \to \pm\infty$, we get

$$\frac{d\langle x^2 \rangle}{dt} = 2D \int_0^t ds\,\phi(s). \tag{1.11}$$

It is certainly not significant, merely interesting, that on time differentiation, (1.11) has an aspect of Newton's equation of motion (Kenkre 1977a) as relating the second time derivative of the mean square displacement (like that of the coordinate of

---

[2] We recommend that the serious student undertake the derivation of Eq. (1.10) using Fourier and/or Laplace transforms as an exercise. It is straightforward but not entirely trivial to do so.

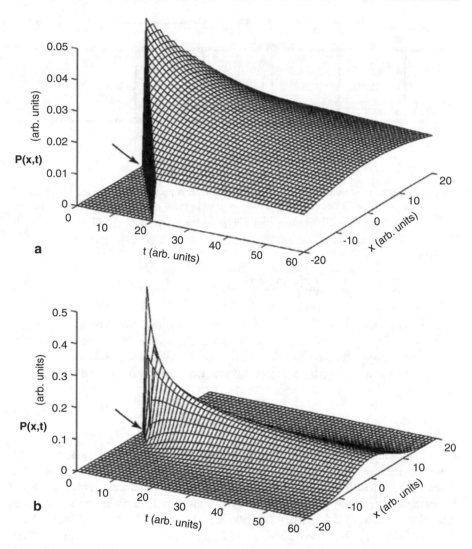

Fig. 1.1 Unified description of coherent (wave-like) and incoherent (diffusive) motion achieved with the memory formalism with exponential memory. Probability density $P(x, t)$ for an initially localized placement as shown by the arrows as a function of position x and time t. See text. In each of the panels, the same value of c is taken, equal to 1 in arbitrary units; α is 0.2 in the upper panel and 3.2 in the lower one, also in arbitrary units. Coherent motion is seen at short times, which is clear in the 'light-cones', and incoherent motion at long times, clear in the Gaussian shape of $P(x, t)$ in space. Adapted with permission from Fig. 1.2 of ref. (Kenkre et al. 1998a); copyright (1998) by the American Physical Society

a particle) to the memory (like the force). Time integration of (1.11) gives, for a localized initial condition,

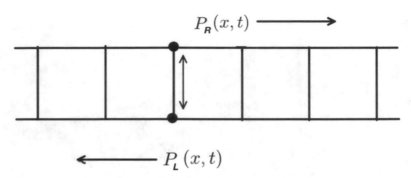

Fig. 1.2 Railway track model in which memories appear. A particle moves to the left (right) if on the lower (upper) track and randomly undergoes transitions at rate $\alpha/2$ from track to track. The result is the appearance of a memory in the evolution of the sum $P(x,t) = P_L(x,t) + P_R(x,t)$. See text

$$\langle x^2 \rangle = 2D \int_0^t dt' \int_0^{t'} ds\, \phi(s). \tag{1.12}$$

The advantage of providing a unification analysis with the help of an expression such as in Eq. (1.12) is simplicity. It treats the moments of the probability distribution rather than the full distribution itself. Observe that substituting the exponential memory for $\phi(t)$ in Eq. (1.12) produces the mean square displacement

$$\langle x^2 \rangle = 2D \left(t - \frac{1 - e^{-\alpha t}}{\alpha} \right) \tag{1.13}$$

for a localized initial condition. Remembering that $D = c^2/\alpha$, one notices that this is a pretty result that passes smoothly from the quadratic form $c^2 t^2$ at short times to the linear form $2Dt$ at long times except for a constant term negligible by comparison to the growing term. The transition time is measured by the characteristic time of the memory function, i.e., $1/\alpha$. We will see, in Chap. 3, how this simple behavior was used (Kenkre and Knox 1974b) to resolve a long-standing puzzle in the literature on excitation and energy transfer which used to go under the name of the Perrin-Förster puzzle (Perrin 1932; Förster 1948; Dexter et al. 1969) to which Davydov (Davydov 1968) had also added a point of debate in the discussion.

It might be fascinating, at least instructive, to notice from the general expression on the right hand side of Eq. (1.12) that, for essentially *any* reasonable memory $\phi(t)$ which decays after a sufficient time, the s-integral on the right will produce a constant for a large enough upper limit of integration. This constant is the integral of $\phi(t)$. It brings about the linear result that $\langle x^2 \rangle$ tends to $2Dt$ at long times, assuming $\phi(t)$ is normalized. At the other extreme of time, i.e., for small enough times, any reasonable $\phi(t)$ can be approximated by its (constant) value at $t = 0$ and produce the quadratic result for $\langle x^2 \rangle$ at sufficiently small times. The unification of the coherent behavior (proportional to t^2) with its incoherent counterpart (proportional to t) is thus seen to hold, not merely for the exponential memory corresponding to the telegrapher's equation, but for *any* reasonable memory function that can be approximated by a constant at short times and that decays for long enough times!

The specific detail in which one extreme transitions into the other will, no doubt, depend on the actual form of the memory function. However, the unification is quite general, i.e., robust. This must be understood as part of the power of the memory formalism.

1.2   An Example of How Memory Functions Arise: The Railway-Track Model

That memory functions are able to address the description of coherence in motion (wave-like versus diffusive behavior) by their very nature is clear from the above. But do they appear naturally in every system? What must we do to encounter memories? Let us examine this important question by considering a simple model of a particle moving on a system of what looks like a railway track, see Fig. 1.1. The particle has probability density $P_L(x, t)$ if it is on the lower track at position x at time t, and with probability density $P_R(x, t)$ if it is on the upper track. If the former, the particle moves to the left (hence the suffix L) and if the latter, it moves to the right (hence the suffix R). In either case the magnitude of the speed with which the particle travels is c. There are random transitions that occur from track to track at rate $\alpha/2$, so that the equations of motion are[3]

$$\frac{\partial P_R(x, t)}{\partial t} = +c\frac{\partial P_R(x, t)}{\partial x} + (\alpha/2)\left[P_L(x, t) - P_R(x, t)\right], \tag{1.14a}$$

$$\frac{\partial P_L(x, t)}{\partial t} = -c\frac{\partial P_L(x, t)}{\partial x} + (\alpha/2)\left[P_R(x, t) - P_L(x, t)\right]. \tag{1.14b}$$

The tracks are simply a vestige of starting the model out in a discrete form before taking the continuum limit; the transitions can take place anywhere on the lines.

The probability density that the moving entity is at position x at time t irrespective of its direction of travel (equivalently irrespective of whether it is on the upper or lower track) is given by the sum $P(x, t) = P_L(x, t) + P_R(x, t)$. To obtain a closed equation of motion for this sum, let us add and subtract the two equations respectively and eliminate the difference $P_L(x, t) - P_R(x, t)$ by taking derivatives. We find the resulting equation of motion for the sum is precisely the telegrapher's equation (1.8) displayed in the previous Section.

---

[3]This model was invented by the author to answer a query by Raoul Kopelman at a conference in the early 1970s whether memory functions required the system to be quantum mechanical or whether they arose in classical systems as well. The model was later generalized by Alden Astwood in collaboration with the author to treat flocking of birds in his Ph.D. thesis. The model is simple enough that it must have surely arisen independently in the literature in other places.

The telegrapher's equation is equivalent to a diffusion equation augmented with $\phi(t) = \alpha \exp(-\alpha t)$. An exponential memory function has thus made its appearance simply because we have looked at partial information of the probability density by eliminating or ignoring the direction of travel. This is a common occurrence: memories often arise when one coarse-grains the description. The reciprocal of α measures the average time between scattering events that change the identity of the track.

1.3   Memories in Quantum Systems

The Fermi Golden Rule is a well-known prescription which makes its appearance as a workhorse in various contexts in quantum physics whenever the description of a transition from a group of states to another group of states is sought. If the system Hamiltonian is $H = H_0 + V$, one's interest is typically in studying transitions from a group of eigenstates of H_0 which we will call ξ, μ, etc, labelled as group 1, to another group labelled as 2. The Rule asserts that the rate of transition is given by the product of $(2\pi/\hbar)|\langle\xi| V |\mu\rangle|^2$ averaged over the group of states ξ the transition is from, and the energy density of states in the group of states μ that the transition is to, there being a conservation of energy requirement of the energies (equivalently eigenvalues of H_0) of the initial and final states. This is taken as a long-time weak-coupling result, i.e., a perturbative formula when the action of V is small relative to the action of H_0, and the time is long enough that microscopic oscillations of occupation probabilities may be neglected.

The subject of the derivation of the Fermi Golden Rule from the microscopic quantum evolution is much more deep and fraught with conceptual traps than most textbooks admit, related as it is to the central problem of nonequilibrium quantum statistical mechanics. However, it should suffice here to state that, if our system consists of two such (macroscopically degenerate) groups of states, which we will call 1 and 2, between which transitions are taking place, we have the probabilities of occupation of the groups P_1 and P_2 satisfying

$$\frac{dP_1(t)}{dt} = F[P_2(t) - P_1(t)], \tag{1.15a}$$

$$\frac{dP_2(t)}{dt} = F[P_1(t) - P_2(t)]. \tag{1.15b}$$

We will call this system a *degenerate* quantum dimer[4] and notice that, for initial occupation of one of the (groups of) states, say 1, the solution for either probability of occupation is an exponential relaxation to equilibrium. The equilibrium value is obviously 1/2.

---

[4]This is meant in the sense of equal energies; the comment is completely devoid of any disapproval on moral grounds.

1.3.1   Memories from Reservoir Interactions

What happens if the transitions are not from one group of states to another but from a single state $|1\rangle$ to a single state $|2\rangle$? The energy density of states in the receiving "group" is 0 for any energy except that of the receiving state in which case it is infinite, in other words described by a δ-function; the Golden Rule formula cannot be used at all. Instead, we must revert to the Schrödinger equation

$$i\hbar\frac{d|\psi(t)\rangle}{dt} = H|\psi(t)\rangle \tag{1.16}$$

where $|\psi\rangle$ is the quantum mechanical state vector with respective amplitudes $c_1(t) = \langle 1|\psi(t)\rangle$ and $c_2(t) = \langle 2|\psi(t)\rangle$ to be in the two sharp equienergetic eigenstates of H_0. Using the same initial occupation of one of the two states assumed along with Eq. (1.15), we can see that the probabilities now obey

$$\frac{d^2 P_1(t)}{dt^2} = 2(V/\hbar)^2[P_2(t) - P_1(t)], \tag{1.17a}$$

$$\frac{d^2 P_2(t)}{dt^2} = 2(V/\hbar)^2[P_1(t) - P_2(t)]. \tag{1.17b}$$

In place of the exponential relaxation to the value 1/2 predicted by Eq. (1.15), we see here sinusoidal oscillations of the probability from 1 to 0 around the average value 1/2: $P_1(t)$ equals $\cos^2(Vt/\hbar)$ while $P_2(t)$ oscillates as $\sin^2(Vt/\hbar)$.

The sharp-states case in the latter example results in oscillations known often as *ringing* in quantum mechanics. The dense-group case in the former example can be said to be representative of the irreversible approach to equilibrium. Can the two be unified?

It is easy to see that memory functions can do the trick. If, for the same initial conditions assumed, we were to consider

$$\frac{dP_1(t)}{dt} = 2(V/\hbar)^2 \int_0^t ds\, e^{-\alpha(t-s)}[P_2(s) - P_1(s)], \tag{1.18a}$$

$$\frac{dP_2(t)}{dt} = 2(V/\hbar)^2 \int_0^t ds\, e^{-\alpha(t-s)}[P_1(s) - P_2(s)], \tag{1.18b}$$

and impose the relation $2(V/\hbar)^2/\alpha = F$, we would find a remarkable unification of the coherent (oscillatory) Eq. (1.17) with the incoherent (decay-type) Eq. (1.15). Typically, the probabilities would oscillate trigonometrically[5] around 1/2 and eventually both relax to the equilibrium values 1/2.

---

[5]Surely, if the degree of coherence happens to be small enough because of the value of $V/\hbar$ relative to that of α, there would be an overdamping phenomenon resulting in oscillations not being manifested. The situation is precisely as what occurs in an overdamped oscillator. In such a

Understanding how an intermediate equation such as (1.18), carrying with it the (exponential) memory function, can arise from microscopic interactions present in the Hamiltonian of the quantum system under consideration is crucial to an understanding of quantum statistical mechanics itself. It will occupy us in some of the immediately following chapters of this book. However, at this point, let us recall that energy differences like $E_\xi - E_\mu$ have been considered sometimes to have an *imaginary* component arising from interactions with a reservoir. Such practice is associated with the names of Weisskopf and Wigner in atomic physics. An eminently usable implementation of this subterfuge of introducing irreversibility into quantum evolution in this somewhat *ad hoc* manner may be found in the discussion given by Wannier (1959) in his book in the chapter on Ohm's Law. It consists of deducing from Eq. (1.16) the equation for the density matrix elements in a representation such as those of the states 1 and 2, and, to describe the reservoir interactions, adding terms that destroy the off-diagonal elements at a rate which we will call α. Thus,

$$\frac{d\rho_{11}(t)}{dt} = -i(2V/\hbar)(\rho_{21} - \rho_{12}), \qquad (1.19a)$$

$$\frac{d\rho_{22}(t)}{dt} = -i(2V/\hbar)(\rho_{12} - \rho_{21}), \qquad (1.19b)$$

$$\frac{d\rho_{12}(t)}{dt} = -i(2V/\hbar)(\rho_{22} - \rho_{11}) - \alpha\rho_{12}, \qquad (1.19c)$$

$$\frac{d\rho_{21}(t)}{dt} = -i(2V/\hbar)(\rho_{11} - \rho_{22}) - \alpha\rho_{21}. \qquad (1.19d)$$

Here, $\rho_{mn} = c_m c_n^*$ are the density matrix elements, the diagonal elements ρ_{11}, ρ_{22} are the respective probabilities P_1 and P_2, and it is important that the decay terms proportional to α appear only in the equations for off-diagonal elements; otherwise the particle itself would begin to disappear.

⊛   It will amply repay the reader, at this stage, to undertake a few simple manipulations of the set of Eqs. (1.19). Subtract the fourth equation from the third; solve the resulting equation for the off-diagonal difference $\rho_{12} - \rho_{21}$, say in the Laplace domain; substitute it in the first and second equations; use the fact that the initial conditions give $\rho_{12}(0) = 0 = \rho_{21}(0)$; and, finally, arrive precisely at the intermediate equation with memory, Eq. (1.18)!

This little known result, that the Wigner-Weisskopf approximation, equivalently the Wannier trick of off-diagonal density matrix element destruction, leads so smoothly to the integrodifferential equation with a memory whose decay rate is α, and whose value at $t = 0$ is proportional to V^2, should whet the appetite of the reader for further investigations of what memory functions can deliver.

---

case, a vestige of coherence would be seen at short times in the curvature of the probabilities at the initial time.

1.3.2 Memories from Spatial Extension

Let us now do away with reservoirs or dense groups of states but extend the spatial extent of our quantum dimer from two sites to infinity. Consider, thus, a particle moving not on a dimer but on a linear chain of infinite sites/states labelled by integers m that take on all values from $-\infty$ to $+\infty$ including 0 of course. Let the Hamiltonian $H = H_0 + V$ be such that H_0 has only one eigenstate per site, each eigenvalue being equal to any other. We will take all these eigenvalues to be 0 without loss of generality. Let us take the interaction matrix elements to be nearest neighbor and let us call them also simply V where the latter is a c-number. The equation of motion for the amplitudes at each site, $c_m(t) = \langle m|\psi(t)\rangle$, is now

$$i\hbar\frac{dc_m(t)}{dt} = V[c_{m+1}(t) + c_{m-1}(t)]. \qquad (1.20)$$

Let the initial occupation be only of a single site, which we will call $m = 0$ without loss of generality, so that $c_m(0) = \delta_{m,0}$.

How do we solve Eq. (1.20)? Translational invariance of the equation suggests discrete Fourier transforms. We multiply the equation by e^{ikm}, sum over all m, and solve the straightforward differential equation for $c^k(t)$ which is simply the sum, with m running over all integers from minus infinity to plus infinity, i.e., equals $\sum_m c_m(t)e^{ikm}$. Then we invert the transform and find out, for the given initial condition, that the amplitude is proportional to the Bessel function J_m of order m such that the probability of occupation of site m at all times is given by

$$|c_m(t)|^2 = P_m(t) = J_m^2(2Vt/\hbar). \qquad (1.21)$$

Displayed in the solution (1.21), are a number of expected features of wave-like propagation of the quantum particle on the linear chain. The probability of the initially occupied site, $J_0^2(2Vt/\hbar)$, decays to zero because the particle moves away from the initial location but does so with simultaneous oscillations compatible with what we have learnt in classical wave physics as a consequence of the Huygens principle. We notice that the characteristic time to leave the original site can be regarded as being proportional to $\hbar/2V$. For this reason we can even state, in some loose sense, that $2V/\hbar$ measures the rate of transfer from site to neighboring site.

Let us calculate the mean square displacement of the quantum particle as we did for the motion of the classical disturbance on an infinite line earlier. For the wave limit in the classical context, we obtained the result that it is proportional to t^2. Let us see if the quantum particle behaves similarly. Calling the intersite distance (what is often called the lattice constant) a, we have

$$\langle x^2(t)\rangle = a^2\langle m^2\rangle = \sum_m a^2 m^2 J_m^2(2Vt/\hbar) = (2Va/\hbar)^2 t^2, \qquad (1.22)$$

in the light of a standard Bessel identity.

We get the same quadratic dependence on time as for the classical wave equation. The difference between the classical and the quantum waves is the absence in the former, but presence in the latter, of dispersion. The classical waves have a single speed whereas each Fourier component of the quantum wave has a different (group) velocity. The effective speed c of the quantum wave arising from the above-given mean square displacement expression is proportional to the average over the tight-binding band of the group velocities characteristics of all locations in the Brillouin zone. This should be quite clear from the tight-binding band result that the group velocity is given by $v_k = (2Va/\hbar) \sin ka$.

Could we associate a memory function with this behavior? Let us use (Kenkre 1978d) a sneaky way[6] of deriving from Eq. (1.21) a memory expression for $\mathcal{W}_{mn}(t)$ or $\mathcal{W}_{nm}(t)$ in an equation like

$$\frac{dP_m(t)}{dt} = \int_0^t ds \sum_n [\mathcal{W}_{mn}(t-s)P_n(s) - \mathcal{W}_{nm}(t-s)P_m(s)], \qquad (1.23)$$

knowing, for sure, that translational invariance will make the memories $\mathcal{W}_{mn}(t)$ as functions of the single variable $m - n$.

Let us first define a new matrix $\mathcal{A}$ such that $\mathcal{A}_{mn} = -\mathcal{W}_{mn}$ for $m \neq n$, and $\mathcal{A}_{mm} = \sum_n \mathcal{W}_{nm}$. Then let us take the discrete Fourier transform

$$\mathcal{A}^k = \sum_l \mathcal{A}_l e^{ikl},$$

where l is an integer equal to $m - n$, and so too of the probabilities. Here and elsewhere, unless otherwise indicated, direct (discrete) space quantities are denoted by subscripts and their (discrete) Fourier transforms by superscripts. Finally, let us take the Laplace transform of Eq. (1.23) with ϵ as the Laplace variable, tildes denoting Laplace transform: for any $f(t)$, we write

$$\widetilde{f}(\epsilon) = \int_0^\infty dt\, f(t)e^{-\epsilon t}.$$

We obtain the connection between the quantities $\widetilde{P^k}(\epsilon)$ and $\widetilde{\mathcal{A}^k}(\epsilon)$ as

$$\widetilde{P^k}(\epsilon) = \frac{P^k(t=0)}{\epsilon + \widetilde{\mathcal{A}^k}(\epsilon)} = \frac{1}{\epsilon + \widetilde{\mathcal{A}^k}(\epsilon)}. \qquad (1.24)$$

---

[6]Perhaps proper scientific practice dictates that we describe the nature of this derivation (Kenkre 1978d) as *indirect*. However, the reader will find, I am sure, that comparison with other independently given derivations, slightly earlier (Sokolov 1976) as well as slightly later (Kühne and Reineker 1979), will indeed justify the terminology used.

The second equality applies to the initial condition that the probabilities are zero for all sites except the one at $m = 0$.

The determination of the elements of the $\mathcal{A}$-matrix is then immediate by taking the reciprocal of the connection relation:

$$\widetilde{\mathcal{A}^k}(\epsilon) = \frac{1}{\widetilde{P^k}(\epsilon)} - \epsilon. \tag{1.25}$$

For an explicit result of the memories for the chain, we need the Laplace-Fourier transform of the square of the Bessel function in light of Eq. (1.21). We invoke the discrete Fourier relation

$$\sum_{m=-\infty}^{+\infty} J_m^2(z)e^{ikm} = J_0\left(2z \sin \frac{k}{2}\right),$$

evaluate the Laplace transform of the zero-order Bessel function, and obtain from Eq. (1.25),

$$\widetilde{\mathcal{A}^k}(\epsilon) = \sqrt{\epsilon^2 + (4V/\hbar)^2 \sin^2(k/2)} - \epsilon. \tag{1.26}$$

Inverting the Laplace and the Fourier transforms, we finally get the memories:

$$\mathcal{W}_{mn}(t) = 2(V/\hbar)^2[J_{m-n+1}^2 + J_{m-n-1}^2 + 2J_{m-n+1}J_{m-n-1}$$
$$- 2J_{m-n}^2 - J_{m-n}J_{m-n+2} - J_{m-n}J_{m-n-2}]. \tag{1.27}$$

Here the argument of each J or J^2 is $2Vt/\hbar$. This is an interesting result. The reader is invited to study the expression for the cases that m and n, respectively, are nearest neighbors and so differ by 1, as well as when they are not nearest neighbor. Identities satisfied by Bessel functions allow one to rewrite the result (1.27) in the much more compact form

$$\mathcal{W}_{mn}(t) = \frac{1}{t}\frac{d}{dt}J_{m-n}^2(2Vt/\hbar). \tag{1.28}$$

The generalization of these results for finite but symmetric lattices of arbitrary dimensionality (Kenkre 1978d) using this same method will be seen to lead to some surprising consequences such as spatially long range transition rates in the presence of incoherence introduced by the Wigner-Weisskopf or Wannier method alluded to earlier. (See, e.g., Sect. 4.4.3.)

⊛ To engage you, the reader, in active learning, may I extend to you an invitation to work out, using methodology similar to that shown here, explicit expressions for memory functions connecting any two sites in a system of an arbitrary number N of sites (provided N exceeds 1)? You should assume that the eigenvalues of H_0

associated with each site are equal so that they may be considered to be zero without loss of generality, and that the matrix element of the intersite interaction between any two sites of the system equals V. You should arrive at the simple result that

$$\mathcal{W}_{mn}(t) = 2(V/\hbar)^2 \cos\left[(Vt/\hbar)\sqrt{N(N-2)}\right]. \tag{1.29}$$

On the basis of the geometry of its intersite connections, we might even give this simple system the weighty name of an *N-hyperhedron*! Notice that, for $N = 2$, the cosine equals 1 and the memory reduces to the simple *constant* expression $2(V/\hbar)^2$ for the ordinary dimer analyzed in (1.17); and that it predicts amusing properties in the opposite limit of very large or infinite N.[7]

1.3.3   A Peek at Richer Memories from Quaum Systems

To complete this introductory tour of memory functions, let me provide the reader here a glance (without showing calculational detail at this stage) at two more candidates that are relatively involved. The first, whose 1-dimensional version we have already encountered, applies to a quantum particle moving among the sites **m**, **n** of a pure crystal (lattice) of arbitrary number of dimensions and composed of N sites, all of equal energy in the unperturbed part of the Hamiltonian. Transfer matrix elements of the Hamiltonian $V_{\mathbf{mn}}$ lead to the motion between sites. These matrix elements are translationally invariant and the system obeys periodic (Born-von Karman) boundary conditions in the appropriate number of dimensions. Thus, we have a generalization of the system considered in the preceding section. The Hamiltonian is

$$H = \sum_{\mathbf{m,n}} |\mathbf{m}\rangle V_{\mathbf{mn}} \langle|\mathbf{n}|, \tag{1.30}$$

and the equations of evolution for the amplitudes c are

$$i\hbar \frac{dc_{\mathbf{m}}}{dt} = \sum_{\mathbf{n}} V_{\mathbf{mn}} c_{\mathbf{n}}. \tag{1.31}$$

Translational invariance allows the evaluation of Fourier transforms in the space of appropriate dimensions: $V^{\mathbf{k}} = \sum_{\mathbf{l}} V_{\mathbf{l}} e^{i\mathbf{k}\cdot\mathbf{l}}$. In terms of them, the memory functions in the Laplace domain are given by

---

[7]The late Bob Silbey and I had entertained ourselves with an inspection of this system in the mid 1970s but never found occasion to make anything substantial from it.

$$\widetilde{W}_{\mathbf{mn}} = -\sum_{\mathbf{k}} \left[\frac{e^{-i\mathbf{k}\cdot(\mathbf{m}-\mathbf{n})}}{\sum_{\mathbf{q}} [\epsilon + i(V^{\mathbf{k}+\mathbf{q}} - V^{\mathbf{q}})^{-1}]} \right]. \tag{1.32}$$

The second example, which we will meet with in Chaps. 4, 6, and 7, is specifically directed at reservoir interactions but not represented through the simple expedient of introducing decay terms. The method used to obtain the memories is a weak-coupling perturbation calculation from an explicitly stated Hamiltonian of the interactions with the reservoir. For a quantum dimer and a boson reservoir, the Hamiltonian is

$$H = V \left(|1\rangle\langle 2| + |2\rangle\langle 1| \right) + \sum_{q} \hbar\omega_q \left(b_q^\dagger b_q + \frac{1}{2} \right) +$$

$$(|1\rangle\langle 1| - |2\rangle\langle 2|) \sum_{q} \hbar\omega_q g_q \left(b_q + b_{-q}^\dagger \right) \tag{1.33}$$

in standard notation. The kets and bras here describe the two site states of an excitation which shuttles between them with matrix element V while interacting with a bath of bosons (phonons) of wavevector q each of which is created by the operator $b_q^\dagger$ and has frequency ω_q, the coupling constant for interaction of the reservoir bosons with the excitation being dependent on wavevector, and denoted by g_q. Note that the interaction with the bosons modulates the site energies of the excitation rather than the intersite matrix elements.

The excitation probability difference between the two sites, $p(t) = P_1(t) - P_2(t)$, obeys

$$\frac{dp(t)}{dt} + 2 \int_0^t \mathcal{W}(t - s) p(s) ds = 0 \tag{1.34}$$

where the memory $\mathcal{W}(t)$ is given by

$$\mathcal{W}(t) = 2(V/\hbar)^2 e^{h(t)-h(0)} + c.c. \tag{1.35}$$

$$h(t) = 2 \sum_{q} g_q^2 \left[n_q e^{i\omega_q t} + (n_q + 1) e^{-i\omega_q t} \right] \tag{1.36}$$

Here the factor n_q is the thermal Bose factor $\left(e^{\hbar\omega_q/k_B T} - 1 \right)$. For a single mode, the memory function is given by

$$\mathcal{W}(t) = 2(V/\hbar)^2 e^{-2g^2(1-\cos\omega t)} \cos(2g^2 \sin\omega t). \tag{1.37}$$

For finite temperatures and multiple vibrational modes, the memory takes on a form that can be more complicated. For optical phonons centered around frequency ω with a narrow width σ, the expression for the memory is still simple (Kenkre 2003):

$$\mathcal{W}(t) = 2(V/\hbar)^2 e^{-2g^2 \coth(\omega/k_B T)(1-\zeta(t)\cos\omega t)} \cos(2g^2\zeta(t)\sin\omega t). \qquad (1.38)$$

Here, T is the temperature, k_B is the Boltzmann constant, and $\zeta(t)$ is the Fourier transform of the phonon density of states, typically given by a Gaussian $\exp(-\sigma^2 t^2)$.

1.4   Overview of Areas Where the Memory Formalism Helps

Applications of the memory formalism appear wherever coherence questions in the motion of an entity are important. We will see this happen prominently in the study of sensitized luminescence (Wolf 1968a; Wolf and Port 1976; Braun et al. 1982; Auweter et al. 1978, 1979) in which energy transfer or the motion of a Frenkel exciton has relevance. Indispensable to the study of these and related processes occurring in organic systems is an excellent book (Pope and Swenberg 1999) that is comprehensive and has emerged as a veritable bible of this field. I strongly recommend that the reader interested in any aspect of this area of science consult the book. The process of exciton dynamics is also important in the study of photosynthesis (Knox 1975). In a photosynthetic antenna, the electronic excitation created by absorption of sunlight travels from whatever location it lands at, to the reaction centers where it is caught and used to cook carbon dioxide and water into sugar (Clayton 1980). The nature of the transport of excitation in complex systems, in particular whether it proceeds coherently or incoherently, fast or slow, has been an active goal of investigations (Robinson and Frosch 1962, 1963; Dexter et al. 1969; Kenkre and Reineker 1982). We will have occasion to examine the basic incoherent theory (Förster 1948) constructed by Förster, (see also his article in (Gordon 1961)) for this process and generalize it to use memory functions.

Sensitized luminescence as a general area involves the doping of molecular crystals and aggregates by impurities, an example being aromatic hydrocarbons such as anthracene with impurities of tetracene placed in them (Wolf 1968a). Light produces excitations that move in the host crystal and are captured by the guest molecules. Memory functions will be found to be important in understanding coherence in the transfer within the host and from host to guest. Similar considerations apply also to the transport of other quasparticles in condensed matter systems, including electrons or holes, often photo-injected, in molecular crystals and aggregates (Schein et al. 1978; Warta and Karl 1985; Reineker et al. 1981; Andersen et al. 1983; Kenkre et al. 1989). Memory functions will be calculated when such electrons, or excitons, interact strongly with molecular vibrations giving rise to polaronic effects (Holstein 1959a,b) in the quasiparticle mobility.

The quite unrelated area of nuclear magnetic resonance microscopy with applications to MRI (magnetic resonance imaging) is studied with the help of the so-called Torrey-Bloch equations obeyed by the spins on the protons making up the water molecules in matter. Memory functions appear in the analysis of the study of probes of confined motion in this context. Remarkably, we will find the formalism of memories applied even to the compaction of sand and other granular materials (Kenkre et al. 1996; Scott et al. 1998; Kenkre 2001a,b) where the memories are not functions of time but of depth in the compaction process. Phenomena such as stress distribution will be studied in this context.

We will meet with the analysis of relations of memory functions to neutron scattering functions, (Van Hove 1954b), to velocity auto-correlation functions, (Kubo 1957), as also to pausing time distributions (Montroll and West 1979) in random walks. We will encounter them in the quantum control (Kenkre 2000) of dynamic localization and in the study of vibrational relaxation (Kenkre and Chase 2017; Ierides and Kenkre 2018). They will appear indispensably even in the central problem of non-equilibrium statistical mechanics (Zwanzig 1964) which is the understanding of the irreversibility of macroscopic phenomena in terms of the microscopic reversibility of the fundamental underlying equations of motion.

Memory functions have also appeared in systems with static disorder as in amorphous molecular aggregates. There are three contexts of such treatment through memory functions in disordered systems. The first consists of postulating memory functions to address specific experiments, as in predicting their effects on transient grating and sensitized luminescence observations. The other example, much deeper, successful and well-known, is by Scher, with his collaborators Lax and Montroll (Scher and Lax 1973; Scher and Montroll 1975). Although the work of Scher et al., and the associated work presented by (Shlesinger 1974), is couched in terms of the so-called pausing time distribution functions $\psi(t)$, a thorough discussion is available (Kenkre and Knox 1974a; Kenkre 1977a; Kenkre et al. 1973) of the equivalence of the $\psi(t)$'s and the memory functions $\phi(t)$'s discussed in the present book.

The second context of memories in disordered systems appears in the analysis of (Haan and Zwanzig 1978), as well as (Gochanour et al. 1979). Wong has given a lucid discussion in (Wong and Kenkre 1982). As the third context, there is also the representation of disorder by a memory description. A publication by Klafter and Silbey (1980) set right some doubts that had been unjustifiably raised about whether a memory description could provide a description of static disorder. Perhaps, of even more practical importance was, rather than a statement of existence, a specific recipe to obtain the memory functions in some usable form of the evolution equation from known facts regarding the disorder. Two instances of such a practical recipe have appeared in the literature. One is the perturbative development in the appendix of the paper of (Scher and Lax 1973). The other, along with its consequences for motion, is the analysis performed more recently by Kenkre et al. (Kenkre et al. 2009) in their investigation and extension of effective medium theories. In that relatively recent analysis, which picks up on the spirit of the work of Haus and Kehr (1987), the authors were able to provide a usable prescription to convert given probability distributions characteristic of the disorder in a given system into explicit memory functions to be used on a corresponding ordered system. Some of these matters will be discussed in Chap. 13 of this book.

My own encounter with memory functions came first when I was a student learning about quantum statistical mechanics from the writings of traditional masters such as van Hove and Zwanzig, and then soon thereafter as a researcher bent on solving vexing coherence issues in Frenkel exciton transport in the very early 1970s. To me, they have been home-grown techniques constructed for specific purposes in the quantum transport of quasiparticles, generally in the field of energy transfer in molecular crystals and aggregates. I have had to twiddle them, and hone them, to suit definite purposes in the arena we were working in. It appears most useful to communicate to the reader whatever I have learned about the subject in that spirit. That is what I have done in this book.

However, I have discovered kindred souls traveling independently on similar journeys, notably Grigolini and his collaborators in the older days, and Yulmetyev in collaboration with Hänggi and other colleagues more recently. The former published a book (Evans et al. 1985) with Evans and Parravicini, that appeared soon after my own with Reineker (Kenkre and Reineker 1982) and has made profound contributions to our subject (Vitali and Grigolini 1989; Bonci et al. 1991) which I want to recommend strongly to the reader. There has been memory function work, earlier in liquids (Lantelme et al. 1979), and more recently in epilepsy and other aspects of neuroscience, in cardiac physics and even in seismology. I know little of those areas (Yulmetyev et al. 2003) but recommend to the reader to learn from that literature how fertile and multifaceted these techniques are, with their association with generalized Langevin equations (Mokshin et al. 2005b) and their connection with the general subject of correlations (Mokshin et al. 2005a). I will restrict my own description in the following pages to the context of the fields in which we developed them, hoping thereby to instruct the reader in their subtleties as we saw them.

The rest of the book is laid out as follows. In its first part, which includes the present Chaps. 1, 2–10 as well, the memory formalism and projection techniques are discussed. The latter is introduced in Chap. 2 in the form presented by Zwanzig in the early 1960s (Zwanzig 1964), and generalized in Chap. 3 to include coarse-graining as was done a decade later (Kenkre 1977a). In that generalized form, they are applied at the end of Chap. 3 to the resolution of a puzzle sometimes associated with the names of Perrin and Förster and used to extend the famous theory of the latter, known to be the basis of the description of energy and excitation transfer in sensitized luminescence and photosynthesis. Usable solutions of the generalized master equations containing memory functions are obtained in Chap. 4 for the often-encountered system of a crystal with focus on Frenkel exciton motion. The relations of memory functions to a variety of other entities such as neutron scattering functions and velocity auto-correlation functions, and pausing time distributions in continuous time random walks, are also described in that chapter. Some of those other entities have been used in condensed matter physics in investigations unrelated to Frenkel excitons: for instance of hydrogen atoms and other light interstitials (such as even muons), and photo-injected charges moving under the action of an electric field.

Applications of the memory formalism to experiments of two classes of the gentle kind will be the subject of Chap. 5. By the word 'gentle' here is meant the kind of experiment in which drastic modifications of the system do not occur during measurement. We will study Frenkel excitons in an ideal crystal: analyze Ronchi rulings to understand triplet motion, and transient gratings (four-wave mixing) to measure the degree of coherence in the transport of singlet excitons. These two phenomena involve the most direct observational probes of exciton dynamics allowing relatively unambiguous conclusions to be drawn about the nature of the motion of those quasiparticles in molecular systems.

The mobility of photoinjected charge carriers, holes as well as electrons, in aromatic hydrocarbon crystals, with specific attention on a long-standing puzzle concerning temperature-independent mobilities in naphthalene will form most of Chap. 6. The mobility is calculated from the integral of the auto-correlation function of the velocity which is in turn obtained from memory functions in the system. Although the study of the temperature dependence of observables does not lend itself to clean elucidation of theoretical concepts, since temperature effects arise not only from phase space considerations but also from reservoir interactions which are typically dirty in concept, the nature of quasiparticle motion enters in a fundamental way in this field.

Projections and memories are then developed in microscopic systems involving vibrational relaxation in Chap. 7 where the Montroll-Shuler equation of that phenomenon (Montroll and Shuler 1957) is studied with a view to generalizing it to coherent situations. Gaining insights into the general problem of the approach to thermal equilibrium of a quantum mechanical system in interaction with a reservoir is an additional goal of this analysis.

A return to the subject of projection operators in Chap. 8 will allow us to learn their use in varied contexts such as the calculation of electrical resistivity, the derivation of the BBGKY hierarchy in statistical mechanics, NMR microscopy, and the quantum control of dynamic localization. A study of these applications will broaden our insights into the technique of projection operators, allowing us to use them for studying NMR signals in confined geometries. The compaction of granular materials will be treated in Chap. 9. We will find in Chap. 10 that memories and projections become useful in nonlinear equations of motion as well, in the dynamics of systems similar to the physical pendulum, and in unexpected contexts such as reaction diffusion systems and the Fisher equation in theoretical ecology.

The next five chapters of the book, devoted to the defect technique developed by Montroll and collaborators, will occupy us in diverse ways in its application within the area of memory equations. The technique will be introduced and explained in Chap. 11 where basic examples will be shown. Applications to the so-called Simpson geometry situations in sensitized luminescence with end detectors, and to coherence effects of exciton transport in molecular crystals will be described. The importance of imperfect trapping of excitations, i.e., the need to refrain from automatically assuming that their absorption occurs at infinite rate, with dramatic repercussions to the interpretation of experiments, will be elucidated in some of this analysis. The v-function concept useful for high concentration of defects will be introduced and a periodic arrangement of defects studied with its help.

By extending the defect method from lattices to the continuum in Chap. 12, we will find it possible to tackle a number of new situations. This includes higher dimensional systems, the use of the Smoluchowski equation, and a theory of coalescence of signaling receptor clusters in immune cells. Whereas the memory function formalism will have been studied thus far in the context of dynamic disorder, Chap. 13 will introduce it from static disorder: memory functions arising from specific distribution functions of disorder will be calculated explicitly and used. A tutorial on how to implement these lessons step by step will be provided in Chap. 14. The defects in all of the systems studied so far in the book involve actual physical interruptions of translational invariance. In Chap. 15 we will develop an approach to non-physical defects that only have a formal (i.e., mathematical) existence. With its help, we will study exciton annihilation in molecular crystals, the neutron scattering function in the context of the stochastic Liouville equation and the transmission of infection in the spread of epidemics.

The final chapter of the book will perform two functions: providing a summary of what we have learnt, and elucidating certain issues that do not directly belong to any specific previous Chapter but are nevertheless relevant in the flow of this material.

1.5   Chapter 1 in Summary

The discussion began with how, when facing the bewildering variety of motions that one witnesses in nature, viewing them in terms of the memory formalism helps unify extremes and provides a description of any intermediate degree of coherence. This power of the formalism stems from the basic nature of memory functions rather than from any particular detail. A damped harmonic oscillator and the telegrapher's equation, with its limits of the wave and the diffusion equations, illustrated the situation. A simple example (the railway track model) showed how memories arise when one chooses to examine only *a part* of a given system. How memory functions can arise in quantum systems was explained next, an attempt being made to involve the reader in calculating results right from the beginning of the book, ensuring thereby an active process of learning. An overview of the various fields of research, ranging from fundamental studies of photosynthesis and sensitized luminescence to applied investigations of NMR microscopy and granular compaction, in which memories help naturally, was given.

Zwanzig's Projection Operators: How They Yield Memories

We have seen in Chap. 1 that memory functions address coherence issues in motion quite naturally, i.e., by their very existence. The important question, then, is whether, in a given system and context in which coherence matters are under investigation, memory functions exist at all; and if they do exist, how to get our hands on them so we can put them to use. In the present chapter, we will achieve two tasks. We will see that, in essentially every real system, memory functions are available for our use: they become apparent to us when we examine the passage of the system description from its microscopic level of atoms, electrons and quantum states to the macroscopic level of everyday perceptions and observations. We will additionally learn the workings of a specific mathematical tool, the *projection operator*, that is defined to facilitate this passage, and how to hone and fine tune it for our purposes.

2.1 Derivation of the Master Equation: A Central Problem in Quantum Statistical Mechanics

A profound question in physics is how microscopic reversibility in the underlying equations of motion, such as Newton's, Hamilton's or Schrödinger's, gives rise to macroscopic irreversibility and approach to equilibrium. Indeed, a central task of the subject of statistical mechanics consists of bridging this gap between evolutions that have such drastically different characters. Large-scale phenomena exhibit decays, approach to equilibrium, irreversibility, and closure in a few variables, although the underlying dynamics at the microscopic level has (Poincaré) recurrences, is therefore quite reversible, and involves a staggeringly larger number of degrees of freedom and corresponding variables describing the evolution of the zillions of constituent particles. A lot of work has been done by various investigators over the years in attempts to resolve this surely paradoxical situation (Uhlenbeck 1955; Dresden 1961). Significant contributions started with the work of Boltzmann. In

© The Author(s), under exclusive license to Springer Nature Switzerland AG 2021
V. M. (Nitant) Kenkre, *Memory Functions, Projection Operators, and the Defect Technique*, Lecture Notes in Physics 982,
https://doi.org/10.1007/978-3-030-68667-3_2

the 1950s, van Hove focused on the passage from the microscopic Schrödinger equation for the wave-vector, equivalently the Liouville-von Neumann equation for the quantum mechanical density matrix ρ,

$$i\hbar \frac{d\rho(t)}{dt} = [H, \rho(t)] \equiv L \rho(t), \tag{2.1}$$

to the Master equation for the probabilities $P_\xi(t)$, which are the diagonal elements $\langle \xi | \rho(t) | \xi \rangle$ of the density matrix in some appropriate representation, say of the eigenstates $|\xi\rangle$ of a part H_0 of the system Hamiltonian $H = H_0 + V$,

$$\frac{dP_\xi(t)}{dt} = \sum_\mu \left[F_{\xi\mu} P_\mu(t) - F_{\mu\xi} P_\xi(t) \right]. \tag{2.2}$$

Several valuable contributions were made at this stage by Van Hove (1954a, 1955). They included (1) pointing out a glaring inadequacy in Pauli's 1927 derivation of the Master equation which entailed making the so-called repeated random phase approximation, (2) assuming conditions on the interaction V, specifically the diagonal singularity conditions on its matrix elements, (3) invoking the thermodynamic limit, and (4) showing that, under a certain "$\lambda^2 t$" limit, one can derive the Master equation (2.2) from the Liouville equation (2.1). The λ is a c-number assumed to multiply V and thereby signify the strength of the interaction. The "$\lambda^2 t$" limit means taking the limit $\lambda \to 0$, and $t \to \infty$, such that the product $\lambda^2 t$ is a finite quantity. Its physical meaning was that the Master equation is a consequence of the microscopic evolution only at long times and on a scale correlated with the weakness of interaction through that slightly special combination of λ and t, only in the thermodynamic limit, and only for weak enough interactions which additionally obeyed the slightly mysterious diagonal singularity conditions. Recognizing the essential role of these restrictions in the derivation of the Master equation was also a somewhat significant contribution of van Hove's work as it made clear that *not all* conditions and environments lead microscopic evolution to give rise to macroscopic irreversibility and the approach to equilibrium.

Close inspection of the van Hove procedure would reveal the fact that the result was at first not the Master equation (2.2) for the probabilities but an equation, that although formally similar to the latter, was *non-local in time*.

This is where the memory functions lay! To reduce that equation to the Master equation, it was necessary to make the Markoffian approximation which actually meant assuming that somehow the memories could be replaced by δ-functions in time. The van Hove derivation used the thermodynamic limit (that the system size is infinite) to justify this approximation.

Many other investigators, among them Prigogine and Resibois (1958), Nakajima (1958), Zwanzig (1961, 1964), Montroll (1962), and Swenson (1963), whose separate analyses all came after van Hove's, contributed towards the derivation of the Master equation with the help of a variety of arguments. However, an intermediate step appeared always to be the generalized master equation (GME) with memory functions. There were debates in the literature of that time as to which parts of the logical flow of the derivation were associated with the conceptual aspects of the reversibility-to-irreversibility physics and which addressed merely a calculational issue. Was the intermediate appearance of the GME (the emergence of memory functions) irretrievably tied to van Hove's (slightly mysterious) $\lambda^2 t$ limit or was it an independent feature? Swenson suggested that in his method it was independent. Zwanzig, with his simple and transparent technique of projection operators, proved that Swenson's suggestion was indeed right, and applicable in general, and showed explicitly how this happened.[1] Although Nakajima had also introduced the technique almost at the same time, and although I have heard it said to me (by my colleague Colston Chandler at the University of New Mexico) that projection operators had been earlier in use in theoretical nuclear physics for other purposes, many of us in statistical physics have permanently stamped the technique with the name of Zwanzig. That is because, through his signature discussions and papers, brief, to the point, and always the epitome of clarity, he cleaned the entire field of extraneous confusions.

The GME, it should not be forgotten, is identical to the Master equation except for the fact that memories appear in a time-non-local equation in the place of transition rates appearing in a time-local equation:

$$\frac{dP_{\xi}(t)}{dt} = \int_0^t ds \sum_{\mu} \left[\mathcal{W}_{\xi\mu}(t-s)P_{\mu}(s) - \mathcal{W}_{\mu\xi}(t-s)P_{\xi}(s) \right]. \tag{2.3}$$

A thorough study of the work mentioned above on the central problem of nonequilibrium quantum statistical mechanics mentioned briefly above uncovers, thus, two points. First, memory functions (or GME's that contains them as an essential component) are unavoidable in the passage from microscopic to macroscopic levels of description of almost *any* system that we might be interested in investigating. It is not necessary to undertake a specialized quest to go in search of them. And second, a large amount of intellectual effort is spent in understanding the process of making those memory functions go away, so that the GME's might validate the Master equation (2.2) that has no memory functions in it.

Allow me to invite you, the reader of this book, to notice that the power of the GME could, indeed *should*, justify a quite different approach. Instead of being in a hurry to get rid of the GME and its memory functions and heave a sigh of relief when you finally land on the Master equation, we might understand the memory functions, pay attention to them, and *use them* to understand coherence issues. A lot of the present book is built on this message.

---

[1] I knew Bob Zwanzig and had the good fortune, when I was a graduate student about to finish my Ph.D., to spend some time in his office at the University of Maryland discussing various aspects of statistical mechanics. I remember his words well. Their sense was that he had introduced projections only to clarify the calculational aspect, and to cut through the conceptual cacophony of the derivation of the Master equation.

In an article I wrote on the occasion of E. W. Montroll's sixtieth birthday festschrift, I accordingly began the article (Kenkre 1977a) with the following passage:

> The generalized master equation (GME) is an entity one meets with on the wayside in one's journey from the microscopic to the macroscopic level of the dynamics of large systems in statistical mechanics. The prominent character in this journey is *not* the GME but is the Pauli master equation (PME) also known as the Master equation. The latter, with its distinctive tendencies as are evident in the H-theorem and its generally built-in irreversibility, possesses the ability to guide the weary traveller safely (!) to the realm of macroscopic phenomena. The importance of the GME is therefore not always appreciated in the normal course of this journey. In fact, usually, the equation is not even allowed to live long. Almost immediately after one makes its acquaintance, a procedure known as the Markoffian approximation is thrust into the GME, destroying its special characteristics and converting it into the sought-after PME. There are, however, researchers, who admire the GME for its own qualities (and not merely for its ability to give birth to the PME), who have recently studied it in its own right and have put to use its potentialities. This article will trace some of this activity concerning this interesting, powerful, but not always appreciated equation.

That passage was written in 1976 and published the next year. The next four decades have given some of us numerous opportunities in diverse areas of physics to carry out the promised program of putting the yoke on every GME we could find, and extract out of its memory structure, useful information and clarification about diverse topics. A function of this book is to demonstrate these applications and rejoice at their continued success, hoping that they will find use in future investigations carried out by the reader.

We make no claim on the profundities in the contributions made by the authors we have mentioned above, from van Hove to Zwanzig. Our job has only been in learning what they have discovered, invented, and perfected, profiting from their techniques and discussions centered around the origin of irreversibility, and the derivation of the Master equation. Our only contribution has been to pick up a largely discarded byproduct of that process, the GME, and to put it to the maximum use possible, on the basis of its memory functions. The projection operator that Zwanzig introduced in the process is a powerful tool with which we will begin our study.

2.2 Memories from Projection Operators that Diagonalize the Density Matrix

The starting point of Zwanzig's derivation is the microscopic Liouville-von Neumann equation (2.1) for the density matrix ρ, the eventual goal is the Master equation (2.2) obeyed by the probabilities of occupation of states such as ξ and μ which are eigenstates of a part H_0 of the total system Hamiltonian H, and the intermediate object we will encounter during this passage is the GME (2.3). This equation looks a lot like the Master equation but has memory functions in it. What we will use for our derivation is a projection operator $\mathcal{P}$ which extracts the diagonal part of any operator on which it acts. This it does by extinguishing all elements of that operator on its right (on which it acts) which are not diagonal in the given representation (of states ξ, μ, etc.) The operator is clearly idempotent, which means that $\mathcal{P}^2 = \mathcal{P}$, it has nothing to do with any time-dependences, thus commutes with operations such as the time derivative, and is linear, which means that $\mathcal{P}(O_1 + O_2) = \mathcal{P}O_1 + \mathcal{P}O_2$.

Armed with these properties of the projection operator, we apply $\mathcal{P}$, and $1 - \mathcal{P}$, where 1 is the identity operator, separately to Eq. (2.1). Here and henceforth, (with the exception of Chap. 6), we suppress showing the $\hbar$ explicitly. The result for the coupled (respective) evolution of the diagonal part $\rho' \equiv \mathcal{P}\rho$ and the off-diagonal part $\rho'' \equiv (1 - \mathcal{P})\rho$ of the density matrix is

$$\frac{d\rho'(t)}{dt} = -i\,\mathcal{P}L\,\rho'(t) - i\,\mathcal{P}L\,\rho''(t), \tag{2.4a}$$

$$\frac{d\rho''(t)}{dt} = -i\,(1-\mathcal{P})L\,\rho''(t) - i\,(1-\mathcal{P})L\,\rho'(t). \tag{2.4b}$$

The equation for the diagonal part of the density matrix contains the off-diagonal part, and vice-versa. This is inevitable. The magic of separating the evolutions, which appears impossible on superficial thought, is now performed by the technique by solving the equation for $\rho''(t)$ formally and substituting the solution in the equation for $\rho'(t)$. The first step presupposes only that we know that the solution of a linear driven equation,

$$\frac{dy(t)}{dt} = -\mathcal{A}y(t) + f(t), \tag{2.5}$$

where $\mathcal{A}$ is a constant operator independent of y, and $f(t)$ is a driving term, is simply given, with the help of the Green function $e^{-t\mathcal{A}}$, by

$$y(t) = e^{-t\mathcal{A}}y(0) + \int_0^t ds\,e^{-(t-s)\mathcal{A}}f(s). \tag{2.6}$$

If $\mathcal{A}$ is time-dependent, $t\mathcal{A}$ in the exponent in the first term of the solution is generalized to $\int_0^t dt'\mathcal{A}(t')$, the exponent $(t-s)\mathcal{A}$ within the integrand in the second term to the integral $\int_s^t dt'\mathcal{A}(t')$, and time ordering of the operators is assumed. We will stick to the simpler expression in Eq. (2.6) unless required by the context.

Replacing y, $\mathcal{A}$, and $f(t)$ first in Eq. (2.5) and then in its solution Eq. (2.6), respectively by $\rho''(t)$, $i\,(1 - \mathcal{P})L$, and $-i\,(1 - \mathcal{P})L\,\rho'(t)$, the equation for the *diagonal* part of the density matrix, that is the first of the pair in Eq. (2.4), can be rewritten in an *essentially closed form*, i.e., almost independently of the off-diagonal part ρ''!

$$\frac{d\rho'(t)}{dt}=-i\mathcal{P}L\rho'(t)-\int_0^t ds\mathcal{P}Le^{-i(t-s)(1-\mathcal{P})L}(1-\mathcal{P})L\rho'(s)-i\mathcal{P}L\,e^{-it\,(1-\mathcal{P})L}\rho''(0).$$

$$(2.7)$$

Why have we used the expression "almost independently" (of the off-diagonal part) above? Because the only place any dependence of the diagonal part of the density matrix on its off-diagonal part comes in, is through the *initial* value of the latter in the last term. It is perfectly possible to have physically important situations in which $\rho''(0)$ is actually zero and, in such cases, the diagonal evolution is indeed independent of the off-diagonal evolution. How is such severing of the connection between the two parts of the density matrix possible? The original equation for the density matrix has interactions fully present between the diagonal and the off-diagonal parts. Understanding the answer to this question is equivalent to understanding how Zwanzig's projection operators clarified the debates in the statistical mechanics literature about whether that separation could be done independently of conceptual assumptions.

The answer actually is quite simple. Whereas the original equation for the entire ρ, which displayed without doubt coupling between its two parts, was time-local, Eq. (2.7) for the diagonal part is closed in itself but requires knowledge of that diagonal part throughout all time! The time rate of $\rho'(t)$ is connected not only to $\rho'(t)$ at the present instant, but to all times s in the past from 0 to the present time t. The projection technique has indeed succeeded in making the evolution of $\rho'(t)$ 'essentially' closed, but the price it has paid is not trivial. There is of course, the additional problem that, typically, the evolution operator in the integral, $e^{-i(t-s)(1-\mathcal{P})L}$, is no light-weight. Difficulties of unraveling it are usually severe and often make necessary brutal approximations, as might be expected, for realistic systems. And yet, the projection technique has unquestionably performed its magic of separating the evolution at least in a formal sense. Quantum mechanics, which takes pride in mixing the diagonal and off-diagonal elements of the density matrix, is thus cheated by the technique, in a manner that is fair and square.[2]

Zwanzig's procedure (Zwanzig 1961, 1964) of showing how the exact consequence of the Liouville-von Neumann equation (2.1) reduces to the GME (2.3) proceeds by proving that the first and the last terms of (2.7) can be neglected, the former exactly because $\mathcal{P}$ diagonalizes and L takes the commutator of H with whatever it acts on, which in that first term is a diagonal operator $\rho'(t)$, and the latter by the assumption (not approximation) that initially we start with a density matrix

---

[2]Although what we are discussing here may not be characterized by great profundity, it is true that many serious practitioners of science, not trained in this particular small area of nonequilibrium quantum statistical mechanics, are not aware of this peculiar way that an evolution of the probabilities of a quantum system can be closed in the probabilities. Indeed, I was present at an event at the Eastern Theoretical Physics Conference held in Rochester in the 1970s when a brilliant theoretical physicist, famous for multiple contributions to our science, shook his head most vehemently at a suggestion made by a junior colleague that such could happen. The latter had been exposed to the projection technique. The former had not. For a small fee along with an oath of secrecy, I might consider naming the parties privately to anyone sufficiently interested.

that is diagonal in the representation chosen. Such initial diagonality is ensured by an initial occupation of a single state ξ which is an eigenstate of H_0. Another example is when the system occupies several such states but initially with random phases between them. This is the allowable "initial random phase assumption" in contrast to the "repeated random phase assumption" made by Pauli in his analysis which van Hove showed (rather dramatically) to be tantamount to requiring that ρ is stationary, i.e., has no time dependence whatsoever.

With the first and the last terms gone from (2.7), taking the diagonal element of the entire equation, we get

$$\frac{d P_\xi(t)}{dt} = -\int_0^t ds \, \langle \xi | \, L e^{-i(t-s)(1-\mathcal{P})L} (1 - \mathcal{P}) \, L \rho'(s) \, | \xi \rangle \tag{2.8}$$

for the time rate of the diagonal element $\langle \xi \, | \, \rho(t) \, | \, \xi \rangle$ of the density matrix, which we also call the probability $P_\xi(t)$ of occupation of the eigenstate $| \xi \rangle$ of H_0. The nature of commutation inherent in L and the appearance of explicitly diagonal operators at several places in the integrand allow us to replace the rightmost occurrence of $(1 - \mathcal{P})L$ by L without approximation. Any further reduction requires an approximation. Putting the c-number λ denoting the strength of the interaction back into the expressions explicitly, writing the $(1 - \mathcal{P})L$ in the exponent as $(1 - \mathcal{P})(L_0 + \lambda L_V)$ where the Liouville operator L is split into a commutator involving H_0 and one involving V, respectively, we make ourselves ready to make the first approximation in the analysis.

This is the *weak-coupling* approximation in which only the lowest-order term in an expansion of the kernel in powers of λ is kept. This is sometimes termed the Born approximation in the manner of similar standard perturbation treatments for the quantum mechanical amplitude. To lowest order in λ (that happens here to be the second), we may write Eq. (2.8) as

$$\frac{d P_\xi(t)}{dt} = -\lambda^2 \int_0^t ds \, \langle \xi | \, L_V e^{-i(t-s)L_0} L_V \rho'(s) \, | \xi \rangle. \tag{2.9}$$

As soon as λL_V is taken out of the exponent to maintain the term of lowest order, the $(1 - \mathcal{P})$ gets replaced by the identity as earlier. Now that we have got rid of the pesky $\mathcal{P}$'s, we can write the L's in terms of the commutators and return to much more familiar expressions. The first step is to rewrite (2.9) as

$$\frac{d P_\xi(t)}{dt} = -\lambda^2 \int_0^t ds \, \langle \xi | \, [V, e^{-i(t-s)H_0} [V, \rho'(s)] e^{+i(t-s)H_0}] | \xi \rangle. \tag{2.10}$$

The second step is to resolve the commutators keeping in mind that H_0 and ρ' are diagonal in the chosen representation but V is not. We can then write as a weak-coupling consequence of the original Liouville-von Neumann equation (2.1) for a density matrix that is initially diagonal in the representation of the eigenstates of H_0, the GME (2.3) with memory functions given by the explicit formula

$$\mathcal{W}_{\xi\,\mu}(t) = 2\lambda^2 |\langle\xi|\,V\,|\mu\rangle|^2 \cos\left[(E_\xi - E_\mu)t\right].$$ (2.11)

In deriving Eq. (2.11), use has been made of the fact that the matrix elements $\langle\xi|\,H_0\,|\mu\rangle$ are nothing other than the unperturbed energies $E_\xi\,\delta_{\xi\,\mu}$, and that, consequently, $\langle\xi|\,e^{-i(t-s)H_0}\,|\mu\rangle$ equals $e^{-i(t-s)E_\xi}\delta_{\xi\,\mu}$.

Return now to the initial discussions in this Chapter in which we pointed out that our goal is to derive the Master equation (2.2) which awaits us with its built-in irreversibility, decay behavior, and enormous success in practical calculations, much that is collected in the compact fact that it satisfies the famous H-theorem of Boltzmann. That means that it guarantees approach to equilibrium and the increase of entropy in macroscopic processes. The trick to pass from the GME (2.11) to the Master equation is to *get rid of the memory functions* that one sees displayed in the GME. The accepted procedure is the Markoffian approximation. With the understanding that one is interested in a time-smoothed description, i.e., one that does not possess too fine a resolution in time, one replaces the memory function $\mathcal{W}_{\xi\,\mu}(t)$ by the product of a δ-function in time and the integral over all time of the memory function, ensuring thereby the simplification of the integro-differential equation into a time-local equation. The latter is the Master equation (2.2) and the integral of each memory $\mathcal{W}_{\xi\,\mu}(t)$ is the transition rate $F_{\xi\,\mu}$ given by

$$F_{\xi\,\mu} = 2\pi\lambda^2 |\langle\xi|\,V\,|\mu\rangle|^2 \delta\left(E_\xi - E_\mu\right).$$ (2.12)

The impressive success of the Zwanzig procedure should now be evident to the alert reader who will have noticed the identity of Eq. (2.12) to the standard, text-book, Fermi Golden Rule prescription for transition rates in the weak-coupling or perturbation approximation, thus to what is expected for the Master equation.

The steps of this classic derivation provided above are self-contained and should be sufficient to rederive the expressions. If further detail is found to be necessary, we refer the reader to several places in the literature (Zwanzig 1961, 1964; Kenkre 1977a) including, in particular, the extremely lucid exposition given by Zwanzig in a conference proceedings book (Meijer 1966) published after his original derivation.

What happens for an initial condition on the system density matrix which does not satisfy the random phase assumption (assumption that $\rho(0)$ is diagonal)? One answer that has been used is that the driven GME must then be used for the evolution wherein the term emerging directly from $-i\mathcal{P}L\,e^{-it\,(1-\mathcal{P})L}\rho''(0)$ in Eq. (2.7), now non-zero because $\rho''(0)$ is non-vanishing, provides the driving. An argument has been used sometimes that there are well-defined situations in which that driving term itself decays to zero for appropriate systems at a rate faster than the memory function does. If this does indeed occur, the GME without the driving term can be considered applicable in the time interval after the decay of the driving term but before the decay of the memory function. Unfortunately, only a small amount of study has been expended on this matter of the neglected initial term. See, e.g., Kenkre (1978b).[3]

We use the next section to familiarize the reader more about projection operators in general and the last section to a reexamination of the Zwanzig result embodied by the present discussion.

2.3   Three Exactly Approachable Examples of Projections and an Exercise

Projection operators have entered the physics scene of the reversibility-irreversibility paradox with an air of a certain mystery and wizardry about them. Let us strip them of these qualities by treating three relatively simple examples and calculating everything without approximation, and then set up an exercise for the ambitious reader bent on mastering the technique. First, we treat the time evolution of a simple complex quantity,[4] resolving the meaning of the projections in Zwanzig's expression exactly. As our second example to familiarize the reader with the technique, we apply projections to extract a single element of a 3-element (Bloch) vector to understand quantum control of dynamic localization. The third example deals with the effects on crystal memories due to reservoir interactions as expressed via the Wigner-Weisskopf procedure. All these examples will illustrate how to use projection operators exactly, i.e., without invoking any approximation. Finally, we will set up an exercise to calculate memories in an open trimer from the von Neumann equation for the density matrix, an exercise whose results will also be used elsewhere in the book.

2.3.1   Evolution of a Simple Complex Quantity

Consider a time-evolving complex variable $z(t) = x(t) + iy(t)$ obeying the simple equation

$$\frac{dz(t)}{dt} = -\Omega z(t) \tag{2.13}$$

where Ω denotes multiplication by the complex number $\alpha + i\beta$.

Let us pretend that our interest lies only in the evolution of the real part of z, viz. $x(t)$, in spite of the fact that the evolution of the real and imaginary parts is intertwined. Let us use Zwanzig's projection operators for the purpose and define $\mathcal{P}$

---

[3]The reader is warned there are special situations in which the initial driving term can change the character of the evolution drastically. This may not be usual but an experimentally relevant realization is discussed in detail for transient grating experiments in the second part of Chap. 5.

[4]I have always wanted to use the two adjectives, simple and complex, together in this manner that surely must be considered legitimate.

as an operator that extracts the real part of whatever it acts on. Since the real part of
a real number is the real number itself, $\mathcal{P}$ is idempotent, and it has little to do with
time derivatives. Thus we can write coupled equations for $z'(t)$ and $z''(t)$ which are
nothing but $x(t)$ and $iy(t)$ respectively, in the manner of Eqs. (2.4). We see that
the evolution of the real part $x(t)$ depends fully on that of the imaginary part $iy(t)$
and yet Zwanzig has taught us how to separate them and get an essentially closed
equation for $x(t)$. Following Zwanzig's procedure, let us take the initial z to be
completely real so that $y(0) = 0$, and therefore write the projected Eq. (2.7) as

$$\frac{dx(t)}{dt} = -\mathcal{P}\Omega x(t) - \int_0^t ds \mathcal{P}\Omega e^{-(t-s)(1-\mathcal{P})\Omega}(1 - \mathcal{P})\Omega x(s). \tag{2.14}$$

What exact manipulations can we justify in the present case? First, we notice that the
first term on the right hand side of Eq. (2.14) does *not* vanish. In the Zwanzig case,
the diagonal part of the result of commuting the Hamiltonian with any diagonal
operator is identically zero. However, the real part of the number obtained by
multiplying a complex number and a real number does not vanish. Indeed that first
term, that we will call the *proper* term of the Zwanzig equation, is simply $-\alpha x(t)$
in our example. But what about the second term that we will call the *memory* term?
    It is wise to study the action of $(1 - \mathcal{P})\Omega$ on a purely imaginary quantity iq.
It produces $i\alpha q$. If this is done a second time, we get $i\alpha^2 q$. And so on. So the
expansion of the exponential in the integrand of the memory term has simply the
result of producing $ie^{-\alpha t}q$. Not to forget that what the exponential in the integrand
is acting on, is indeed a purely imaginary quantity since it starts with $(1 - \mathcal{P})$. So
the action of the integrand is, we know exactly, to multiply iq by $e^{-\alpha t}$. We have got
rid of the pesky $\mathcal{P}$'s in the exponential without making any approximation!
    That iq on which the exponential acted is $i\beta x(s)$ and so, after the action of
the exponential full of the uncomfortable projection operators $\mathcal{P}$'s we have, as
the full integrand, $\mathcal{P}\Omega$ acting on $i\beta x(s)e^{-\alpha(t-s)}$. In other words, we have been
able to resolve the complicated expressions involving the $(1 - \mathcal{P})$'s through exact
calculations. The simple result is

$$\frac{dx(t)}{dt} = -\alpha x(t) - \int_0^t ds \beta^2 e^{-\alpha(t-s)} x(s). \tag{2.15}$$

A differentiation of Eq. (2.15) leads to the interesting evolution of the real part
of the complex number z:

$$\frac{d^2 x(t)}{dt^2} + 2\alpha \frac{dx(t)}{dt} + (\alpha^2 + \beta^2)x(t) = 0 \tag{2.16}$$

Is the result correct? It is simple to check. If we had not bothered to use the
Zwanzig procedure, we could have simply separated Eq. (2.13) into its real and
imaginary parts, expressed y and its derivatives in terms of x, and arrived at the
final result (2.16). That the real part of $z(t)$ is identical to the displacement of a

damped harmonic oscillator of frequency $\sqrt{\alpha^2 + \beta^2}$ and damping coefficient 2α is unimportant to this demonstration.

What have we learnt through this exercise? We have seen how to seek exact rules of simplification of terms involving projection operators and use them wherever possible. In realistic systems this will be possible in part but typically we will then have to take the help of weak-coupling approximations as done by Zwanzig (shown in the previous Section) in the derivation of the Master equation from the Liouville-von Neumann starting point.

2.3.2   Projection Operators for Quantum Control of Dynamic Localization

Dynamic localization is a fascinating phenomenon discovered by Dunlap during the analysis of the motion of a quantum particle such as an electron moving along a 1-dimensional chain of sites via nearest-neighbor matrix elements, and simultaneously subjected to a time-sinusoidal electric field (Dunlap and Kenkre 1986). It is related to, but is by no means the same as, the Stark ladder effect in solids, and involves the periodic occurrence of localization as one varies the frequency of the electric field. It has attracted a great deal of attention in the literature including clever experimentation by Raizen and collaborators (Madison et al. 1998) who have verified the effect in an optical lattice using cold sodium atoms. The phenomenon has been analyzed in various ways (Dunlap and Kenkre 1988a,b) and the present author has used projection techniques in one attempt (Kenkre 2000) to clear a specific confusion of an incorrect identification of the effect with another that appeared in the literature. Here in this section, our interest lies neither in the effect itself nor in that clarification of the confusion but only in a pedagogical examination of the use made of projection operators.

Consider a quantum particle that shuttles back and forth between just two states $|1\rangle$ and $|2\rangle$ via interstate matrix elements V and as the result of an applied field causes a time-dependence described by $\pm\mathcal{E}(t)/2$ to appear in the site energies. The evolution is governed by the standard Liouville-von Neumann equation (2.1) for this simple system, the Hamiltonian being

$$H = \begin{pmatrix} \mathcal{E}(t)/2 & V \\ V & -\mathcal{E}(t)/2 \end{pmatrix}. \tag{2.17}$$

To understand our projections better, we will rewrite the evolution equation as one for the so-called Bloch vector of three (real) quantities made from the four elements of the density matrix: $p = \rho_{11} - \rho_{22}$, $q = i(\rho_{12} - \rho_{21})$, and $r = \rho_{12} + \rho_{21}$:

$$\frac{d}{dt}\begin{pmatrix} p \\ q \\ r \end{pmatrix} + \begin{pmatrix} 0 & -2V & 0 \\ 2V & 0 & -\mathcal{E}(t) \\ 0 & \mathcal{E}(t) & 0 \end{pmatrix}\begin{pmatrix} p \\ q \\ r \end{pmatrix} = 0. \tag{2.18}$$

We have put $\hbar = 1$ as earlier and we will henceforth call the 9-element square matrix appearing in Eq. (2.18) as Λ. Our goal happens to be to derive a closed equation for the single element $p(t)$ of the Bloch vector without reference to the others. Therefore, we will apply projections to Eq. (2.18) and design the operator to extract simply the *first* element of the Bloch vector, ignoring the other two. The motivation behind displaying this case is simply to increase the reader's familiarity with the varied use of projections.

Linearity, time-independence, and idempotency of the projection operator $\mathcal{P}$ defined thus as

$$\mathcal{P} \begin{pmatrix} \eta_1 \\ \eta_2 \\ \eta_3 \end{pmatrix} = \begin{pmatrix} \eta_1 \\ 0 \\ 0 \end{pmatrix} \tag{2.19}$$

for any η, with the respective elements as shown, is obvious. We have, as an immediate consequence of the Zwanzig procedure for an initial condition that q and r initially vanish, $q(0) = 0 = r(0)$,

$$\frac{d}{dt} \begin{pmatrix} p(t) \\ 0 \\ 0 \end{pmatrix} + \mathcal{P}\Lambda(t) \begin{pmatrix} p(t) \\ 0 \\ 0 \end{pmatrix} = \mathcal{P}\Lambda(t) \int_0^t ds\, e^{-\int_s^t dt'(1-\mathcal{P})\Lambda(t')}(1-\mathcal{P})\Lambda(s) \begin{pmatrix} p(s) \\ 0 \\ 0 \end{pmatrix}. \tag{2.20}$$

Because the Hamiltonian is time-dependent through the time dependence of $\mathcal{E}$, we have had to generalize earlier expressions to include a non-difference kernel (which is thus a function of s and t but not simply of $t - s$). Also, needless to say, time-ordering is implicit in the expressions displayed.

As in the example of the decay of the complex number treated in the previous section, we must now seek to obtain exact results, if we can, to simplify the projection operator expressions in the exponential of the integrand. If we succeed in such exact simplification, we will have actually gained from the use of the technique. If not, we will have to resort to some kind of weak-coupling approximation.

A bit of inspection shows, that for any column vector η,

$$\Lambda(t) \begin{pmatrix} \eta_1 \\ \eta_2 \\ \eta_3 \end{pmatrix} = \begin{pmatrix} -2V\eta_2 \\ 2V\eta_1 - \mathcal{E}(t)\eta_3 \\ \mathcal{E}(t)\eta_2 \end{pmatrix}. \tag{2.21}$$

This has three clear consequences. First, the "proper" term of Eq. (2.20), that happens to be the term immediately preceding the equality sign, vanishes identically. Second,

$$(1 - \mathcal{P})\Lambda(s) \begin{pmatrix} p(s) \\ 0 \\ 0 \end{pmatrix} = 2V \begin{pmatrix} 0 \\ p(s) \\ 0 \end{pmatrix}, \tag{2.22}$$

which means that the action of $(1 - \mathcal{P})\Lambda$ on any column with only the first element non-vanishing is to switch that element to the second place, place a zero in the first and the third elements, and multiply by $2V$. And third, when $(1 - \mathcal{P})\Lambda(t')$ is applied to any column vector of the form

$$\begin{pmatrix} 0 \\ \eta_2 \\ \eta_3 \end{pmatrix},$$

i.e., with its first element vanishing, gives

$$\begin{pmatrix} 0 \\ \eta_3 \\ -\eta_2 \end{pmatrix},$$

when applied once. More importantly, when applied twice in succession, it yields a c-number times

$$\begin{pmatrix} 0 \\ \eta_2 \\ \eta_3 \end{pmatrix},$$

thus returning to the original column vector! Such a structure in the action of the operators is what we were looking for. As a result, the exponential operator in the projection operator expressions in (2.20) produces the sum of a cosine series and a sine series, thereby allowing an exact *analytic* evaluation.

We have thus shown that the diagonal element difference $p = \rho_{11} - \rho_{22}$ exactly satisfies the integrodifferential equation

$$\frac{dp(t)}{dt} + 2 \int_0^t ds\, \mathcal{W}(t, s) p(s) = 0. \tag{2.23}$$

The memory $\mathcal{W}(t, s)$ is not a function of the difference $t - s$: a consequence of the time dependence of $\mathcal{E}(t)$ in the Hamiltonian. It is given explicitly by

$$\mathcal{W}(t, s) = 2V^2 \left[\phi_c(t)\phi_c(s) + \phi_s(t)\phi_s(s) \right]. \tag{2.24}$$

where the functions $\phi_c(t) = \cos \left[\int_0^t ds\, \mathcal{E}(s) \right]$ and $\phi_s = \sin \left[\int_0^t ds\, \mathcal{E}(s) \right]$ are obtained from the time variation of $\mathcal{E}(t)$ in the Hamiltonian.

Our pedagogical examples, one showing the application of projections to the evolution of a complex number and the other to the extraction of one element of a Bloch vector, are complete. Although these examples are rather simple, we hope they are useful in learning the craft of projection techniques. Careful inspection of how the structure of the problem is exploited (differently in the two cases) to gain analytic insights into the simplification of projection expressions will repay the reader interested in applying the technique to new problems.

2.3.3   Projections for Interactions with a Reservoir

In the last chapter we have been introduced to the Wigner-Weisskopf idea of the treatment of reservoir interactions (Wannier 1959) by adding decay terms for the off-diagonal elements of the density matrix as in Eqs. (1.19). We have also calculated there, exactly, through an interesting indirect method, the memory functions that arise in a pure (i.e., without interactions with any reservoir) and translationally invariant chain, as in Eq. (1.27) or (1.28). Let us attempt to combine the two here and calculate the memory functions that arise for the motion of a quantum particle moving among the sites m of a chain in the presence of interactions with a reservoir. We start with Eq. (1.20) expressed in terms of the evolution of the elements ρ_{mn} of the density matrix and add a decay term proportional to α to the equation of each off-diagonal element. Keeping in mind that from this chapter onwards we do not display $\hbar$ explicitly for notational simplicity, we have the governing equation which sometimes goes under the name of the simple SLE (stochastic Liouville equation):

$$\frac{d\rho_{mn}(t)}{dt} = -iV[\rho_{m+1\,n}(t)+\rho_{m-1\,n}(t)-\rho_{m\,n+1}(t)-\rho_{m\,n-1}(t)]-\alpha(1-\delta_{m\,n})\rho_{m\,n}.$$
(2.25)

Let us express Eq. (2.25) as a von Neumann equation of a more general form

$$\frac{d\rho(t)}{dt} = -iL\rho(t) = -i(L_c + L_i)\rho(t),$$
(2.26)

the Liouville operator being split into two additive parts, L_c and L_i. The two parts stem, respectively, from intra-crystal interactions and interactions with the reservoir. Here, the suffix c represents 'coherent' interactions and i represents 'incoherent' (reservoir) interactions. Specifically, L_c acting on any operator means taking the commutator of the crystal Hamiltonian H in the absence of the reservoir with that operator, in other words, $L_c O = [H, O] = [H_0 + V, O]$, and

$$\langle m|L_i O|n\rangle = -(1 - \delta_{m\,n})\alpha O_{m\,n}.$$
(2.27)

In the projected equation (2.7) resulting from the Zwanzig formalism, we now have the expression

$$e^{-it\,(1-\mathcal{P})L} = 1 + (-it)\left(L_c'' + L_i''\right) +$$

$$\frac{(-it)^2}{2!}\left[(L_c'')^2 + (L_i'')^2 + L_c''L_i'' + L_i''L_c''\right] + \dots$$
(2.28)

where $\mathcal{P}$ is the standard diagonalizing operator (that diagonalizes the operator it acts on in the representation of m,n, etc.), and the notation $O'' = (1 - \mathcal{P})O$ has been used for any operator O.

Equations (2.26) and (2.27)) show explicitly that L_i'', when acting on an off-diagonal operator, merely multiplies it by $-i\alpha$. Therefore, Eq. (2.28) yields, for any off-diagonal operator O'',

$$e^{-it(1-\mathcal{P})L}O'' = \left[1 + (-it)\left(L_c'' - i\alpha\right) + \frac{(-it)^2}{2!}\left(L_c'' - i\alpha\right)^2 + \ldots\right]O''.$$
(2.29)

This crucial result shows that L_c'' and L_i'' commute in the expansion of the left-hand side of Eq. (2.28). It therefore takes the simple and useful form

$$e^{-it(1-\mathcal{P})L}O'' = e^{-\alpha t}e^{-it(1-\mathcal{P})L_c}O''.$$
(2.30)

It is clear from Eq. (2.27) that L_i contains within it the action of the off-diagonalizing operator $(1 - \mathcal{P})$ and consequently makes no contribution to the first L and the last L in the kernel of the memory term of (2.7) for this system. The consequence is the elegantly simple relation

$$\mathcal{W}_{mn}(t) = e^{-\alpha t}\mathcal{W}_{mn}^c(t).$$
(2.31)

The relation is a statement of the fact that the memory in the presence of interactions with the reservoir equals simply the product of the exponential factor $e^{-\alpha t}$ and the memory $\mathcal{W}_{mn}^c(t)$ in the absence of the reservoir, provided the reservoir interactions can be completely described by the simple procedure of adding a decay to the off-diagonal elements of the density matrix!

This simple yet powerful result (Kenkre 1978a) which is thus obtained by identifying and exploiting *exact* properties of projection operators for this system, is true not only for the infinite chain but for any case in which a quantum particle on a pure lattice, infinite or finite, or in any number of dimensions, including the simple dimer displayed in Eq. (1.19), interacts with a reservoir in the simple Wigner-Weisskopf manner. It was been explicitly verified via an independent calculation (Reineker and Kühne 1980) involving a term-by-term evaluation of terms in an expansion of the exponential operator in the Zwanzig kernel.

In light of the coherent linear chain result of Eq. (1.27), the prescription (2.31) provides us with the exact memory function for an infinite linear chain with nearest neighbor interactions for *arbitrary* degree of coherence as

$$\mathcal{W}_{mn}(t) = 2(V/\hbar)^2 e^{-\alpha t}[J_{m-n+1}^2 + J_{m-n-1}^2 + 2J_{m-n+1}J_{m-n-1}$$

$$- 2J_{m-n}^2 - J_{m-n}J_{m-n+2} - J_{m-n}J_{m-n-2}].$$
(2.32)

Solutions of the propagators emerging from this memory are obviously of high relevance in our subsequent analysis. They will be derived in Chap. 4 and used in subsequent discussions of experiments and observations.

2.3.4 Exercise for the Reader: The Open Trimer

Consider a quantum particle moving on the three degenerate sites 1, 2, and 3, of an open trimer with nearest neighbor matrix elements V such that the particle moves between sites 1 and 2, and 2 and 3, but not directly between 1 and 3 (hence the terminology *open*.) Therefore, the amplitudes $c_{1,2,3}$ obey (we put $\hbar = 1$ as always)

$$i\frac{dc_1}{dt} = Vc_2, \tag{2.33a}$$

$$i\frac{dc_2}{dt} = V(c_1 + c_3), \tag{2.33b}$$

$$i\frac{dc_3}{dt} = Vc_2. \tag{2.33c}$$

⊛ The assigned exercise is to proceed via projection operators using equations such as Eq. (2.7) and the particular structure of the Hamiltonian, and to obtain results which allow the derivation of the GME obeyed by the probabilities $P_m = |c_m|^2$ for $m = 1, 2, 3$, with explicit evaluation of the memory functions. You should show that, in addition to the expected, spatially local, memories

$$\mathcal{W}_{12}(t) = \mathcal{W}_{21}(t) = 2V^2 \cos\left(\sqrt{2}Vt\right),$$

between sites connected by the interaction matrix element V of the Hamiltonian, a memory function also develops between sites 1 and 3 unconnected by V, and that it is given by

$$\mathcal{W}_{13}(t) = \mathcal{W}_{31}(t) = 2V^2 \sin^2\left(\sqrt{2}Vt\right).$$

Note with care that the local (former) memory starts at $t = 0$ finite and oscillates as a cosine similarly to many other microsocopic memories you might have encountered; but that the non-local memory rises from a zero value initially. We will have occasion to comment on the significance of these features later on in the book.

Your proof should follow the establishing of rules of what certain combinations of the projection operators and the Liouville operator do in the manner illustrated in the two examples explained above. You should also independently check the result: directly, without using the projection operators. If, after attempting to do this exercise, you want a key to the calculations, you will find the projection operator derivation in an appendix in (Kenkre 1978d) and the indirect derivation, i.e., the check, in (Kenkre 1977c). As explained in detail in the former of these references, as well as in Chap. 4 of the present book, these are important findings of relevance to spatially long-range memories and transfer rates arising for strong coupling from short-range interactions.

2.4 What Is Missing from the Projection Derivation of the Master Equation

Let us now return to the Zwanzig derivation of the memory in the context of the passage from the Liouville equation to the Master equation and the resolution of the question of how macroscopic irreversibility emerges from the underlying microscopic reversibility.

The fact that the Markoffian approximation on the cosine memory produces the Fermi Golden Rule, which is a well-known and much used fixture of physics in varied contexts, is certainly satisfactory. However, one may voice one point of dissatisfaction. The derivation has shown that the memory function is a cosine. How can a single cosine decay in time? Unless it decays, how can we expect to be able to make a Markoffian approximation on it? Does it make sense to replace such a trigonometric function which oscillates forever by a δ-function in time even if such a procedure comfortably hands over to us, on a platter, a Fermi Golden Rule expression? What we really need is surely a *sum* of cosines, that, perhaps in a Fourier transform-like spirit, could indeed decay. Where is such a sum to be found in the derivation that we have provided? We will take this up in the next chapter.

2.5 Chapter 2 in Summary

Chapter 1 showed that looking at *part* of a system naturally gave rise to memories. This happens also in the quite different context of the study of the central problem of statistical mechanics, viz. understanding how reversibility in the microscopic equations of motion gives rise to irreversibility in the macroscopic equations of motion. Memories arise as an intermediate feature in this voyage from the microscopic to the macroscopic levels of description. Projection operators are particularly designed to perform this passage. After learning the basic concept from the Zwanzig technique of looking at only the diagonal part of a density matrix, the chapter analyzed four examples constituting a mix for simple and involved systems, the final one of which was assigned as an exercise, with solution provided in the literature: a time-dependent complex quantity, the Bloch vector appearing in the control of dynamic localization, the modification of a memory in a lattice due to reservoir interactions, and the open quantum trimer. Attempting the associated exercises should familiarize the reader with analytic calculational details of the projection technique.

Building Coarse-Graining into Projections and Generalizing Energy Transfer Theory

3

We have seen in the last chapter that the Zwanzig procedure of projection operators succeeds in a powerful and efficient manner at arriving first at the generalized master equation (GME) and then, when the Markoffian approximation is made, at the Fermi Golden Rule for transition rates. This would be a fine way of understanding the passage from microscopic reversibility to macroscopic irreversibility and approach to equilibrium had it not been for the fact that it involves making the Markoffian approximation on *single cosines*. Since a single cosine never decays, this procedure seems flawed. How shall we solve this problem?

An answer that we provide to this question, and thereby complete the Zwanzig procedure by augmenting its projection operator, is the first order of the day in the present chapter. With the answer in hand, we will then generalize the theory of excitation transfer provided by Förster and by Dexter to cover short time and coherent aspects of energy transfer and Frenkel exciton transport.

3.1 The Need to Coarse-Grain

The general expectation, or belief, in statistical mechanics is that the primary factor that brings in irreversibility from the reversible equations of evolution at the micro scopic level is the enormously large number of degrees of freedom in macroscopic systems. Although it is certainly odd to make a Markoffian approximation on a single cosine, it is clear that, if only we had to handle a sum of *many* cosines, we could have the sum yielding us functions decaying in time, in the precise manner a sum does in Fourier transforms. And then it would be appropriate to make the Markoffian approximation on such a decaying function.

Surely, if our interest is in the evolution of macroscopic observables, we should not even seek evolution equations for the occupational probabilities of the microscopic states ξ and μ that appear, say in Eq. (2.3) of the last chapter. We

V. M. (Nitant) Kenkre, *Memory Functions, Projection Operators, and the Defect Technique*, Lecture Notes in Physics 982, https://doi.org/10.1007/978-3-030-68667-3_3

should, instead, focus attention on *groups* of such states, introducing thereby *coarse-graining* into our description. If we call such a group M, and the fine-grained state $|\xi\rangle$ as identical to $|M, m\rangle$, where m labels the states within a group M, we might attempt summing Eq. (2.3) over m. This would mean that, from Eq. (2.3), we seek an equation for the macroscopic probability $P_M(t) = \sum_m P_\xi = \sum_m P_{M,m}$. It would make the left hand side of Eq. (2.3) $dP_M(t)/dt$. Perhaps this would provide a step towards making our GME macroscopic.

3.2   Constructing the Coarse-Graining Projection Operator

The procedure does not work smoothly. What the m-sum produces in the first term on the right hand side of Eq. (2.3) is $\sum_{N,n} \sum_m \mathcal{W}_{M,m\,N,n}(t-s) P_{N,n}(s)$. If it is justified in the given system to assume that the m-sum $\sum_m \mathcal{W}_{M,m\,N,n}$ produces a quantity that is *independent of* n, we could call that quantity $\mathcal{W}_{MN}$ and transfer the n-sum to the probabilities $P_{N,n}$. That first term on the right hand side of Eq. (2.3) would then give $\sum_N \mathcal{W}_{MN}(t-s) P_N(s)$ as desired. A similar situation would occur in the second term which would become $\sum_N \mathcal{W}_{NM}(t-s) P_M(s)$. But are we guaranteed a priori that for every system we consider, the property of n-independence of the m-sum holds?

To resolve this problem and yet obtain macroscopic evolution equations, let us replace Zwanzig's projection operator by a generalized version that not only diagonalizes but also performs a coarse-graining operation at the same time.

Let us retain the diagonalization property of Zwanzig's projection operator but define $\mathcal{P}$ as having the additional property that it coarse-grains by summing the diagonal elements within groups, and take the opportunity provided to impose our own desired dependence of the diagonal elements within the groups. It is easiest to explain the detail for the case when the microscopic states form a product of a space of groups M, N, with a space of states within the group m, n.[1] Then,

$$\langle M, m|\mathcal{P}O|N, n\rangle = \delta_{m,n}\delta_{M,N}\left[\sum_{m'}\langle M, m'|O|M, m'\rangle\right]\left(\frac{q_m}{\sum_{m'} q_{m'}}\right). \qquad (3.1)$$

Notice what the three factors are designed to do in the right hand side of Eq. (3.1). The Kronecker δ symbols ensure diagonalization, and the contents of the square brackets represent the summation (coarse-graining) within the group M. Because the coarse-graining cleans out the m-dependence of the diagonal elements within

---

[1]When they do not, other expressions may be written such as the one given in Kenkre and Knox (1974a), Kenkre (1975a), and Kenkre (1977a). One of them, for instance, is $\langle\xi|\mathcal{P}O|\mu\rangle = Q_\xi \delta_{\xi,\mu}\left[\sum_{\xi\epsilon M}\langle\xi|O|\xi\rangle\right]\left[\sum_{\xi\epsilon M} 1\right]^{-1}$, with the requirement that Q_ξ summed within the grain or group M must equal the number of states g_M within the grain.

a group, we have an opportunity to impose our own substitute dependence q_m.[2]
The choice we make would not affect the coarse-grained probabilities in a truly
exact calculation but, particularly when we are forced to make a weak-coupling
perturbation later on in the development, our choice, while otherwise arbitrary, has
the power to help (or complicate) the calculations practically. The idempotency
requirement on the projection operator ($\mathcal{P}^2 = \mathcal{P}$) imposes the division by $\sum_{m'} q_{m'}$.

This freedom of designing the third factor in (3.1) will appear in almost every
application of projections, as we will see in a later chapter. Typically, we might use
q_m as equal to unity or a thermal factor within the m-space such as $e^{-\beta \epsilon_m}$ where $\beta = 1/k_B T$ with ϵ, k_B and T as, respectively, the within-group energy, the Boltzmann
constant, and the temperature.

With our generalized definition of the projection operator, the evaluation method
given to us by Zwanzig for the $\mathcal{P}$ that only diagonalizes, explained in Chap. 2 of
this book in detail, proceeds unscathed even when we make the weak-coupling
approximation. We obtain a GME in the space of *groups (of states)* M, N:

$$\frac{dP_M(t)}{dt} = \int_0^t ds \sum_N [\mathcal{W}_{MN}(t-s)P_N(s) - \mathcal{W}_{NM}(t-s)P_M(s)]. \tag{3.2}$$

The memories for the transitions from the group N to the group M are given by

$$\mathcal{W}_{MN}(t) = 2 \sum_{m,n} \frac{e^{-\beta \epsilon_n}}{Z} |\langle M, m|V|N, n\rangle|^2 \cos[(\epsilon_m - \epsilon_n)t + (E_M - E_N)t], \tag{3.3a}$$

$$\mathcal{W}_{NM}(t) = 2 \sum_{m,n} \frac{e^{-\beta \epsilon_m}}{Z} |\langle M, m|V|N, n\rangle|^2 \cos[(\epsilon_m - \epsilon_n)t + (E_M - E_N)t], \tag{3.3b}$$

the *only* difference between these two equations being in the label of the microen-
ergy in the exponent: it is n in one case but m in the other. That difference produces
a detailed balance relationship between the integrals of the memories, respectively
F_{MN} and F_{NM}, if the group energies E_M and E_N are not identical to each other:

$$F_{MN}(t) = 2\pi \sum_{m,n} \frac{e^{-\beta \epsilon_n}}{Z} |\langle M, m|V|N, n\rangle|^2 \delta[(\epsilon_m - \epsilon_n) + (E_M - E_N)], \tag{3.4a}$$

---

[2]Not to forget here is that the projection maps O into another quantity of the nature of O defined
on the same space as the latter.

$$F_{N\,M}(t) = 2\pi \sum_{m,n} \frac{e^{-\beta\epsilon_m}}{Z} |\langle M, m|V|N, n\rangle|^2 \delta[(\epsilon_m - \epsilon_n) + (E_M - E_N)].$$

(3.4b)

The partition function of the microstates, $\sum_n e^{-\beta\epsilon_n}$, is here denoted by Z. As written above, Eqs. (3.3) and (3.4) correspond to the textbook statement that one performs a thermal average for initial states and a sum over the final states.

Our understanding of the memories obtained is enhanced by recasting the second of the Eqs. (3.3) in the form[3]

$$\mathcal{W}_{N\,M}(t) = 2 \sum_{m,n} \frac{e^{-\beta\epsilon_n}}{Z} |\langle M, m|V|N, n\rangle|^2 \cos[(\epsilon_m - \epsilon_n)t - (E_M - E_N)t].$$

Notice the change from $\beta\epsilon_m$ to $\beta\epsilon_n$ in the thermal factor and the consequent change of sign in combining the ϵ-difference with the E-difference.

The reader might want to consult Kenkre and Chase (2017) for a recent careful explanation of how these manipulations produce detailed balance in expressions for the relevant transition rates in non-degenerate systems. Application to an explicit realization of the microstates as the states of a boson bath will be found both in a recent paper (Ierides and Kenkre 2018) and in early derivations (Kenkre and Rahman 1974; Kenkre 1975c).

In all these cases, the memory function has the over-all appearance of a *Fourier transform*. Concentrating on the essentials and absorbing the weight factors within a function $Y(\omega)$, we see that the memory is of the form

$$\mathcal{W}(t) = \int d\omega \, \cos \omega t \, Y(\omega)$$

with the ω-integration being essentially the ξ- or μ- summation within grains. The quantity $Y(\omega)$ has, within it, the product of the matrix element squared of the interaction V, and the density of states in the grain of the states over which the summation is. An oscillatory $\mathcal{W}(t)$ would arise from a singular $Y(\omega)$, and a decaying $\mathcal{W}(t)$ typically from a non-singular $Y(\omega)$. How the GME could become a Master equation, with its irreversibility and approach to equilibrium, should now become eminently clear. The secret lies in how the coarse-graining changes the character of $Y(\omega)$. A small system with well-spaced discrete states has a density of states that is a clearly recognizable sum of δ-functions (in energy) and which is thus quite singular. Since, as macroscopic observers, we are not of the size of

---

[3]This rewriting requires the assumption that the matrix element of V remains unchanged on switching m and n, which is, of course, always valid if the state space is a product of M-space and m-space as mentioned earlier.

electrons or atoms,[4] we usually cannot avoid taking the thermodynamic limit that was suggested by van Hove in his derivations. The limit brings those states closer together, the corresponding Poincaré recurrence periods are pushed to infinity, our perception is of a continuum of states and, provided other factors are not singular, $Y(\omega)$ becomes non-singular, resulting in approach to equilibrium. Non-dissipative systems will be represented by either a squared V matrix element that continues to have a singularity even in the limit or by states that, in the limit, bunch into one or more localized groups (at locations in energy space) leading to a singular $Y(\omega)$. Such situations, exemplified by systems exhibiting, e.g., zero-phonon lines (Fitchen 1968), could result in persistent correlations or oscillations that survive even in the thermodynamic limit. A relevant discussion awaits you in Chap. 4.

As mentioned elsewhere, all this is based on general ideas developed by many workers within quantum statistical mechanics from the days of van Hove. What is perhaps new in our own construction is the idea of looking at $\mathcal{W}(t)$ as a Fourier transform and of connecting thereby the dissipative properties of the memory to system properties lying within the interaction V and such factors as the density of states.

The early elucidation of how our generalization of the Zwanzig projections to include coarse-graining makes the onset of irreversibility much more palatable (than when the Markoffian approximation is thrust into a lonely cosine) may be found in the Appendix[5] to Kenkre (1977a). The motivation for that analysis came in part from a pressing need for the generalization (Kenkre and Knox 1974a) of the Förster theory of excitation transfer (Förster 1948) that was used widely in photosynthesis and similar contexts. An explanation of the issues involved, the problems and their resolution, will form the main part of what remains in the present chapter. But first, in the next section, let us inspect an implementation of the technique at various *different* levels of coarse-graining in the same system (a ferromagnet in interaction with a bath) that was provided in Kenkre (1975a).

I have no doubt that other investigators must have reported similar coarse-graining implementations in their studies independently, often without knowledge of each other's work. I have uncovered two such instances, (Emch 1964, 1965) that seems to me to be the first, and also to have emphasized the importance of time-smoothing in addition to coarse-graining, and (Sewell 1967) who did refer to the earlier work of Emch. I have chosen to give details of my own procedure here because, while perhaps not particularly sophisticated in form, they are quite simple to understand and implement.

---

[4]Unless we can seriously emulate Antman during one of his rare trips into the microscopic realm.

[5]You might find this old article fun to read. The Appendix is titled "Poor Man's Version of the Explanation of the Origin of Irreversibility". It is presented along with the self-shielding statement "Here, within the relative safety of an appendix, I shall venture to present, *very* briefly, ... a private picture."

3.3   Implementing a General Coarse-Graining Plan at Various Levels of Description

The approach to equilibrium of a simple model of a ferromagnet interacting with a reservoir of bosons was studied by Goldstein and Scully (1973) via standard methods. How the Zwanzig projection technique could be used in that study was shown by Wang (1973) who rederived the results of the earlier study but did not exploit any independent features of the GME. Those two papers provide an excellent illustrative study to which to apply (Kenkre 1975a) the coarse-graining generalization of the Zwanzig operator explained in the preceding section. In taking a brief look at that application in the following, we will have an opportunity to examine how coarse-graining can be applied at *various* different levels within the same given system.

Consider a ferromagnet made up of N spins treated under the mean field approximation and described in customary notation by spin operators $\vec{S} = \sum_{i=1}^{N} \vec{s}_i$, with $\vec{S}$ and $\vec{s}_i$ having respective Cartesian components labelled by superscripts x, y, and z. Let us abbreviate as customary, the expressions $S^x \pm iS^y$ as $S^\pm$. With μ as the magnetic moment of each spin, B as the applied magnetic field, and J as the interaction parameter, the ferromagnet Hamiltonian is

$$H_f = -2JS^z S^z + 2\mu B S^z.$$

A bath of bosons each with annihilation operator b_α and energy ϵ_α has the Hamiltonian

$$H_b = \sum_\alpha \epsilon_\alpha b_\alpha^\dagger b_\alpha.$$

The system-bath interaction with coupling constants λ_q is given by

$$V = \sum_\alpha \lambda_\alpha (S^+ b_\alpha + S^- b\dagger_\alpha + S^+ b\dagger_\alpha + S^- b_\alpha)$$

which we will write in terms of obvious symbols as

$$\sum_\alpha \lambda_\alpha (V_\alpha^{+-} + V_\alpha^{-+} + V_\alpha^{++} + V_\alpha^{--}).$$

Let us use b here for the bosons instead of a used in the three publications mentioned to maintain consistency with later usage in the current book.[6]

The total Hamiltonian is

---

[6]Not to mention the fact that the word 'boson' begins with the letter 'b'.

$$H = H_f + H_b + V. \tag{3.5}$$

This model, studied in Kenkre (1975a), incorporates a trivial generalization of the one studied in Goldstein and Scully (1973) and in Wang (1973) in that, in the latter two papers, the last two terms in each of the parentheses pair were omitted. Their neglect is not necessary and appears to have been made by those authors in those studies through force of habit in optics investigations where, for analytic tractability, the so-called counterrotating terms are often dropped.

For all details, the reader should consult the original publication (Kenkre 1975a). All that is important to emphasize here is that, with the help of the coarse-grained description, it is possible to study the system with various GME's at various *different* levels of description. First, it is simple to derive, in the manner of Zwanzig, a GME at the very microscopic level where the evolving probabilities are $P_{\{s\},\{n\}}(t)$ that the ferromagnet is in a state with all the constituent spins and the bath has n bosons differentiated by their energies ϵ_α. Use is made here of the projection operator $\mathcal{P}_1$ defined as

$$\langle\{s\}, \{n\}|\mathcal{P}_1 O|\{s'\}, \{n'\}\rangle = \delta_{\{s\},\{s'\}}\delta_{\{n\},\{n'\}}\langle\{s\}, \{n\}|O|\{s\}, \{n\}\rangle,$$

which is devoid of any coarse-graining; the analysis is essentially identical to that by Wang (1973).

But observe that one can also derive GME's with a more macroscopic flavor, directly, by using a different projection operator $\mathcal{P}_2$ which eliminates the bath variables but imposes their thermal distribution for calculations. It is given by an expression such as in Eq. (3.1) with q's chosen to be thermal factors involving the boson energies. The resulting evolution equation (GME) is for the probability of each of the constituent spins having one of the possible eigenvalues without reference to the bath.

Another coarse-graining description can be directly given if what is of interest is $P_m(t)$, the probability that the ferromagnet has *total* magnetization m without any reference to the state of individual spins or of the bath state. One uses a $\mathcal{P}_3$ that goes yet further in the coarse-graining operation by not only summing over bath states but also over individual spin configurations compatible with a given magnetization. If you feel ambitious, you might want to work out the form and the consequences of $\mathcal{P}_3$ explicitly and check your results against those given in the publication mentioned.

The probability of m is connected in this GME to probabilities differing by one unit as a result of the given interaction and the memory functions involve the product of two bath functions with trigonometric functions of the product of the ferromagnet energy differences and time. The memory functions contain bath contributions that are given by

$$\eta_c(t) = 2\sum_\alpha \lambda_\alpha^2[2\bar{n}(\epsilon_\alpha) + 1]\cos\epsilon_\alpha t, \qquad \eta_s(t) = 2\sum_\alpha \lambda_\alpha^2 \sin\epsilon_\alpha t,$$

in terms of the thermal Bose factors $\bar{n}(\epsilon_\alpha) = (e^{\beta \epsilon_\alpha} - 1)^{-1}$. You might discern a similarity of these bath functions with an ingredient of one of the last of the memory functions in the preceding chapter, viz., Eq. (1.36). This is expected since the reservoir consists of a set of bosons in either case. Clarification of how the interaction strengths λ will change drastically the character of the bath functions will be found in a subsequent analysis published by Ierides and Kenkre (2018) that will be explained in Chap. 7 of this book.

The analysis in the original publication (Kenkre 1975a) also includes further discussions on coarse-graining, derivation of moment equations, application to a two-spin system and use of the memory feature of the GME obtained, as also cross-connections with the analysis of Goldstein and Scully (1973) and Wang (1973). While interesting in their own right, they are not germane to our present discussion. Detailed expressions have not been displayed here to avoid distraction since the results are not used in this book except in concept. The take-home message is that, whatever one's need for a coarse-graining description, there is a projection operator for that.[7] It should be particularly appreciated, however, that the procedure of building coarse-graining into projections can be used, in the manner shown, not only to eliminate reservoirs from the description but also to remove from consideration the detail of finer structures belonging to the given system. Such a procedure assists one to focus attention *directly* on more macroscopic features.

3.4   Challenges in Energy Transfer Theory in the 1960s

It is exciting to work in an area of science when there are points of disagreement, divergence of interpretation, even some healthy confusion in the field. Such was the state of affairs in the field of energy transfer in the late 1960s and early 1970s. The physical system under debate was a molecular aggregate that might be a crystal and therefore possess translational invariance but might be also a two-state system (a dimer), an n-mer, a solution, a mixed crystal, or even a spatially complex structure such as a photosynthetic unit. The entity whose motion was under question was a Frenkel exciton. Details of how this quasiparticle generally behaves was known from familiar reviews (Knox 1963; Dexter and Knox 1965; Davydov 1971). Suffice it to note that, if the m-th molecule in the aggregate is electronically excited while the other molecules are unexcited, one can assert that a Frenkel exciton occupies the m-th site. Frenkel exciton dynamics, or energy transfer through its motion, is the field of study of how this exciton moves from one location to another. The field has always been of great relevance to the study of sensitized luminescence, to the physics of molecular, particularly organic, crystals such as of the aromatic

---

[7]Quite what Steve Jobs might have said were he to have had occasion to focus on statistical physics rather than mac apps.

hydrocarbons, and to processes in the biological realm as for instance in the context of photosynthesis as reviewed by Knox (1975).[8]

There was a large established community of researchers who felt that experiment and theory in this field were in fine shape. The transfer of excitation was considered established to arise from interactions of the dipole-dipole type for singlet excitons (relatively short-lived quasiparticles that gave off blue light by radiation) and of the exchange kind (much longer-lived ones that gave off red light when they emitted radiation). The theory for the motion was due to Förster for singlet excitons (Förster 1948) and due to Dexter for triplet excitons (Dexter 1953). Both theories were based on calculating Fermi Golden rule rates for the transitions. Particularly useful as part of the Förster theory was a highly appreciated prescription to obtain these rates directly from the emission spectrum of the donor molecule and the absorption spectrum of the acceptor molecule without the need for detailed model assumptions and calculations.

This perception of satisfaction with the theory, wide-spread in certain circles as it certainly appeared, was by no means universal. A number of researchers cast doubt from time to time on the use of the Fermi Golden Rule in these contexts and raised the question of whether the analysis of Perrin (1932) should provide, instead, the underlying mode of transfer. In other words, would a Master equation be most appropriate for the description or was the finer level Schrödinger equation called for? (Robinson and Frosch 1962, 1963; Davydov 1968; Powell and Soos 1975). Although many debates existed about such issues as the magnitude of the exciton diffusion constant, whose estimation performed at respectable laboratories around the world differed sometimes by *several* orders of magnitude for the same species in the same systems at the same temperature, perhaps the most important question brandied around was about the *coherence* of the exciton. Is the exciton coherent or incoherent? Does it move as a quantum mechanical wave might, as would an electron in a metal, or is its motion like that of a random walker?[9] Perhaps better put, what is the quantitative extent of the coherence of the exciton?

Investigators did not, at that time and for many more years, appear unified even on what the word 'coherent' does, or should, precisely, mean. Some attached the coherence question to spin dynamics since the excitations did also have magnetic properties; thus, optically detected magnetic resonance phenomena, questions of lineshapes of various kinds, and allied topics also muddled the issue. Discussions about whether the transfer was cold (post-relaxation among the vibrational manifolds) or hot (pre-relaxation or during relaxation) were also often mixed with coherence analysis.

---

[8]There have been reports of observations of coherence in energy transfer in photosynthesis as recently as within the last year. See e.g. observations of coherent behavior in Venus A206 dimers at room temperatures (Kim et al. 2019) and in other systems of relevance to photosynthesis (Cao et al. 2020).

[9]Needless to say, this question of whether waves were appropriate or not had nothing to do with wave-particle duality; that would be a quantum versus classical issue.

The debate was presented sometimes as related to the point of contention about whether excitation transfer rates should be proportional to the reciprocal of the third or sixth power of the intermolecular separation. The source of this R^3/R^6 puzzle was that the dipole-dipole interaction matrix element, established to be responsible for the motion of singlet excitations, was inversely proportional to the cube of the separation. The question was whether the rate of transfer should vary as the matrix element or its square: the former was related to a frequency of oscillation and the latter to the Fermi Golden Rule transition rate.

Among the clearest representatives of the debate was an exchange that appeared in Physica Status Solidi between Davydov (1968) and Dexter et al. (1969). The former was an unmitigated attack on the use of the Fermi Golden Rule and the latter was a cogent reply. While the reply was succinct and successful in clarifying the issue, the authors claimed among other points that an oscillatory excitation transfer was not actual transfer but *only phase* evolution!

3.5 Generalization of the Förster-Dexter Theory of Excitation Transfer

The astute reader will already realize that the development of the GME approach, as explained in the preceding sections, is eminently suited to address these issues that had surfaced in the field of energy transfer. We have seen how the limit of no reservoir, or of fine-grained states, is connected to the opposite limit of the effects of a reservoir in a system with dense groups of states. It should be clear from the previous sections how the incoherent Fermi Golden Rule transitions appear *emergently* as the density of states increases, and how the oscillatory, microscopic, Perrin-like behavior passes into macroscopic irreversible behavior as the system grows in complexity. There should be no doubt in the reader's mind that oscillatory transfer between sharp states is not 'mere phase evolution' as claimed in Dexter et al. (1969) or any less real than transfer between coarse groups obeying the Golden Rule is; only that in *most* realistic situations the latter is more applicable than the former.

Further thought will reveal that the hot/cold issue is an independent problem having to do with the relative time scales of thermalization and transfer.[10] We will see below how the GME theory solves the R^3/R^6 puzzle effortlessly. We have already seen that our Fourier transform prescription for memory functions provides an easy understanding of how underlying microscopic reversible equations lead to macroscopic irreversible behavior. We will see below that the Fourier transform also allows us to generalize the spectral prescription of Förster's theory to obtain memories from observed spectra with little calculation.

Let us begin with this spectral generalization.

---

[10]Indeed, the interested reader will find it addressed in a different manner in the literature (Tehver and Hizhnyakov 1975; Kenkre 1977b).

3.5.1   Extending the Spectral Prescription to Calculate Memories

When appropriate conditions are satisfied, Förster's spectral prescription allows one to do away with model assumptions and calculations whose validity may be doubtful as a result of inevitable approximations that one might be forced to invoke. The prescription replaces them with a recipe which takes as inputs, *observed* absorption and emission spectra of the molecules, and smoothly outputs transition rates for excitation transfer. The origin of the prescription becomes clear when we examine the matrix element of the dipole-dipole transfer interaction between donor and acceptor states. Within it, one can identify the matrix element for absorption of light to create the excitation on the acceptor and that for emission of light from the donor to eject it radiatively.[11] *If the bath can be assumed to be the same* for the spectral process on the one hand and the motion process on the other, the spectral prescription holds.

Förster's prescription for the transfer rate F for a singlet excitation to move from a donor molecule to an acceptor molecule, the two separated by a distance R, is (Förster 1948)

$$F = \frac{C_1}{R^6} \int_{W=0}^{\infty} dW \, \frac{\bar{\epsilon}(W)\bar{A}(W)}{W^4}, \tag{3.6}$$

in general, where $\bar{\epsilon}$ is the extinction coefficient or absorption spectrum, and $\bar{A}$ is the emission spectrum. For like molecules, the expression reduces to

$$F = \frac{C}{R^6} \int_{W=0}^{\infty} dW \, \frac{\bar{\epsilon}(2W_0 - W)\bar{\epsilon}(W)}{(2W_0 - W)W}, \tag{3.7}$$

on invoking Levschin's law of mirror correspondence between absorption and emission spectra. Here, the constants C and C_1 are proportional to each other and contain factors representing the speed of light, the Avogadro's number and the refractive index as well. They may be found in the original papers (Förster 1948; Kenkre and Knox 1974a) and have not been displayed here to eliminate distraction from what is under emphasis.

Comparison of the relation between Eqs. (3.3) and (3.4) should make at once clear that our GME extension of the spectral prescription to derive the memory function from spectra is

$$\mathcal{W}(t) = \frac{C_1}{\pi R^6} \int_{-\infty}^{+\infty} d\left(\frac{\Delta W}{\hbar}\right) \cos\left(\frac{\Delta W}{\hbar}t\right) \int_{W=0}^{\infty} dW \, \frac{\bar{\epsilon}\left(W + \frac{\Delta W}{2}\right)\bar{A}\left(W - \frac{\Delta W}{2}\right)}{\left(W + \frac{\Delta W}{2}\right)\left(W - \frac{\Delta W}{2}\right)^3}, \tag{3.8}$$

---

[11] This does not mean that the transfer process consists of actual emission followed by subsequent absorption, that being a separate physical process that can also occur. The identification we speak of can be done, e.g., in Eq. (9') of Förster (1948) or Eq. (23') of Kenkre and Knox (1974a).

in the general case, with a reduction, in the case of like molecules, to

$$\mathcal{W}(t) = \frac{C}{\pi R^6} \int_{-\infty}^{+\infty} d\left(\frac{\Delta W}{\hbar}\right) \cos\left(\frac{\Delta W}{\hbar}t\right) \int_{W=0}^{\infty} dW \, \frac{\bar{\epsilon}\left(W + \frac{\Delta W}{2}\right) \bar{\epsilon}\left(W - \frac{\Delta W}{2}\right)}{\left(W + \frac{\Delta W}{2}\right)\left(W - \frac{\Delta W}{2}\right)}.$$

(3.9)

As is well-known, Förster's prescription asks that one calculate an overlap integral of the spectra in the manner shown. Each of our recipes, (3.8) and (3.9), is identical to its respective Förster counterpart, except for two changes. The first is the introduction of a multiplicative factor $1/\pi$. The second, an additional integration of the product of a cosine and the overlap integral, redone after a shift of the spectra. Expressed differently, one calculates F as in Förster's prescription but calls it $F(0)$, recalculates it by shifting the two curves $\bar{A}(W)/W^3$ and $\bar{\epsilon}(W)/W$ (sometimes termed 'modified spectra' after the division by the appropriate powers of W) on the frequency axis, the former by $\Delta W/2$ and the latter by $-\Delta W/2$, and redoes the integral, naming it the 'generalized overlap integral' $F(\Delta W)$. The final computational step consists of performing the Fourier transform of this quantity, to yield the memory $\mathcal{W}(t)$. Surely, it is related to the transfer rate as given by Förster, through $\int_0^\infty dt\, \mathcal{W}(t) = F$, as it should be.

There is an alternate, equivalent, manner (Kenkre 1975c) to express our generalization of the spectral prescription. One may assert that, except for proportionality constants, the memory is the real part of the product $I^e(t)I^a(t)^*$ where the two time-dependent functions in the product (that can be identified with what Lax (1952) termed "characteristic functions") are the respective Fourier transforms of the modified spectra, sometimes expressed as $\omega^{-1}\sigma(\omega)$ and $\omega^{-3}I(\omega)$ respectively. Here, σ is called the absorption cross section and I the emission probability at the frequency ω. This result, given by Kenkre (1975c), makes contact with the spectral work of a number of investigators (Soules and Duke 1971; Lax 1952).

As an illustrative example, consider a "modified absorption cross section"

$$I^a(\omega) = C_0 L_{\alpha_0}(\omega) + C_s L_{\alpha_s}(\omega - \omega_s)$$

that possesses both a zero-phonon line and a sideband. If mirror symmetry applies, the "modified emission spectrum" will be given by

$$I^e(\omega) = C_0 L_{\alpha_0}(\omega) + C_s L_{\alpha_s}(\omega + \omega_s).$$

Here, $L_{\alpha_0}(\omega) = \alpha/(\omega^2 + \alpha^2)$ is the Lorentzian with width 2α, the zero of ω being taken at the zero-phonon line. Following our generalization of the spectral prescription, we get the memory to be

$$\mathcal{W}(t) = \text{const}\left[C_0^2 e^{-2\alpha_0 t} + 2C_0 C_s e^{-(\alpha_0 + \alpha_s)t} \cos \omega_s t + C_s^2 e^{-2\alpha_s t} \cos 2\omega_s t \right].$$

We plot the spectra and the memory in Fig. 3.1 as given by these expressions describing the hypothetical system considered. We suppress the oscillations in the

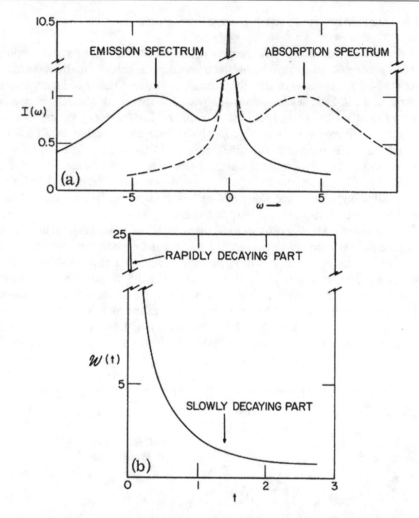

Fig. 3.1 Spectral features reflected in the memory function for transfer in a hypothetical system. Modified absorption (dashed line) and emission (solid line) spectra plotted in the upper panel for exaggerated relative values of parameters (see text) showing zero-phonon peaks and sidebands. The resultant memory is shown in the lower panel by suppressing oscillations arising from the Stokes shift to make clearer the multiple-time-constant nature of the memory. All units are arbitrary. We see that the memory has a rapidly decaying part arising from the sidebands and a longer-living part coming from the sharp zero-phonon peak. Adapted with permission from fig. 1 of Ref. Kenkre (1975c); copyright (1975) by the American Physical Society

memory caused by the Stokes shift ($\omega_s \neq 0$) and note that the slowly decaying component arises from the zero-phonon peak. More instructional insights can be drawn from this example as has been done in Kenkre (1975c) to which we refer the reader for further details. This example will be of use in the next chapter when we compare the GME theory to other approaches to the theory of excitation transfer.

3.5.2 Obtaining Realistic Memory Functions

As a first realistic application of the generalization of the Förster prescription to predict dynamics from spectral inputs, let us outline the calculation by Kenkre and Knox (1974a) for the case of transfer from an anthracene donor to an anthracene acceptor when in cyclohexane solution. Figure 3.2 shows the generalized overlap integral $F(\Delta W)$ with its argument in wavenumbers. The memory function obtained from its cosine transform is in Fig. 3.3. Note how the structure in $F(\Delta W)$ is reflected via the transform in the time behavior of the memory function $\mathcal{W}(t)$. The connection of spectral features and the memory function evolution is seen even more clearly in Fig. 3.4 where three further examples of the GME generalization are shown, following Kenkre and Knox (1976). Insets show the spectra and the main figure plots the corresponding memory functions.

While the above are all room temperature examples of spectra and memories, we show here one instance of low temperature spectra and expected coherent memory, as provided by the F_3^+ center in NaF. It illustrates effects of sharp zero-phonon lines (Stiles Jr and Fitchen 1966). In this example, see Fig. 3.5, the absorption spectrum has been assumed mirror symmetric since only the excitation spectrum is known in the multiphonon region. A persistent component of the memory calculated from the spectra is observed at 0.8 ps which is a relatively long time in comparison to the decay during the first 20–30 fs which is similar to the high-temperature examples.

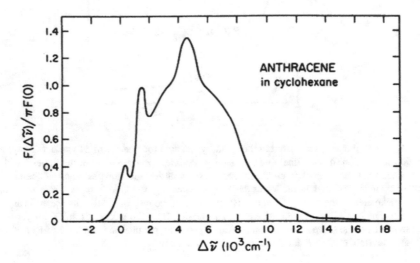

Fig. 3.2 Generalized overlap integral (see text) of the anthracene emission and absorption spectra (modified) in cyclohexane solution. Details of the structure of the shape in this figure are reflected in the cosine Fourier transform of the memory function shown in Fig. 3.3. Reprinted with permission from fig. 1 of Ref. Kenkre and Knox (1974a); copyright (1974) by the American Physical Society

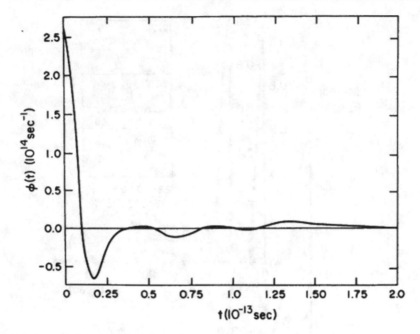

Fig. 3.3 Memory function $\mathcal{W}(t)$ for anthracene in cyclohexane computed as the cosine Fourier transform of the curve shown in Fig. 3.2. Reprinted with permission from fig. 2 of Ref. Kenkre and Knox (1974a); copyright (1974) by the American Physical Society

   The (generalized) spectral prescription that our GME theory inherits from its close connection with the Förster analysis is particularly powerful because it avoids problems associated with most assumptions that have to be made during the construction of models of reservoirs. Whatever Nature shows us through absorption and emission spectra is input into our calculation. However, this can have its dangers in those cases where the observed spectra do not reflect the interactions present during excitation transfer. An unavoidable instance is encountered when the spectra are broadened inhomogeneously. What this means is that, because different locations in the sample have differing environments, the observed structure of the spectra might be a superposition of contributions from distant locations in the sample which are not involved in the transfer process at all! The real spectra may be sharp and representative of coherent transfer. Yet the observed spectra may show broadening which would be then misinterpreted as representative of less coherence. This is a feature of any spectral prescription and not a peculiarity of the GME formalism. However, it does mean that there are situations in which we must supplement or replace the spectral prescription by model calculations for obtaining memories. More details of such model calculations will be presented further on in the book but we refer the reader to Kenkre and Rahman (1974) for the first such analysis in our own work. It dealt with the excitonic polaron model introduced in this field by Grover and Silbey (1971).

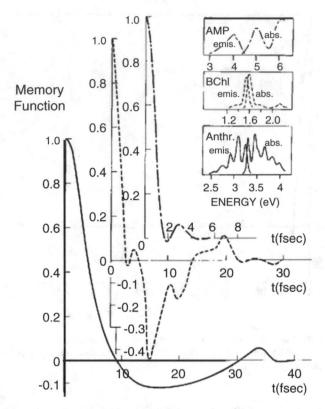

Fig. 3.4 Three Memory functions $\mathcal{W}(t)$ computed from the Kenkre-Knox generalization of the spectral prescription of Förster for three systems: from bottom to top in the main figure, anthracene, bacteriochlorophyll, and adenosine monophosphate (AMP); inputs into the calculation are the respective observed spectra, shown in the inset with arbitrary ordinate units. Ordinate scale for the memories is normalized so that the initial value of the memory, $\mathcal{W}(0)$, equals 1. Reprinted with permission from fig. 1 of Ref. Kenkre and Knox (1976); copyright (1976) by Elsevier Publishing

3.6   Transfer Rates: Resolution of the R^3 Versus R^6 Puzzle

Since the matrix element for the transfer interaction V in the case of singlet excitations is inversely proportional to the cube of the intermolecular distance R for a dipole-dipole source of transfer, the much debated R^3 versus R^6 puzzle can be reworded in terms of the interaction: should the transfer rate be expected to vary as V or V^2? The GME theory can solve the puzzle in the simplest fashion not requiring any particular detail or shape of memory. We show this below by first remarking that a broken plot of the transfer rate had been displayed by Förster in one of his later publications, the extremes in the R^{-3} and R^{-6} regions corresponding to large and small interaction respectively. We shall now show that the extremes as well as a

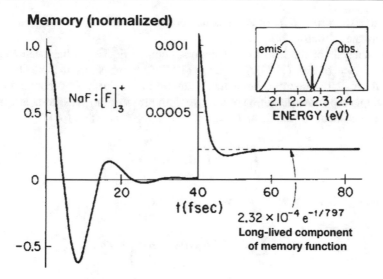

Fig. 3.5 Memory function $W(t)$ with spectral insert to show high coherence effects of zero-phonon lines. Notice the long-lived component (although of very small magnitude) shown in the right of the main figure. See text. Adapted with permission from fig. 2 of Ref. Kenkre and Knox (1976); copyright (1976) by Elsevier Publishing

smooth quantitative connection between them is provided by the GME theory for almost *any* memory function arising from the basic interactions.

Let us start with the assumption that the excitation motion occurs via

$$\frac{dP_m(t)}{dt} = \int_0^t ds\, \phi(t-s) \sum_n [F_{mn} P_n(s) - F_{nm} P_m(s)], \qquad (3.10)$$

in which, for simplicity, we have taken the memories to be given by a single time function multiplying all the transition rates: $W_{mn}(t) = F_{mn}\phi(t)$. We know that R^3 is associated with ringing (oscillations) and R^6 with what has been called 'true' transfer and that it had been suggested in the past that the two could never be compared even in principle. Can we, nevertheless, ignore the suggestion and introduce a new definition of transfer rate that could cover both extremes? The solution in Kenkre and Knox (1974b) was to define the rate as the reciprocal of the time taken by the exciton to traverse a unit distance, i.e., to reach the neighboring site. We imagine the exciton placed initially[12] on one of the sites and call it the origin ($m = 0$), calculate the time it takes for the exciton mean square displacement

---

[12]Whether it is easy in an experiment to place the exciton initially in such a way on a site is irrelevant to this discussion. No violation of a physical principle occurs in such a placement. The consequences of the mathematics involved clarify for us the issue. This statement is made because a lot of time and effort has been sometimes misspent in related discussions.

(msd) to rise from 0 to a^2 (a is the lattice constant), and call the reciprocal of that time the rate of transfer w.

The analysis of the msd from equations such as Eq. (3.10) has been available in detail since Kenkre (1974). In Kenkre (1977a), Expressions such as (1.11), and particularly (1.12), connect the memory function to the msd via a double integral. If we set up the above definition of the rate w in terms of those expressions, we can write, as was done in Kenkre and Knox (1974b), the expression for w,

$$\int_0^{1/w} dt \int_0^t ds \, \phi(s) - \frac{1}{A}. \tag{3.11}$$

Here, A is generally given by m^2 weighted by the F's,

$$A = \left[\sum_n P_n \sum_m F_{mn}(m-n)^2 \right] / \sum_n P_n, \tag{3.12}$$

and, for nearest neighbor interactions $F_{mn} = F(\delta_{m,n+1} + \delta_{m,n-1})$, reduces to $A = 2F$.

Let us take F proportional to V^2, the square of the relevant matrix element for nearest-neighbor transfer as it would be from the Fermi Golden rule. It is easy to see that, for almost *any* reasonable memory function $\phi(t)$, the rate w will transition smoothly from being proportional to V in the coherent regime to V^2 in the incoherent regime. This is shown by the solid line in Fig. 3.6 which is a plot from Kenkre and Knox (1974b) based on an exponential memory,

$$\phi(t) = \alpha e^{-\alpha t}. \tag{3.13}$$

The form (3.13) reduces Eq. (3.11) to the implicit expression for the transfer rate w as

$$\frac{\alpha}{w} + e^{-\alpha/w} = 1 + \left(\frac{\alpha}{2V}\right)^2. \tag{3.14}$$

Notice that the rate becomes respectively proportional to V^2 and V in the incoherent and coherent extremes. It is straightforward to confirm that this limiting behavior does not require the exponential form of the memory and is obtained whenever $\phi(t)$ is non-pathogical at the origin and decays at long enough times.

⊛   The reader is encouraged to derive Eq. (3.14) without help and to explore other forms that one may arrive at if one assumes decaying memories other than the exponential in Eq. (3.13).

In Fig. 3.6, u denotes V in essence, the segments marked a, b, c refer to portions of a similar explanatory plot in a publication by Förster in the book Gordon (1961), and the reader should consult (Kenkre and Knox 1974b) for all detail. Thus, we have a unified description of the rate for arbitrary degree of coherence which bridges

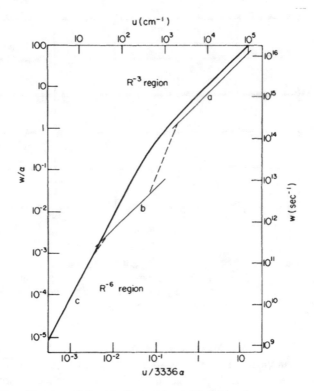

Fig. 3.6 The transfer rate as defined in the text (scaled to the incoherence parameter α) showing a smooth passage from the incoherent region of low interaction to the coherent region of strong interaction for transfer. The latter, denoted by u scaled as shown, is in effect the intersite matrix element V. This plot, which shares some features in appearance to one appearing in a Förster publication to be found in the book (Gordon 1961) (but not the bridging of the coherent and incoherent limits), resolves the puzzle of fast and slow rates by unifying them within the same description. It removes the reservations that had been expressed in the earlier literature that the two rates could not be compared to each other. Reprinted with permission from fig. 1 of Ref. Kenkre and Knox (1974b); copyright (1974) by the American Physical Society

the coherent R^3 limit with the R^6 counterpart. Indeed, it is easy to obtain from this theory a continuously varying exponent of the intermolecular separation R as a function of the interaction strength (see Fig. 3.7). The plot is of the exponent n in expressing the transfer rate as proportional to R^n for a dipole-dipole interaction. The exponent is given by

$$n = \frac{d \ln w}{d \ln R} = 6 \left(\frac{w}{\alpha} - \frac{1}{1 - e^{-\alpha/w}} \right). \tag{3.15}$$

An explicit application of this development to observations may be found in Kenkre and Knox (1974b) where the authors considered a complex of five bacte-

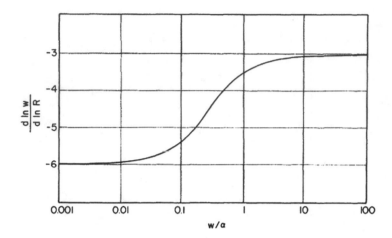

Fig. 3.7 The exponent $n = \frac{d \ln w}{d \ln R}$ of the intermolecular separation R for the case of dipolar interaction, as a function of the transfer rate w expressed in units of the incoherence parameter α. See Eq. (3.15) in the text. Reprinted with permission from fig. 2 of Ref. Kenkre and Knox (1974b); copyright (1974) by the American Physical Society

riochlorophyll (BChl) molecules whose spectra had been studied by Philipson and Sauer (1972). From the low-temperature spectrum, the average pairwise interaction strength was deduced to be $V = 125 \text{cm}^{-1}$. The incoherence parameter α was extracted using the BChl-α monomer spectrum at room temperature, assuming in the process a Stokes shift comparable with that observed in bacteriopheophytin. The value of α was, thus, evaluated to be $2.3 \times 10^{14} \text{sec}^{-1}$. The transfer rate was found from the figure to be $w = 0.020\alpha = 4.8 \times 10^{12} \text{sec}^{-1}$. It was then possible to read off the effective exponent of the intermolecular separation. The result was an $R^{-5.9}$ dependence of the transfer rate on the intermolecular distance R.

Lest one conclude that the assumption of the governing GME as being separable in the form given in Eq. (3.10) is a poor one, it deserves to be stated that the msd it leads to is *precisely* the one that the crystal memory for a system in interaction with a bath, (2.32) does. This may be surprising on two counts. The latter memory has terms connecting not only nearest neighbors but links a given site to all sites in the chain, in spite of the fact that the interaction is nearest-neighbor at the Schrödinger level. Also, it can be easily shown (see, e.g., Silbey 1976) that for a nonnegligible degree of coherence, the separable approximation compatible with (3.10) leads to negative probabilities at large enough times. In spite of this undesirable result for probabilities, its prediction for the mean square displacement is always accurate.[13]

---

[13] After bouts of a recurring argument in which Bob Silbey would look down upon my colleagues and me for using the approximate equation, he put an end to the argument one day by confiding in me that Bob Zwanzig had confirmed this peculiarity that the msd remained accurate even though

In ending this chapter, let us emphasize the fact that the GME formulation of the memory formalism naturally makes four contributions to the resolution of the vexing conceptual and practical issues in the theory of energy transfer. First, it unifies the description of the coherent or ringing dynamics at the microscopic level with that of the incoherent or irreversible transition dynamics at the macroscopic level in an *emergent* manner. Second, it generalizes Förster's spectral prescription to the coherent domain and thereby provides a way of obtaining memory functions for many systems rather directly. Third, it allows one to calculate memory functions explicitly through perturbation schemes from assumed Hamiltonians. And fourth, by its very nature, it allows us to solve the puzzle and clear the confusion centered around the so-called fast (R^{-3}) and slow (R^{-6}) rates for excitation transfer.

3.7   Chapter 3 in Summary

It was pointed out that, while the projection operator introduced by Zwanzig effortlessly gives rise to memory functions and, through a lucid scheme of approximations, leads from the microscopic von Neumann equation to the macroscopic Master equation, the memory functions involved are always cosines of time arguments which by themselves always oscillate. To make memories decay in time and thereby be able to describe irreversible dynamics, it is necessary to build the coarse-graining operation *within the projection operator*. This technique, in the simple but useful form originated by the author, was explained. The recipe was applied in the context of the unrelated phenomenon of ferromagnetism to elucidate how a succession of projection operators can be fashioned for use at different levels of description. The situation around 1970 in the field of energy transfer, with its various interesting conflicts, was described. The theory developed in the first part of the chapter was shown to result in a generalization of the Förster theory of excitation transfer in molecular aggregates as given by Kenkre and Knox in 1974, with explicit recipes for calculating memory functions from absorption and emission spectra. Memories were displayed for several different molecules. The unification of the so-called fast and slow rates addressing questions of exciton coherence was also provided on the basis of memory functions by presenting how, by their very nature, memories help solve coherence problems. This gave a clear resolution of the Perrin-Förster-Davydov puzzle of R^3 versus R^6 transfer rates, where R is the intermolecular distance between a donor and a receptor molecule.

---

the probabilities went negative. In his inimitable manner said Silbey to me, "When Bob Zwanzig speaks, I listen."

Relations of Memories to Other Entities and GME Solutions for the Linear Chain

4

Now that we have understood how the memory formalism originates from the Hamiltonian and how it resolves, at least in broad outline, long-standing coherence issues, let us turn first to the relations it bears to other theoretical approaches to excitation transfer, then to other entities in statistical mechanics and condensed matter physics, and finally let us seek explicit solutions of the GME in perhaps the most relevant system, the infinite linear chain or the 1-dimensional crystal. These three form the content of the present chapter.

4.1 Relation of the Memory Formalism to SLE Theories of Excitation Transfer

Attempts at the unification of the description of exciton dynamics, developed on the basis of the memory formalism and described in the preceding chapter, were predated by at least two other avenues of investigation carried out approximately at the same time. One was conducted at the University of Stuttgart and the other at MIT. With two of his students, Strobl and Reineker, Haken had introduced a stochastic approach to what they called the "coupled coherent and incoherent motion of excitons". The other was authored by Silbey, first with his student Grover and later with his student Rackovsky. Reineker stayed in the field and was a prime mover who analyzed many problems of exciton transport dealing with lineshapes, resonance phenomena and coherence matters. So did Silbey who led many of the theoretical efforts on this side of the pond.

There were studies by Katja Lakatos-Lindenberg and her colleagues on the basis of the above two approaches, reviews of the experimental and theoretical situation by several leading chemical physicists (Hochstrasser 1966; Robinson 1970; Soos 1974), the injection of percolation and metal-insulator transition concepts into the exciton field by Kopelman et al. (1975) and several other analyses as well (Aslangul

© The Author(s), under exclusive license to Springer Nature Switzerland AG 2021
V. M. (Nitant) Kenkre, *Memory Functions, Projection Operators, and the Defect Technique*, Lecture Notes in Physics 982,
https://doi.org/10.1007/978-3-030-68667-3_4

and Kottis 1974; Soules and Duke 1971; Munn 1973, 1974; Fayer and Harris 1974; Shelby et al. 1976; Zewail and Harris 1975a,b; Harris and Zwemer 1978).[1] Let us focus on the two theoretical studies mentioned here first, the Stuttgart investigation and the MIT analysis. Because of the form of the equations of motion that came out of the analysis of both groups, we will call both together as the Stochastic Liouville Equation (SLE) approach.

4.1.1   The Haken-Reineker-Strobl Form of the SLE Approach

The Stuttgart approach (Haken and Reineker 1972; Haken and Strobl 1973) had the power of simplicity and focus. They described their method as that of the description of *coupled coherent and incoherent* motion of excitons. If they suspected on the basis of experiments that there was such a blend of behavior in the motion, they introduced parts of the Hamiltonian in the von Neumann equation for the exciton density matrix that had the corresponding characteristics, applied stochastic methods to the incoherent or fluctuating part, assumed simple white noise for the stochastics, and derived an equation which combined within itself quantum mechanical elements of a dynamical kind, side by side with terms arising from the stochastic treatment. That equation has the same general form as Eq. (2.25) introduced into the exciton field by Avakian et al. (1968) but prominently augmented by a hopping Master equation part:

$$\frac{d\rho_{mn}(t)}{dt} = -iV[\rho_{m+1\,n}(t) + \rho_{m-1\,n}(t) - \rho_{m\,n+1}(t) - \rho_{mn}(t)]$$

$$-(1 - \delta_{mn})(\alpha\rho_{mn} + \bar{\alpha}\rho_{nm}) + \delta_{mn}\gamma_1[\rho_{m+1\,m+1}(t) + \rho_{m-1\,m-1}(t) - 2\rho_{mm}(t)].$$

$$(4.1)$$

While the V and γ_1 terms have been shown here as nearest-neighbor ones for simplicity, the Stuttgart forms have general arbitrary-range forms. The last term denotes the (usually phonon-assisted) additional hopping channel of motion. The off-diagonal density matrix element destruction shown in the term previous to the last has not only the Wigner-Weisskopf or Avakian term with rate α but an additional term, with a different rate $\bar{\alpha}$, that links to, not $\rho_{mn}(t)$, but the elements with indices flipped: $\rho_{nm}(t)$.

A variety of observations were addressed with the help of the SLE, with a good deal of success, including electron spin resonance and optically detected magnetic resonance lineshapes. An excellent reference to learn this approach is from the writing of Peter Reineker which forms the second half of the book (Kenkre and Reineker 1982).

---

[1] Two experimentalists in this field who were my contemporaries and engaged in intense debates on exciton coherence made path-breaking innovations in experimental science: Mike Fayer at Stanford and (the late) Ahmed Zewail at Caltech. The latter won the 1999 Nobel prize in Chemistry.

4.1.2 The Grover-Rackovsky-Silbey Form of the SLE Approach

The MIT approach to the problem was formulated by Silbey (Grover and Silbey 1971; Rackovsky and Silbey 1973) in a quite different manner, its essential characteristic being polaron methodology along the lines of Holstein's theory (Holstein 1959a,b). That work happens to be appropriate to strong exciton-phonon interactions. The strong nature of the interactions necessitates a so-called dressing transformation in such systems. After this transformation is carried out, a particular perturbation technique is undertaken (Silbey 1976) which surprisingly leads approximately to the same form of the evolution equation as arose from the Stuttgart approach based on stochastic considerations. The differences *in form* are very small although the physics content leading to the form are different. With the same simplification of nearest neighbor interactions used for simplicity, the equation can be written as

$$\frac{d\rho_{mn}(t)}{dt} = -i\widetilde{V}[\rho_{m+1\,n}(t) + \rho_{m-1\,n}(t) - \rho_{m\,n+1}(t) - \rho_{m\,n}(t)]$$

$$-(1 - \delta_{mn})\alpha\rho_{mn} + \delta_{mn}\gamma_1[\rho_{m+1\,m+1}(t) + \rho_{m-1\,m-1}(t) - 2\rho_{m\,m}(t)]. \tag{4.2}$$

Relative to Eq. (4.1) that represents the Stuttgart approach, this Eq. (4.2) from MIT has the following differences. The V in the former is replaced by $\widetilde{V}$ in the latter, in order to denote that one refers here not to the bare bandwidth but to the dressed bandwidth, drastically reduced via an exponential factor as a result of strong interactions with phonons; and there is no $\bar{\alpha}$ or the flipped density matrix element in Eq. (4.2). Because of the striking similarity in form, the two have been regarded together often in the literature as

$$\frac{d\rho_{mn}(t)}{dt} = -iV[\rho_{m+1\,n}(t) + \rho_{m-1\,n}(t) - \rho_{m\,n+1}(t) - \rho_{m\,n}(t)]$$

$$-(1 - \delta_{mn})\alpha\rho_{mn} + \delta_{mn}\gamma_1[\rho_{m+1\,m+1}(t) + \rho_{m-1\,m-1}(t) - 2\rho_{m\,m}(t)], \tag{4.3}$$

and termed the SLE. In this combined form one finds the equation displayed without the tilde on the V and omitting the additional decay of the flipped density matrix peculiar to the Stuttgart approach. We will use this form in the book, call Eq. (4.3) the SLE (stochastic Liouville equation) and point out that the original Stuttgart expression has a factor 2 multiplying the rates which it is not necessary to show here.

4.1.3 Relation of the SLE to the GME Approach

One way to explore the relation of the SLE approach to the memory-function (GME) approach is to calculate consequences such as the mean square displacement arising from the SLE and compare them to the corresponding consequences of the

GME, or indeed, to cast the SLE's in the GME form and examine the memory functions to which they correspond. Both these avenues were undertaken, the first in publications such as Kenkre (1974) and Kenkre and Knox (1974a) and the second in Kenkre (1978a). Let us analyze the relations between the two approaches here by considering (1) the presence or absence of multiple timescales in the memory function, (2) the polaron transformation and particularly the subsequent perturbation scheme graphically, and (3) by discussing the very important issue of emergence versus coexistence.

Multiple Time Scales in the Memory

It is at once clear that the memory function in the SLE approach, whether of the Stuttgart or the MIT variety, consists of two parts with widely different time constants (Kenkre 1975b). Indeed, one part is a δ-function in time (the one multiplied by γ_1). Thus, putting the arbitrary-range interactions back into the SLE (as their authors obviously always meant to), we see that the memory functions $\mathcal{W}_{mn}(t)$ in a GME corresponding to the SLE are given by (Kenkre 1975c):

$$\mathcal{W}_{mn}(t) = \mathcal{W}^c_{mn}(t)e^{-\alpha t} + \gamma_{mn}\delta(t), \tag{4.4}$$

the memory $\mathcal{W}^c_{mn}(t)$ in first term arising from coherent interactions on the chain, the multiplicative exponential factor from the Wigner-Weisskopf or Avakian destruction of the off-diagonal elements of the density matrix, and the additive δ-function term from the (typically phonon-assisted) hopping interaction. The first term is precisely as in Eq. (2.31).

What this means without doubt, is that the SLE approach is applicable to a situation where two drastically different mechanisms responsible for such time behaviors are present in the system. If such is the case for the system under investigation, the SLE approach is forearmed and already well equipped to describe the system. If, however, such is not the case for the given system, the SLE is less suitable for the analysis than it would be if it did not have built into it such a two time-constant memory. This is made clear by Fig. 4.1 where the main figure shows (on an arbitrary scale) a two-time constant memory as clearly characteristic of the SLE whereas the inset shows a single time-constant memory. If the system is such that the spectral prescription of Förster (see the end of the previous chapter) applies, and the situation is as described in the inset, there is no freedom left to the transport description to accommodate the multiple time-constant nature shown in the main figure. If the given system is similar to the case displayed in Fig. 3.5, the two-time constant memory, consequently the SLE approach, is highly appropriate and can be said to have arisen from the sharply differentiated spectrum shown in the inset of Fig. 3.5. However, it also follows that, if the system is represented by spectra similar to the other cases displayed, such as in Fig. 3.2, the memory is not multiple time-constant at all, being as given in Fig. 3.3. The differentiation in time behavior, that the SLE demands to be present in the two parts of the memory, might then be counter to the nature of the system examined.

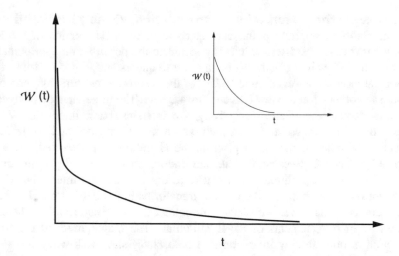

Fig. 4.1 Multiple-time-constant memory which would justify an SLE approach. This is a schematic illustration of a situation shown in Fig. 3.1. Inset, by contrast, shows a single time-constant memory function where the SLE approach would not be appropriate. Units, obviously, are arbitrary. Adapted with permission from figs. 1 and 2 of Ref. Kenkre (1975b); copyright (1975) by the American Physical Society

It is, no doubt, possible to replace the δ-function in the second part of the SLE memory by a more gradual time function, an attempt of this kind having been indeed suggested (Kenkre 1975b,c).[2]

The SLE approach is, thus, excellent where the separation into different time-constant memory parts is warranted for the particular system being studied, but not universally. In systems with no multiple time-constants in their memory there is a danger that the application of the SLE might introduce spurious effects.

The mention of spectral connections made here does not presuppose that the motion is necessarily due to dipole-dipole interactions of relevance to singlets. The importance of the general connection between spectra and motion has been discussed by various authors known for their significant contributions to the subject of energy transfer such as Hochstrasser and Prasad (1972) and Soules and Duke (1971).[3]

---

[2]Along with a somewhat cumbersome terminology involving the GSHRS equation distinguished from the GSHRS formalism. However, in a sense, the reason for constructing the SLE in its two-part form, thus its focus, are then certainly lost.

[3]The latter authors have also commented on the important question of whether the donor and the acceptor are coupled to the same or different baths, and the issue of inhomogenous broadening of spectra.

Differences in the Perturbation Approximation, Visually Displayed

There is another important point, characteristic of the MIT version of the SLE approach that deserves discussion: it is relevant to the perturbation approximation employed by Silbey and collaborators. In order to understand it, it is useful to view a graphical representation (Kenkre 1975c) of the polaron transformation employed by them and others. Let us view it as our Fig. 4.2 here. The three intersite interaction quantities depicted in it are, respectively, the intersite (bare) interaction V, the reduced dressed interaction $\tilde{V}$ after the polaron transformation, and the left-over interaction $\mathcal{V}$. The interaction as given in the Hamiltonian is depicted in the top left panel of Fig. 4.2. Because the intrasite interaction shown by lightning arrows is too strong to treat perturbatively (relative to the "horizontal" interaction V, not dependent on the bath state), the polaron transformation is carried out. It renders the system as in the lower left panel. Subsequent to the transformation, there are no intrasite matrix elements of the Hamiltonian. The major effect of the strong interaction is now already in the dressed bath states shown as wavy lines. The remaining effect is to make the left-over interaction not horizontal, i.e., to make it connect different bath states on the different sites.

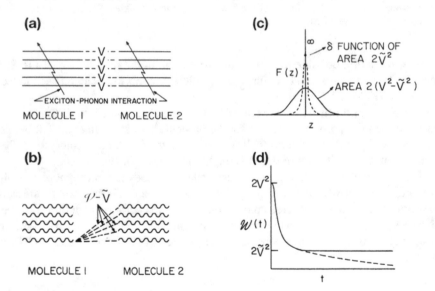

Fig. 4.2 A graphical depiction of the polaron model employed by Silbey and collaborators. The bare system before the polaron transformation is shown in (**a**). We see it is independent of the vibrational state depicted by the straight horizontal lines but there is a (strong) interaction within the vibrational manifold shown by the lightning lines. After the polaron transformation that removes the strong intra-site interaction, the system is viewed as in (**b**), the wavy lines representing the dressed states. The leftover interaction, in orders of which a perturbation expansion is then to be carried out, is $\mathcal{V} - \tilde{V}$ as shown. See text for further explanation. Modified with permission from fig. 3 of Ref. Kenkre (1975c); copyright (1975) by the American Physical Society

Typically, one now carries out a straightforward perturbation calculation in the left-over interaction. In the work of Holstein and many other authors this leads to expressions for the transition rate for the polaron to move between the sites. Precisely the same is done in the GME theory, but what is calculated is the memory function rather than the transition rate. The technical steps of the latter analysis are pointed out explicitly, and the results displayed, in an early paper (Kenkre and Rahman 1974), analyzed in considerable physics detail in Kenkre (1975c), and studied afresh in a recent publication aimed at the investigation of GME's for vibrational relaxation rather than motion (Ierides and Kenkre 2018).

The perturbation development followed in the MIT (SLE) approach does not follow the above path. The left-over intersite interaction after the polaron transformation is *further* split into a horizontal part (the same for all bath states) and used to transform the site-specific dressed states (shown as wavy lines) into dressed k-states which are their Bloch combinations. What is finally left-over is used in a Fermi golden rule rate expression in k-space.

Some of this is graphically suggested in the right top panel of Fig. 4.2. When the polaron transformation is performed, the $F(z)$ whose Fourier transform the memory $\mathcal{W}(t)$ is, with examples displayed in Fig. 3.2 and the insets of Fig. 3.5, equals, in essence, the sum of the δ-function (in z) of area $2\widetilde{V}^2$ and the broad bell-shaped curve of area $2\left(V^2 - \widetilde{V}^2\right)$ both shown in solid line. The δ-function in z arises from the *zero-phonon lines* in spectra that arise in systems such as the present one in which harmonic oscillators are an ingredient. It is well-known (Fitchen 1968) that the zero-phonon lines in such systems cannot be broadened by linear interactions. In real molecular aggregates, nonlinear interactions or other mechanisms exist and cause broadening.

However, unless such mechanisms are added externally to the stated Hamiltonian, the corresponding memories must have a non-decaying part. The dashed line in the upper right panel of Fig. 4.2 shows the effect of such a mechanism *external* to the Hamiltonian used in the system: the δ-function has now become broadened into a Gaussian or Lorentzian shape. The consequent effect on the memory function (the decay) is seen in the lower right panel. This was carefully argued in Kenkre (1975c) and is the origin of the difference of the calculational methods used often along with the GME on the one hand and the (MIT) SLE on the other. What effectively happens in the latter procedure is that the decay of what was the δ-function memory artificially equals the integral of the other part of the memory!

Blind approximation techniques, as opposed to systematic developments in orders of a small dimensionless parameter, appear, unavoidably, at various places in physics. They offer embarrassing alternatives in their use because it is not possible to determine their validity a priori. One might argue (as in the MIT approach) that, after diagonalizing the strong intrasite interaction among the bath states via the polaron transformation, another (Bloch) transformation to remove the horizontal interactions is good practice: one should come as close as possible to the final eigenstates of the system before applying a procedure such as the Golden Rule. One might, on the other hand, also argue against that approach. The supporting statement would be the observation that the intrasite interaction is the stronger one by far in the system. This demands that the perturbation procedure needs to be carried out from the localized (dressed) states to preserve the nature of the system dynamics.

There is also the matter that the horizontal interaction that is removed in the second diagonalization step is a thermal average over the bath states and therefore has no dependence on the individual state. An artificiality may be introduced, thus, in the method in constructing the Bloch states for each combination of equal-energy localized states. This viewpoint was the basis of calculations of models of the model performed and reported in Kenkre (1975c). The calculations showed that our procedure (called generally the GME procedure) gave more accurate results. It was difficult to resolve the conflict given the complexity of the problem.

The issue of which approximation procedure is better must be thus left to the choice of the investigator. However, two developments in the published literature deserve mention. The application of both these theoretical schemes to experiments on (not excitons but) photo-injected charge transport in naphthalene showed, after careful study, that our GME methodology is preferable. It leaves out the second (Bloch) diagonalization before doing the perturbation. It was shown to succeed when compared to experiment (Kenkre et al. 1989) while the MIT procedure for that problem (Silbey and Munn 1980) was shown to be inadequate. The GME theory with our perturbation scheme was also appreciated by independent theorists who applied it successfully to observations on a different material, alpha-sexithiophene (Wu and Conwell 1997). Our theory has been accepted as appropriate in modern surveys of the field such as in the book by Pope and Swenberg (1999).

Equally significant is the demonstration (Kenkre et al. 1996), through a numerical procedure, that the GME approach produces impressive agreement with (numerically) exact calculations.[4] I believe that further investigations of the latter kind would benefit our understanding of the problem.

Emergence of Incoherence from Coherence Versus Their Coexistence

Perhaps the most important ingredient in the relations among the three approaches to exciton transport is the *raison d'être* of each, which happens to vary sharply. The GME approach was constructed in order to understand how incoherence *emerges* from the fundamental coherent dynamics, exemplified by how Perrin transfer evolves into Förster transfer. Coarse-graining provided the conceptual answer and the Fourier summation, as explained in Chap. 3, gave the explicit detail of how memories that were coherent at the microscopic level acquired decays through that summation.

Neither of the SLE approaches asked this question of emergence. The Stuttgart approach assumed the existence of what they called "coupling" of the coherent and

---

[4]Lest too much be read into this, note that what the GME method was compared to in that particular investigation, was not the MIT procedure discussed here. Instead, it was a semiclassical approximation that was shown to be relatively crude. All that demonstration does in the present context is to strengthen our faith in the accuracy of the GME method. It is to be emphasized that the latter method does not undertake the second diagonalization made in the MIT procedure.

incoherent motion and presented a powerful method of analyzing them together, treating the latter one stochastically. The MIT approach was initially not directed at coherence at all, its target being the elucidation of how to treat strong interactions with vibrations. The characteristic of that method is that the assumed strength of these interactions is so large that a perturbation in the *given* interaction is doomed to fail. The polaron transformation, termed in some circles as the "displaced oscillator transformation" (because the strong interaction displaces the equilibrium configuration of the representative oscillator), is the primary element of the MIT approach, supplemented by perturbation choices particular to the authors of that approach.

One acquires an immediate insight from the Stuttgart method into how separate channels, envisaged as coherent and incoherent, if they exist and are coupled in a given system, affect observations. The nuances of the dressing transformation and their consequences for transport are what one learns from the MIT SLE. An understanding of the emergence of incoherence from coherence as a result of bath interactions is what the GME approach is directed at, being thus the only one among the three that is closely connected to formal questions of the central problem of nonequilibrium statistical mechanics.[5]

4.2 Equivalence of the GME to Continuous Time Random Walks

Around the time that the GME theory of exciton transport was being developed at the University of Rochester, an unrelated project was also going on in its proximity under the authorship of Harvey Scher who addressed time of flight experiments in photo-copying machines. Those experiments were of interest to Xerox in their Webster research lab. Scher's work had been in its preliminary phases in collaboration with Lax and in later stages with Montroll and used a theoretical tool invented years earlier by Montroll working with Weiss and others. The tool, called continuous time random walks (CTRW), was characterized by "pausing time distribution functions" $\psi(t)$, the essential concept being that a random walker might *pause* at sites without immediately hopping to another site. I felt certain these CTRW's were equivalent to GME's with simple relations existing between the latter's memory functions $\phi(t)$ and the former's $\psi(t)$. There was, however, a difference of opinion expressed by influential investigators. A simple proof of the equivalence settled the issue unequivocally. It was published in Kenkre et al. (1973) in a simplified form and in Kenkre and Knox (1974a) and Kenkre (1977a) in its

---

[5]The GME as a tool is made for problems that require the computation of probabilities, or observables that depend only on probabilities. If off-diagonal elements of the density matrix are required, one must step back and use the SLE. It is tempting to fall into the erroneous trap of regarding everything one meets with as a nail if one happens to have acquired a hammer. I know of investigators for whom memories and projections are a panacea. Fortunately, physics is richer than that and demands that we remain flexible and move from tool to tool to perform our tasks.

general version. The simple case was like that of Eq. (3.10) where the space part is separable from the time part whereas, in the general version, the product form does not apply. We explain the equivalence below.

Let us recast the CTRW originated by Montroll and collaborators (Montroll and Weiss 1965; Montroll and West 1979) in a simple form. The elementary Chapman-Kolmogorov equation governs the probabilities of occupation of states ξ, μ at set intervals of time. The Montroll generalization to continuous time, allows, as stated above, that the walker may pause at sites.[6] The time evolution of the probability of occupation of state ξ is

$$P_\xi(t) = P_\xi(0) \left[1 - \int_0^t ds\, \psi(s) \right] + \int_0^t ds\, \psi(t-s) \sum_\mu Q_{\xi\mu} P_\mu(s). \tag{4.5}$$

Here $Q_{\xi\mu}$ is the probability to transition from μ to ξ, ψ is the pausing time distribution such that $\psi(s)ds$ is the probability that a hop (transition) will occur between time s and $s + ds$. The Q's summed over all states give 1 since, in the absence of the ψ's, a transition is definite to occur to some state.

It is quite straightforward to show that the GME as given in Eq. (2.3), which is repeated here for convenience,

$$\frac{dP_\xi(t)}{dt} = \int_0^t ds \sum_\mu \left[\mathcal{W}_{\xi\mu}(t-s) P_\mu(s) - \mathcal{W}_{\mu\xi}(t-s) P_\xi(s) \right],$$

is equivalent to Eq. (4.5). All one need perform is simple manipulations with the Laplace transforms. The memory functions $\mathcal{W}$ are related to the pausing time distribution ψ and the probabilities Q via a multiplying factor which is a function of the Laplace variable:

$$\widetilde{\mathcal{W}}_{\xi\mu}(\epsilon) = Q_{\xi\mu} \left(\frac{\epsilon \tilde{\psi}(\epsilon)}{1 - \tilde{\psi}(\epsilon)} \right). \tag{4.6}$$

One can rewrite the GME we have used in the product form $\mathcal{W}_{\xi\mu}(t) = R_{\xi\mu}\phi(t)$ as

$$\frac{dP_\xi(t)}{dt} = \int_0^t ds\, \phi(t-s) \sum_\mu \left[R_{\xi\mu} P_\mu(s) - R_{\mu\xi} P_\xi(s) \right], \tag{4.7}$$

where

---

[6]Surely this reminds the reader of the well-known fairy tale in which a man went to sleep on a mountain slope for many years while the rest of the world moved on in time. Accordingly, I called the walker's behavior *Rip Van Winkle* in Kenkre (1977a) after the name of the man in the fairytale.

$$\tilde{\phi}(\epsilon) = \frac{1}{\alpha} \left(\frac{\epsilon \tilde{\psi}(\epsilon)}{1 - \tilde{\psi}(\epsilon)} \right). \tag{4.8}$$

Here the rates R in the GME are given simply by multiplying the corresponding Q by any constant of the dimensions of the reciprocal of a time,

$$R_{\xi\mu} = \alpha Q_{\xi\mu}. \tag{4.9}$$

We call the constant α, note that its only function is to make $\phi(t)$ have dimensions reciprocal to that of t just like $\psi(t)$. It may be convenient some times to take α to be the reciprocal of the moment of the pausing time distribution

$$\frac{1}{\alpha} = \int_0^\infty dt\, t\psi(t). \tag{4.10}$$

However, this is not necessary since α actually drops out of the product $R\phi(t)$; in a number of applications α, as defined in Eq. (4.10), vanishes, the moment being infinite. We mention this in particular because interesting theories of transport have been based on pausing time distributions where some of the moments, including the first, are infinite (Scher and Montroll 1975; Shlesinger 1974).

The equivalence of Eq. (4.5) to the GME is shown simply first by Laplace-transforming the former and noticing the splitting

$$\frac{\epsilon}{1 - \tilde{\psi}(\epsilon)} = \epsilon + \left(\frac{\epsilon \tilde{\psi}(\epsilon)}{1 - \tilde{\psi}(\epsilon)} \right).$$

Conversely to Eq. (4.8), the memories are obtained from the pausing time distribution via

$$\tilde{\psi}(\epsilon) = \frac{\alpha \tilde{\phi}(\epsilon)}{\epsilon + \alpha \tilde{\phi}(\epsilon)}. \tag{4.11}$$

The way to go from a general GME that need not be of a product form to the equivalent CTRW has also been shown explicitly (Kenkre and Knox 1974a; Kenkre 1977a).[7] The result is that, from the memory functions, i.e. $\mathcal{W}$'s, one constructs the *state-dependent* pausing time distributions

---

[7] Related results in the area of transient nucleation and dynamics with internal states were published about the same time or a little later, independently, by Shugard and Reiss (1976), and by Landman et al. (1977), respectively. How they emerge from the expressions in Kenkre and Knox (1974a) has been shown in footnote 36 of Kenkre (1977a).

$$\tilde{\psi}_\xi(\epsilon) = \frac{\sum_\mu \widetilde{W}_{\mu\xi}(\epsilon)}{\epsilon + \sum_\mu \widetilde{W}_{\mu\xi}(\epsilon)}, \qquad (4.12)$$

and the generalized probability functions

$$\tilde{Q}_{\xi\mu} = \frac{\widetilde{W}_{\xi\mu}(\epsilon)}{\epsilon + \sum_\mu \widetilde{W}_{\mu\xi}(\epsilon)}. \qquad (4.13)$$

These then go into the general form of the CTRW which is not of the product form:

$$P_\xi(t) = P_\xi(0)\left[1 - \int_0^t ds\, \psi_\xi(s)\right] + \int_0^t ds \sum_\mu Q_{\xi\mu}(t-s)P_\mu(s). \qquad (4.14)$$

★ The above recipes constitute a complete connection between the CTRW and GME formulations. One can freely move from one to the other as taste and convenience demand. Here is an open exercise for the interested. Calculate a few pausing time distributions ψ for given cases of the memory function ϕ and vice-versa; check, in particular what ψ is for a δ-function memory (ϕ) and for an exponential memory function.

Continuous time random walks and memory functions, whose precise equivalence has been demonstrated above, form two of the commonly used formalisms of description. Two more are nowadays often employed: fractional derivatives and the approach of time-dependent transport coefficients. To learn the fascinating field that is the former, I recommend reading (Mannella et al. 1994; Mainardi 1997; West et al. 2012). Their relationships with the formalism of memory functions has also been established (Giuggioli et al. 2009; Kenkre and Sevilla 2007), and presented in review form in a recent book (Kenkre and Giuggioli 2020).

4.3   Connection of Memories to Other Entities

Memory functions are very similar to what has been sometimes called "force-force correlation functions." It should be clear that they must be closely related to standard correlation functions in statistical mechanics. They are, indeed. Here let us explore what kinship memories bear to velocity auto-correlation functions and neutron scattering functions. The former are of importance wherever mobility or conductivity of charges such as electrons or holes is studied. The latter find application in the field of the motion of hydrogen atoms in metals or related dynamics of light interstitials such as muons in solids.

In order to understand the precise source of the connection, it would help to recall that a typical linear correlation function in response theory (Kubo 1957) in near-equilibrium statistical mechanics has the form

$$\langle O_1(t)O_2 \rangle = Tr \, \rho_e e^{itH} O_1 e^{-itH} O_2$$

where ρ_e is the equilibrium density matrix of the system (usually thermal), from which state the stimulus is applied (for instance an electric field). The system Hamiltonian is H which means that, normally, ρ_e is of the canonical form $e^{-\beta H}/Tr \, e^{-\beta H}$. The two operators O_1 and O_2 are those whose correlation we describe. And $\beta = 1/k_B T$ where T is the temperature and k_B is the Boltzmann constant. For convenience, let us consider the symmetrized correlation function $C(t) = \frac{1}{2}\langle O_1(t)O_2 + O_2 O_1(t) \rangle$ and observe, from cyclic permutation within the trace, that it can be written as

$$C(t) = \sum_\xi O_1^\xi \langle \xi | e^{-itH} K e^{itH} | \xi \rangle \qquad (4.15)$$

if the operator O_1 has eigenstates labeled by ξ with corresponding eigenvalues O_1^ξ. Rewriting in this fashion makes clear that K is formally nothing but a density matrix obeying the von Neumann equation

$$i\frac{dK(t)}{dt} = [H, K(t)],$$

with the initial value $K(0) = \frac{1}{2}[O_2\rho_e + \rho_e O_2].$[8]

So, in evaluating correlation functions such as $C(t)$ in linear response theory, we encounter the diagonal elements of a density matrix in the representation of states ξ. This is the basic connection. For the case of velocity auto-correlations, each of the operators O is the velocity operator v, and a simple relation can be derived if the representation of the ξ is taken to be site-localized states. for the case of the neutron scattering function, the O's are e^{ikx} and e^{-ikx} where x is the position operator, and again, the connection is immediate. Details and what can be done with this connection, follow.

4.3.1 To Velocity Auto-correlation Functions

A role of special importance in Kubo's linear response theory (Kubo 1957) is played by the velocity autocorrelation function $\langle v(t)v \rangle$ whose integral is proportional to the mobility or conductivity of a system, in which charges move under the action of an applied electric field. The notation of the time dependence of the operator and of the angular bracket refers to standard practice. The former means that $O(t)$ is the Heisenberg expression $e^{itH} O e^{-itH}$, and the latter stands for multiplication by the

---

[8]The idea and the notation come from early work (Kenkre and Dresden 1971, 1972) on the application of projection operators to the calculation of correlation functions. More on this will be found in Chap. 8.

equilibrium density matrix ρ_e followed by taking the trace. The field must be not too strong so that linear response theory applies (Kubo 1957). In order to establish a relation with memory functions in the GME, we begin with the relationship given by Scher and Lax (1973),

$$\frac{1}{2}\langle v(t)v + vv(t)\rangle = \frac{1}{2}\frac{d^2}{dt^2}\langle x(t)x + xx(t)\rangle, \tag{4.16}$$

where x is the position operator and the correlation in the right hand side is often called the *generalized* mean square displacement. If we consider a particle moving among discrete sites labeled by m, n, etc., and call $P(m, t \mid n, 0)$ the probability that it would be at m at time t if it occupied n at time 0, we can use the *approximate* result given in Scher and Lax (1973) that either side of Eq. (4.16) is equal to

$$\frac{1}{2}\frac{d^2}{dt^2}\sum_{m,n}(m - n)^2 P(m, t \mid n, 0)\langle n|\rho_e|n\rangle. \tag{4.17}$$

If we are dealing with a translationally invariant system such as a molecular crystal, $\langle n|\rho_e|n\rangle$ equals $1/N$, where N is the number of sites in the crystal. It is then at once possible to use the relationship between the mean square displacement and the Fourier transform of the memory function to obtain

$$\frac{1}{2}\langle v(t)v + vv(t)\rangle = -\frac{1}{2}\left[\frac{\partial^2 \mathcal{W}^k(t)}{\partial k^2}\right]_{k=0}. \tag{4.18}$$

Here $\mathcal{W}^k(t)$ is the discrete Fourier transform $\sum_l \mathcal{W}_l(t)e^{ikl}$ with $l = m - n$.

Equation (4.18) is a powerful relation (Kenkre 1974, 1978d) between the velocity auto-correlation function and the memory function. It was used by Reineker et al. (1981) to fit the Stuttgart SLE to data on the temperature dependence of the mobility of photo-induced charge carriers in naphthalene, the basis being the blend of the general relation (4.18) and the two-term expression (4.4) for the memory function corresponding to the SLE. It will also prove useful to us in our detailed analysis of the mobility problem from the GME in Chap. 6 without the use of the Stuttgart SLE. However, the validity of (4.18) is limited to systems to which the validity of the Scher-Lax approximation explained above applies. These are narrow-band situations for which the prescription of (4.16) can be reduced to one in which $P(m, t \mid n, 0)$ appears rather than a more general counterpart as we will see below.

What if the band is not narrow and the equilibrium density matrix ρ_e cannot be approximated by the identity matrix? The way of going beyond the Scher-Lax approximation has been given in Kenkre et al. (1981). The exact consequence of the Scher-Lax relation (4.16) turns out to involve the mean square displacement but calculated not from a localized initial condition at n. Instead, the initial density matrix to be taken for the calculation has elements given by

$$\langle a \mid \rho(0) \mid b \rangle = -\frac{1}{2} \langle a \mid \rho_e \mid b \rangle \left(\delta_{a,n} + \delta_{b,n} \right).$$

The approximation of Scher and Lax, valid for narrow-band materials, uses, instead,

$$\langle a \mid \rho(0) \mid b \rangle = \delta_{a,n} \delta_{a,b}.$$

The approximate relation (4.18) requires the assumption that the equilibrium density matrix ρ_e, whose canonical form at temperature T is $\exp(-H/k_B T)$ divided by the normalizing trace factor $Tr \exp(-H/k_B T)$, is diagonal in the site representation. If used without care, this requirement leads one swiftly to an absurd result. If ρ_e were really diagonal in the site representation, it would commute with the position operator x; the latter is, obviously, also diagonal in the site representation. At the very first stage of the Kubo linear response theory, one would then have the autocorrelation function identically zero.

The analysis in Kenkre et al. (1981) shows that the approximate relation (4.17), valid only for narrow-band materials, must be generally replaced by

$$\frac{1}{2} \langle v(t)v + vv(t) \rangle = \frac{1}{2} \frac{d^2}{dt^2} \sum_{m,n} (m-n)^2 p_m^{(n)}(t). \tag{4.19}$$

In this expression, $p_m^{(n)}(t)$, which replaces $P(m, t \mid n, 0)\langle n|\rho_e|n\rangle$ from the approximate relation, means the probability of occupation of m, starting out not from a localized initial condition at n but as explained above. The initial condition used for calculating the mean square displacement is obtained from the equilibrium density matrix by erasing all its elements except those in the nth row and the nth column. The approximate relation of Scher and Lax corresponds to further erasing all the elements except the single one at the intersection of the nth row and the nth column. For the case when the bandwidth is not too narrow with respect to $k_B T$, these remaining elements in the row and column produce a correction factor. For narrow-band materials, the Scher-Lax expression is valid.

4.3.2   To Neutron Scattering Functions

Important narrow-band[9] physical systems exist in which the transport of microscopic (quasi)particles constituting them fails to conform to extreme limits. Examples range from metal hydrides and solid electrolytes to molecular crystals and even muons in solids. An excellent (and traditional) experimental probe used in these

---

[9]Lest the phrase "narrow-band" cause confusion, note that by the expression we mean here that the bandwidth is much smaller than the thermal energy $k_B T$. The usage "narrow-band" here has nothing to do with the coherence issue which happens to be associated with the ratio of the bandwidth to a scattering rate α.

situations is that of neutron scattering. The basic theory was given by Van Hove (1954b) and later reviews may be found in expositions such as by Springer (1972).

Several investigations have been reported by Brown and the present author in this subject. Among them are: an application (Brown and Kenkre 1983) of the SLE of the Stuttgart kind, and our GME formalism, to investigate coherence effects on the scattering lineshape to address issues raised in Sköld (1978); a study (Brown and Kenkre 1985) of the *coupling* between hopping and tunneling transport channels of atoms moving in solids to check the validity of a treatment by Casella (1983); an extension of the Kutner-Sosnowska formulae on general non-Bravais lattices near the incoherent motion limit (Brown and Kenkre 1986); and the analysis of scattering lineshapes for non-degenerate systems as required for $NbO_x H_y$ (Brown and Kenkre 1987).

The interesting method of calculation (Kenkre and Brown 1985) that underlies these applications will be reserved in this book for Chap. 15 where it is more appropriate to discuss. Here, with the help of a consequence of that calculation, let us mention without discussion the result for the simple system described by the motion parameters V and α to understand the effect of coherence on scattering lineshapes; and then pay attention to the *coupling* of tunneling and hopping interactions of atoms (or comparable quasiparticles) as they move through a crystal, an approximate treatment of which was given by Casella (1983). This choice of the material comes from the fact that the governing equation in both instances of dynamics has already been the focus of description in this chapter.

The central role in the neutron scattering experiments on such systems is played by the scattering correlation function

$$I(k, t) = Tr \, \rho_e e^{-ikx} e^{itH} e^{ikx} e^{-itH}, \tag{4.20}$$

where ρ_e is the equilibrium density matrix, x is the position operator of the particle that scatters the neutrons, and k is the momentum transfer to the target; x and k are generally vectors and kx is their dot product.[10] The Fourier transform of $I(k, t)$ is the scattering function $S(k, \omega) = \int_0^\infty dt \, I(k, t) e^{i\omega t}$.

Coherence Effects on the Lineshape

In order to address in a simple and quick manner coherence effects on the motion of light interstitials or hydrogen atoms in solids, it is convenient to start from the *simple* SLE, i.e., the Avakian equation (2.25) as governing the motion. The scattering correlation function in Eq. (4.20) is found to have the Laplace transform

---

[10] A notational alert! So far in the book, k has represented the dimensionless quantity reciprocal to the site index m. In Eq. (4.20) it is, however, dimensioned and reciprocal to the (dimensioned) position x. Given that in most contexts $x = ma$ where a is the lattice constant on a chain, k in Eq. (4.20) is obtained by dividing the dimensionless k by a. Having given careful (and agonizing) consideration to defining two different symbols for the two different k's, I have decided to avoid the consequent clutter. I rely on the reader's alertness, and indulgence, to determine by inspection (it is quite easy) which use is meant in a given context.

$$\frac{1}{\sqrt{(\epsilon + \alpha)^2 + [4V \sin(ka/2)]^2} - \alpha}$$

where a is the lattice constant. The scattering lineshape is calculated explicitly as shown in Brown and Kenkre (1983):

$$S(k, \omega) = \frac{\sqrt{(1/2)\left(\sqrt{f^2 + (2\alpha\omega)^2} + f\right)} - \alpha}{\sqrt{f^2 + (2\alpha\omega)^2} + \alpha^2 - 2\alpha\sqrt{(1/2)\left(\sqrt{f^2 + (2\alpha\omega)^2} + f\right)}}. \tag{4.21}$$

where f equals the expression $\alpha^2 + (V^k)^2 - \omega^2$, the band energy being denoted by V^k, i.e., the discrete Fourier transform of V_{m-n}.

The calculated lineshape exhibits all the expected features such as broadening and motional narrowing. We refer the reader to the original reference for details and plots. The investigation of the anomalous variation of the halfwidth of the lineshape reported in Sköld (1978) has been successfully carried out in Brown and Kenkre (1983).

Coupling Between Hopping and Tunneling Channels of Motion

Consider a more complex situation in which, as envisaged by the SLE approach discussed earlier, there are two channels for the motion of the particles, a hopping channel (typically phonon-assisted) and a band or tunneling channel (typically phonon-hindered.) The underlying equation is the full SLE (4.3) incorporating both terms. If one neglects as in Casella (1983), the coupling of the tunneling and hopping motions, one writes $I(k, t)$ as the product of the correlation functions due to the individual processes involving band and hopping motion respectively, normally denoted by suffixes b and h. The assumed product property means a convolution of the scattering functions:

$$S(k, \omega) = \int_{-\infty}^{\infty} d\omega' \, S_b(k, \omega - \omega') S_h(k, \omega'), \tag{4.22}$$

It is amply clear from our analysis of the SLE, appropriate to the coexistence of hopping and tunneling motion, that the scattering function due to hopping motion alone would be

$$S_h(k, \omega) = \frac{1}{\pi} \left[\frac{\gamma^0 - \gamma^k}{\omega^2 + 2(\gamma^0 - \gamma^k)^2} \right], \tag{4.23}$$

superscripts k denoting discrete Fourier transforms over space. This assumes that the hopping is done via rates γ as in a Master equation obeyed by the probability $P_m(t)$,

$$\frac{dP_m(t)}{dt} = \sum_n \left[\gamma_{mn} P_n(t) - \gamma_{nm} P_m(t) \right]. \tag{4.24}$$

Using the theory in Kenkre and Brown (1985), it is possible to calculate the neutron scattering lineshape from the SLE in both cases: when there is no correlation between the two channels, hopping and band, as assumed in the Casella limit; and exactly, i.e., without neglecting such a correlation that naturally exists as a consequence of the equation. The respective limits given by Brown and Kenkre (1985) are

$$\pi S(k, \omega) = Re \left[\frac{1}{\left[(i\omega + \gamma^0)^2 + V^2(k) \right]^{-1/2} - \gamma^k} \right] \tag{4.25}$$

for the Casella approximation, and

$$\pi S(k, \omega) = Re \left[\frac{1}{\left[(i\omega + \gamma^0 - \gamma^k)^2 + V^2(k) \right]^{-1/2}} \right] \tag{4.26}$$

from the exact calculation in which correlations between the two channels are retained. Here, $V(k)$ denotes the k-dependent band energy of the moving particle, given in a tight-binding model by $2V \cos ka$.

The slight difference in the expressions has profound consequences. By inspecting plots of the lineshape versus the frequency (specifically against the dimensionless ratio $\omega/V(k)$), one can at once see that broadening followed by narrowing and finally again broadening occurs for the exact result as one steps through increasing values of the ratio γ^0/V while the Casella approximation produces only broadening and no narrowing.

4.4 GME Solutions for a 1-Dimensional Crystal

We have seen in Sect. 1.3.2 that the general procedure to analyze the Schrödinger equation in a linear chain is to employ discrete Fourier transforms by multiplying the amplitudes c_m by e^{ikm} and summing over all sites m, to solve the resulting equation for the transform of the amplitude for any k, and to invert the transform. The same method applies in the extreme incoherent case.

4.4.1 The Incoherent Case and the Diffusion Limit

In some places in the last chapter, our notation used M, N as site indices and m, n to represent internal states of vibration or a reservoir. In the rest of the book, unless where explicitly indicated otherwise, we will return to the use of lower case letters for sites.[11] This is in keeping with most usage in the literature. With this notational change, in the incoherent case, the exciton obeys the Master equation for the probabilities $P_m(t)$,

$$\frac{dP_m(t)}{dt} = \sum_n [F_{mn} P_n(t) - F_{mn} P_n(t)]. \tag{4.27}$$

Let us begin the analysis for a system of a finite ring of N sites obeying periodic boundary conditions, and then take the limit of infinite N. If the transition rates obey translational invariance, i.e., if F_{mn} depend only the difference between the site labels and so can be expressed as F_{m-n}, the discrete Fourier transform results in the solution of Eq. (4.27) as

$$P_m(t) = \sum_n \Pi_{mn}(t) P_n(0),$$

the probability propagator being given by

$$\Pi_{mn}(t) = (1/N) \sum_k e^{-ik(m-n)} e^{-tA^k}. \tag{4.28}$$

Here, A^k is the discrete Fourier transform of the translationaly invariant A_{mn}, defined in terms of the rates F through

$$A_{mn} = -F_{mn} \text{ whenever } m \neq n, \text{ and } A_{mm} = \sum_n F_{nm}. \tag{4.29}$$

As the size of the crystal, i.e., N, becomes larger, the allowed k-values come closer together, their interval being proportional to $1/N$. In the limit of the infinite crystal, $N \to \infty$, and the k-summation becomes a continuous integral:

$$\Pi_{mn}(t) = (1/2\pi) \int_{-\pi}^{\pi} dk \, e^{-ik(m-n)} e^{-tA^k}. \tag{4.30}$$

What happens if the interactions in the linear chain are only nearest-neighbor, so that the Master equation (4.27) is actually of the form

$$\frac{dP_m(t)}{dt} = F[P_{m+1}(t) + P_{m-1}(t) - 2P_m(t)] \tag{4.31}$$

---

[11] Chapter 7 will force an exception on us in that we will have to use M, N as site indices again and use m, n to represent internal states of vibration.

with a single F? The integral is now identifiable as proportional to a well-known and well studied special function, the propagator (4.28) becoming

$$\Pi_{mn}(t) = I_{m-n}(2Ft)e^{-2Ft},$$

where I is a modified Bessel function of order $m - n$. Familiar properties of the Bessel functions describe correctly the transport: The self-propagator decays to zero starting from 1 at the initial time, whereas the other propagators start at 0, rise to a maximum, and decay to zero as time goes on. The physics of the evolution of the initially localized quasiparticle at a single site is simply its passage from the initial site to infinity in both directions. The mean square displacement, defined as $a^2 \sum_m m^2 P_m(t)$ for the initial location at the origin, is immediately obtained from standard Bessel function identities, as being $2Fa^2t$, i. e., rising linearly with time. Here a is, of course, the lattice constant (the distance between nearest neighbor sites).

The continuum limit of these results should be familiar. If a is taken to vanish, and F is supposed to tend to infinity such that Fa^2 is a finite quantity that we call the diffusion constant D, Eq. (4.31) reduces to the diffusion equation (1.6). The propagator for the probability *density* is then the well-known Gaussian function, the mean square displacement retaining its value $2Dt$.

4.4.2 The Case of Arbitrary Degree of Coherence

We have already encountered the fully coherent case in Chap. 1. The amplitude propagator to be used with the Schrödinger equation (1.16) is $(-i)^{m-n} J_{m-n}(2Vt)$ i.e., proportional to the ordinary Bessel function.[12] The *probability* propagator is then proportional to the square of the Bessel function,

$$\Pi_{mn}(t) = J_{m-n}^2(2Vt),$$

as shown in Chap. 1.

Let us now derive the intermediate propagator valid for an arbitrary degree of coherence, in other words applicable whatever the value of V/α. All the intermediate results and expressions are already available in the earlier part of the book. To obtain the probability propagator for Eq. (2.25) for any degree of coherence, we start first with the memory function that we have derived for the extreme (purely) coherent case for an infinite chain, i.e., Eq. (1.27), and multiply it by the factor $\exp(-\alpha t)$ to obtain Eq. (2.32) as per the prescription we have derived in (2.31). The $\mathcal{A}^k$ in Eq. (1.26) corresponding to this situation with bath interactions

---

[12]Of course we have put $\hbar = 1$ as we have already been doing. Simplicity rather than wisdom is the motto in this regard. I realize this will endear me to some readers and make me a subject of scorn for other readers. Surely, that is what happens in real life as well.

is now replaced by

$$\tilde{\mathscr{A}}^k(\epsilon) = \sqrt{(\epsilon + \alpha)^2 + (4V)^2 \sin^2(k/2)} - (\epsilon + \alpha). \tag{4.32}$$

This is the consequence of the well known Laplace shift rule that, if the Laplace transform of a function $f(t)$ is $\tilde{f}(\epsilon)$, that of $f(t)\exp(-bt)$ is $\tilde{f}(\epsilon+b)$. The Laplace-inverse of the Fourier transform of the probability propagator (see Eq. (1.24)) is then, for initial localization at the site 0,

$$\tilde{\Pi}^k(\epsilon) = \frac{1}{\sqrt{(\epsilon + \alpha)^2 + (4V)^2 \sin^2(k/2)} - \alpha}. \tag{4.33}$$

Bessel function identities allow the inversion of the Laplace transform into the time domain

$$\Pi^k(t) = J_0\left[4Vt\sin\left(\frac{k}{2}\right)\right]e^{-\alpha t}$$
$$+ \int_0^t du\,\alpha e^{-\alpha(t-u)} J_0\left[4V\sqrt{t^2 - u^2}\sin\left(\frac{k}{2}\right)\right]. \tag{4.34}$$

Equation (4.34), obtained by Kenkre and Phatak (1984), will be used extensively in Chap. 5 in the application to the interpretation of Ronchi ruling and transient grating observations and the measurement of the degree of coherence of Frenkel excitons in aromatic hydrocarbon crystals at various temperatures. Other equivalent expressions include that given in Kenkre (1978d) and the one shown in Wong and Kenkre (1980), and others which are much more difficult to implement as explained in Kenkre and Phatak (1984). Inversion of the Fourier transform implicit in Eq. (4.34) immediately yields the time dependence of propagators in real space on noticing that the inverse discrete Fourier transform of $J_0\left[z\sin^2(k/2)\right]$ equals $J_m^2(z/2)$. These propagators in real space,

$$\Pi_m(t) = J_m^2(2Vt)e^{-\alpha t}$$
$$+ \int_0^t du\,\alpha e^{-\alpha(t-u)} J_m^2\left(2V\sqrt{t^2 - u^2}\right), \tag{4.35}$$

will find use in the description of Ronchi ruling experiments in the first part of Chap. 5, even more directly in the analysis of transient grating experiments in the second part of that chapter, as also of sensitized luminescence experiments in Chap. 11. The self-propagator, i.e., the $m = 0$ case of Eq. (4.35) is plotted for various degrees of coherence in Fig. 4.3 as shown.

This solution that we have obtained to the simple GME for a 1-dimensional infinite crystal, i.e., a linear chain, can also be said to be the solution to the equation of motion that has appeared under different guises in diverse contexts. We see

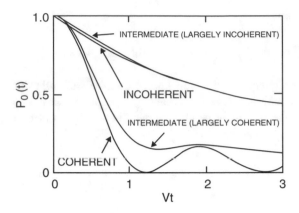

Fig. 4.3 Self-propagator of the probability of an exciton (the probability of the initially occupied site) in an infinite 1-dimensional chain with nearest-neighbor transfer interactions V and degree of incoherence α, plotted as a function of the dimensionless time Vt. The four curves are for four different values, as displayed, of V/α in decreasing order (∞,1, 0.1, and 0, respectively). This quantity is essentially the ratio of the mean free path to the lattice constant.The extreme coherent case shows oscillations which are completely absent from the extreme incoherent case. Adapted with permission from fig. 1 of Ref. Kenkre and Phatak (1984); copyright (1984) by Elsevier Publishing

the equation in Wannier's book on *Solid State Theory* that he published in 1959 to explain Ohmic conduction; in the description of triplet excitons by Avakian et al. given in 1968 (Avakian et al. 1968); in the review by Silbey in 1976 (Silbey 1976); and in our book on exciton dynamics in 1982 (Kenkre and Reineker 1982). The practical usefulness of the expressions we have derived above, in Eqs. (4.34) and (4.35) for reciprocal and direct spaces respectively x, should be immediately obvious on inspecting counterparts for the same that have appeared in the literature elsewhere, in one place, so much less transparently (see Kenkre and Phatak (1984) for a discussion).[13]

---

[13]For instance, one of those earlier expressions actually used for numerical computations was

$$P_m(t) = \frac{\alpha}{2\pi} \int_{-\bar{\kappa}}^{\bar{\kappa}} dk\, e^{ikm} \frac{\exp\left[t\left(-\alpha + \sqrt{\alpha^2 - 16V^2 \sin^2(k/2)}\right)\right]}{\sqrt{\alpha^2 - 16V^2 \sin^2(k/2)}}$$

$$+ \frac{1}{8\pi^2} \int_{-\pi}^{\pi} \int_0^{\pi} dl\, e^{ikm}\, [8V \sin l \sin(k/2)]^2 \left[\frac{e^{t(-\alpha + i4V \sin(k/2)\cos l)}}{16V^2 \sin^2(k/2)\sin^2 l - \alpha^2}\right]$$

where the first term is present conditionally and determines the value of $\bar{\kappa}$. Given that this expression contains two integrations and its connection with the extreme limits is by no means transparent, it is clear that our result Eq. (4.35), from Kenkre and Phatak (1984), is much preferable to use.

4.4.3   Long-Range Transfer Rates as a Consequence of Strong Intersite Coupling

The conscientious reader of this book will have done the assigned exercise in Sect. 2.3.4 and seen for herself that strong intersite transfer can produce memories (consequently, transfer rates) that have a longer spatial range than the matrix elements in the Hamiltonian. For instance, in an open trimer where there are matrix elements V connecting nearest neighbor sites, a memory also develops between the end sites. This is surely a consequence of the exact result requiring higher order terms (in a perturbation series sense) than the lowest, something that becomes important for strong intersite coupling. Let us now comment on interesting cases of this effect.

As a general formula applicable for quasiparticle motion on any translationally invariant system of N sites (with periodic boundary conditions), and matrix elements of arbitrary range V_{m-n}, the bath rate for the destruction of off-diagonal elements of the density matrix being α, it is possible (Kenkre 1978d) to write down the memories in the Laplace domain as

$$\widetilde{W}_{mn}(\epsilon) = -\sum_{k}\left(e^{-ik(m-n)}/\sum_{q}\left(\epsilon + \alpha + i(V^{k+q} - V^{q})\right)^{-1}\right). \tag{4.36}$$

Here V^{k} is the discrete Fourier transform of the interaction matrix elements, hence the same as the band energy at quasimomentum k. Because the integral of the memory $\int_{0}^{\infty} dt\,\mathcal{W}(t)$ equals the rate F, we have, as a consequence,

$$F_{mn} = -\sum_{k}\left(e^{-ik(m-n)}/\sum_{q}\left(\alpha + \frac{(V^{k+q} - V^{q})^{2}}{\alpha}\right)^{-1}\right). \tag{4.37}$$

Equations (4.36) and (4.38) describe a ring of N sites (or higher-dimensioned counterparts) obeying periodic boundary conditions. This means that k (and q) have allowed values decided by $e^{ikN} = 1$. These discrete allowed values pass on to a continuum in the limit that N becomes infinite, the summations in (4.36), (4.38) become integrals, and the rates F_{mn} for the infinite chain show the explicit long-range character expressed as (Kenkre 1978d)

$$F_{mn} = \left[(-1)^{|m-n|+1}\sqrt{\alpha^{2} + 8V^{2}}\,P_{1/2}^{|m-n|}(\beta)\Gamma\left(\frac{3}{2}\right)\right]\left[\sqrt{\beta}\Gamma\left(|m - n| + \frac{3}{2}\right)\right]^{-1}. \tag{4.38}$$

Here, $\beta = (1/\alpha)\sqrt{(\alpha^2 + 8V^2)/(\alpha^2 + 8V^2)}$, and $P_{1/2}^m$ is a Legendre function of fractional order.[14]

For the reader interested in working out the calculations (there is little physics insight one can gain without performing calculations), I recommend deriving the Master equation in the extreme incoherent limit for the opposite limit of size, the smallest interesting example of a ring, viz., one of 4 sites. Calculate the allowed values of k and q, and show that, although in the Hamiltonian there are only nearest-neighbor interactions, rates also develop between opposite corners of the ring. Obtain the Master equation, investigate whether it predicts the probability of occupation of any initially fully occupied site, say 1, to obey

$$\frac{dP_1}{dt} = \xi \left(\frac{2V^2}{\alpha}\right)(P_2 + P_4 - 2P_1)) + (1 - \xi)\left(\frac{2V^2}{\alpha}\right)(P_3 - P_1). \qquad (4.39)$$

as claimed in Kenkre (1978d). Here $\xi = \alpha^2/(\alpha^2 + 8V^2)$ is the factor that equals 1 in the weak-coupling limit $V/\alpha \to 0$ and controls the presence or absence of the new long-range effect we have elucidated here. For weak coupling, the second term of Eq. (4.39) vanishes. As part of the exercise, it is suggested you explore[15] what changes, qualitative as well as quantitative, occur as you vary the coherence ratio V/α.

4.5   Chapter 4 in Summary

Relations of the memory function (GME) theory of exciton transport to other theories and entities in statistical mechanics and condensed matter, as well as solutions of the GME theory for a specific system of special interest to experiments, the linear chain, were the content of this chapter. Our theory is built as a generalization of the theories of Förster as well as Dexter but looks quite different from those developed by Haken and Reineker, and by Grover and Silbey. It was shown that the approaches can be unified by establishing relations among them by inspecting them all in terms of memory functions. Their conceptual differences from the GME theory were also analyzed. It was argued that each method has its own particular purpose and use, separate from the others. The connection between memory functions in probability equations and correlation functions encountered

---

[14]Surprisingly, I have not come across elsewhere this interesting expression for transfer rates on a chain for strong coupling interactions, a very common entity, for the last four decades since it was derived. In a modest manner, the expression hints at what happens when one does not operate under the famous $\lambda^2 t$ limit of van Hove and yet makes the Markoffian approximation to pass from the GME to the Master equation.

[15]Some of the details of the solution for the exercise may be found in Kenkre (1978d) where the time-dependence of the opposite corner site is also shown graphically. Attempt to draw conclusions about the system beyond those set out in that publication.

more usually in statistical mechanics within the framework of linear response was analyzed. The two examples selected for this comparison were velocity correlation functions used in mobility or conductivity theory (Kubo 1957), and scattering functions used in neutron diffraction (Van Hove 1954b). Respectively, they find application in charge mobility physics and in the study of the movement of hydrogen atoms and other interstitials in metals. The next content of the chapter was explicit solutions of the GME extended to have the simple Wigner-Weisskopf incorporation of incoherence in them. These solutions will be useful in subsequent chapters where the effect of coherence is studied on transient grating, Ronchi ruling, and sensitized luminescence observations. Solutions for the full SLE case (with hopping terms added) were also provided. Finally in this chapter, partially nonlocal memories and transfer rates were shown to arise from strong intersite coupling. This chapter thus connects memory functions, primary to the book, to more familiar entities in statistical mechanics and condensed matter physics.

Direct Determination of Frenkel Exciton Coherence from Ronchi Ruling and Transient Grating Experiments

5

This chapter addresses the direct determination of transport coherence of Frenkel excitons in aromatic hydrocarbon crystals, both singlet and triplet, by applying the GME theory explained in the previous chapter.

The first kind of direct experiment used for decades to determine motion characteristic of Frenkel excitons was based on sensitized luminescence. In Chap. 11 of the book we will see, however, a subtle kind of problem that sensitized luminescence observations naturally run into, in spite of the fact that they are the traditional method for measurement and have been repeated in countless instances for numerous systems. Indeed, the problem compelled the late Dankward Schmid and me to conduct, in collaboration with Paul Parris, a protracted and careful analysis of various experiments, culminating in a paper titled "Investigation of the appropriateness of sensitized luminescence to determine exciton motion parameters in pure molecular crystals" (Kenkre et al. 1985a).

It turns out that the preferred direct experiment is by far the one that we discuss in the present chapter. The essential idea is to create an initial distribution of excitons within the crystal and watch the signal coming from the distribution as the distribution homogenizes itself over time as a result of the exciton motion. The methodology for creating a simple spatial distribution is different for Ronchi ruling experiments treated in the first half of this chapter versus for transient grating (four-wave mixing) observations, treated in the second half. It is also different in the two cases for measuring the signal as the time evolution to uniformity occurs. Clever experimental means, which are fundamentally different, are employed in both cases and we analyze the processes in turn.

Coherence of a Frenkel exciton is a concept that has been explored, discussed, sometimes misrepresented and confused, more than most in this field. The purpose of the analysis in the rest of this chapter is to take a precise and economical viewpoint, shun philosophy, insist on quantitatives, and focus on the extraction of coherence information from reported and doable experiments. Such experiments

© The Author(s), under exclusive license to Springer Nature Switzerland AG 2021
V. M. (Nitant) Kenkre, *Memory Functions, Projection Operators, and the Defect Technique*, Lecture Notes in Physics 982,
https://doi.org/10.1007/978-3-030-68667-3_5

using *Ronchi rulings* have been carried out on triplet excitons in Ern et al. (1966), Ern (1969), and Ern and Schott (1976). Counterpart *transient grating* experiments on singlet excitons were introduced by Fayer in Salcedo et al. (1978), analyzed in Agranovich and Hochstrasser (1983) and performed by him with several collaborators (Salcedo et al. 1978; Rose et al. 1984). Several other researchers also made important contributions to our understanding in this area by performing those experiments on inorganic materials (Tyminski et al. 1984; Lawson et al. 1982; Morgan et al. 1986). On the basis of some of those experiments, and of the theory we have developed that we will expound in detail in the present chapter, one can state without ambiguity the primary results.

The degree of coherence, as revealed by these experiments, is measured by the ratio of the mean free path (how far the exciton travels on the average between scattering events) and the distance between nearest neighbor sites of the crystal, the lattice constant. The analysis shows that in anthracene, the singlet exciton moves on the average 20 inter site distances before getting scattered at 20 K; about twice that far at 10 K; and more than 10 times that at 1.8 K showing very coherent motion then. Triplet excitons at room temperatures and a bit lower are quite incoherent, the relevant ratio being a fraction like 0.1 and never rising to much more than twice that even at 118 K, the systems being anthracene, naphthalene and 1,4-DBN. The stark difference between the triplet situation at room temperature and singlet behavior at low temperatures is expected: triplets move much slower, as is well known. As scattering mechanisms are probably of the same order of magnitude in both systems, the difference is not surprising. Explicit tables will be displayed at the end of the description of the theoretical analysis.

For the interpretation of the experiments, we will consider the simple SLE without the additional hopping terms, i.e. Eq. (4.3) without its last term. If there is any reason to suspect the existence of the additional hopping channel, it is straightforward to include that term and obtain the solution. (See Sect. 15.1.1, in particular Eq. (15.8), for the explicit derivation.) The simplicity of the equation we use here will allow us to avoid distractions and focus on the essential issues. The simple SLE we use has appeared earlier in this book as (2.25) and it has the memory given by Eq. (2.32) in the GME that corresponds to it. Its probability solutions are obtained from the propagators in Eq. (4.35) in real space and from Eq. (4.34) in k-space.

What does this equation say about the dynamics of the exciton?

If you place the exciton initially in any of the Bloch states k, (eigenstates of the system in the absence of α), the amplitude of the quasiparticle rotates in phase at frequency proportional to $V \cos k$, (k regarded here as dimensionless as in the rest of the discussion), and scatters equally at a rate proportional to α from that initial state into all the other Bloch states. The ratio of the two inverse time constants, respectively V and α, is thus an excellent measure of the degree of coherence as the exciton transitions from being coherent to incoherent.

If, instead, you place the exciton initially at a single site state m, (Fourier transform of the Bloch state), the quasiparticle moves as a wave (described by Bessel functions as we have seen) exhibiting oscillations in space, with scattering events happening, on the average, in time of the order of $1/\alpha$. The average of the velocity with which the initially localized exciton moves (for no scattering) is the average of the group velocity over the tight-binding band, i.e., proportional to Va where the lattice constant a has been reintroduced. Since scattering events occur on the time of $1/\alpha$, the product of these two quantities is the mean free path $\Lambda = Va/\alpha$.

Dividing the mean free path Λ by the lattice constant a, one gets the natural measure of coherence (addressing the question of how many intersite distances the exciton travels before being scattered, on the average). The result of the division is the same quantity V/α as the coherence measure as we obtained in the picture natural to Bloch states.

It should thus be clear that stating the value of the ratio V/α is equivalent to stating the degree of coherence of the exciton.[1] Coherence viewed thus does not depend on the basis states assumed (in contrast to claims one finds in some assertions in the literature) and it is not necessary to meander among multiple approaches to the concept. All we need is an experimental scheme that can probe the ratio V/α for a given system. The experiments analyzed in this chapter have this ability.

5.1   Ronchi Rulings for Measuring Coherence of Triplet Excitons

The basic idea for this class of experiments came out of the fertile mind of scientists such as the late Vladimiro Ern and his collaborators working in the 1960s (Ern et al. 1966; Ern 1969) in the research laboratory at Dupont. It has been carefully explained in later reviews such as Ern and Schott (1976). The possible effects of coherence in such experiments have been set out in three papers I wrote with him and Alain Fort (Kenkre et al. 1983a,b; Fort et al. 1983).[2] The essential features of that development are explained in this section and end in the tabular presentation of coherence and mobility parameters for triplet excitons in three aromatic hydrocarbon crystals.

The underlying theoretical result that we will use throughout here is Eq. (4.34) from the last chapter, which we multiply by an exponential decay factor, and rewrite

---

[1] Surely, one can encounter situations when, for a fixed V, coherence may be approached more conveniently by considering the magnitude of the scattering time $1/\alpha$ relative to the exciton lifetime, thus by varying the dimensionless quantity $\alpha\tau$.

[2] Vladi Ern invited me to Strasbourg to develop the theory of exciton coherence in Ronchi ruling experiments in the spring of 1983. I learnt much about the design of exciton experiments from the clever mind of that enormously inventive scientist. He was an entertaining story-teller as well as researcher and I had a great time working with him and his very friendly and kind colleague Alain Fort even as they fed me French baguettes and palmier sweets, every day, as compensation. Alain reminded me the other day (when I succeeded in finding him again four decades after our collaboration) that I had strongly urged him to publish a paper with Bob Knox. Apparently I had promised him that I would refer to their result as the Fort-Knox Golden rule. Ah, the foibles of youth!

here as[3]

$$
P^k(t) = e^{-t/\tau} J_0 \left[4Vt \sin\left(\frac{k}{2}\right) \right] e^{-\alpha t}
$$
$$
+ e^{-t/\tau} \int_0^t du\, \alpha e^{-\alpha(t-u)} J_0 \left[4V\sqrt{t^2 - u^2} \sin\left(\frac{k}{2}\right) \right]. \tag{5.1}
$$

The equation describes the time evolution of an initially excited kth mode of the exciton distribution and the exponential factor $e^{-t/\tau}$ represents the radiative (and radiationless) decay with lifetime τ. The coherence is represented by the parameter V/α. We treat here the simplest case when the governing equation is the GME for a crystal with a radiative decay term appended to it. Except for a short discussion in the first half of the chapter, where we have addressed anisotropy investigations by Ronchi ruling experimentalists, the observations are in essence 1-dimensional. Consequently, we have restricted our attention to a linear chain. It is straightforward to extend the number of dimensions of the system to more as explained later below. The solution (5.1) makes the assumption common to normal derivations of the GME that the initial condition is one of random phases in the m-representation. Departures from this assumption are analyzed carefully in the second part of the present chapter.

If one shines light on a crystal and thereby creates triplet excitons in a spatial region small enough so that a substantial fraction of the excitons created can move out of that region within their lifetime, the delayed fluorescence signal arising from their mutual annihilation decays in time on account of their motion as well as of the finiteness of their lifetime. The faster the motion, the more the depletion of the population from the initially illuminated region. The illumination is made through a periodic linear array of alternating opaque and transparent strips (called the Ronchi ruling) placed under the crystal. The delayed fluorescence signal is collected from the entire crystal. Our point of departure for the analysis is the GME augmented by a decay term $-P_m(t)/\tau$ that describes the radiative and radiationless decay with lifetime τ, a term $\mathcal{I}_m(t)$ from the source of illumination, and a term $-\gamma' P_m^2(t)$ that represents the signal from the mutual annihilation of excitons (producing the delayed fluorescence):

$$
\frac{dP_m(t)}{dt} = \int_0^t ds \sum_n [\mathcal{W}_{mn}(t-s)P_n(s) - \mathcal{W}_{nm}(t-s)P_m(s)]
$$
$$
- \frac{P_m(t)}{\tau} + \mathcal{I}_m(t) - \gamma' P_m^2(t). \tag{5.2}
$$

---

[3]It is straightforward to replace this expression by another corresponding to the underlying equation of motion, e.g., the diffusion equation if the motion is totally incoherent and the continuum limit is taken, the Master equation if the first of these assumptions is appropriate but not the second, or the full SLE with the additional "phonon assisted" Master terms if that is what is required.

In keeping with the philosophy and technique of the experiment, the illumination intensities are assumed kept small enough so that the last term can be neglected in the computation of P_m (but of course kept in the calculation of the signal). There are two parts of the experiment, the first (buildup) part in which the delayed fluorescence builds up to its saturation value under the action of a constant illumination source, and a second (decay) part in which the source is shut off and the fluorescence decays from the saturation value. During the buildup part, no excitons are initially in the system. The solution then is obtained in terms of the GME propagator $\Pi_{m-n}(t)$. At the simplest level of description, our choice for the propagator including the effect of the decay term is Eq. (5.1).

5.1.1 Additive Terms and Expressions for the Signal

There are no excitons initially in the system, i.e., $P_m(0) = 0$, and the source term may be written down as the product of the intensity of illumination multiplied by the appropriate $S_0 \rightarrow T_1$ absorption coefficient, the Heaviside step function $\Theta(t)$ and the ruling geometry function g_n,

$$I_n(t) = i_0 g_n \Theta(t).$$

The computation of the expressions for the signals is done by taking discrete Fourier transforms and solving the time-dependent equations. A step-by-step derivation may be found in the original paper (Kenkre et al. 1983a). Let us denote the normalized results for the build-up and decay stages of the experiment by $\Phi^b(t)$ and $\Phi^d(t)$, respectively. Careful experimentation dictates that one focus attention on the difference of these signals and their values in the absence of the motion, equivalently in the absence of the ruling (or as the ruling period becomes infinite.) Denoting these differences by $\Delta\Phi^b(t)$ and $\Delta\Phi^d(t)$, respectively, we obtain

$$\Delta\Phi^b(t) = (1 - e^{-t/\tau})^2 E_b(t), \tag{5.3a}$$

$$\Delta\Phi^d(t) = e^{-2t/\tau} E_d(t), \tag{5.3b}$$

with the most important quantities of the theory being,

$$E^b(t) = \left[\frac{1}{1 + \sum_{l=1}^{\infty} A_l}\right] \left[\sum_{l=1}^{\infty} A_l \left(\left(\frac{\int_0^t ds\, e^{-s/\tau} \Pi^{\eta_l}(s)}{(1 - e^{-t/\tau})\widetilde{\Pi}^{\eta_l}(1/\tau)}\right)^2 - 1\right)\right], \tag{5.4a}$$

$$E^d(t) = \left[\frac{1}{1 + \sum_{l=1}^{\infty} A_l}\right] \left[\sum_{l=1}^{\infty} A_l \left(\Pi^{\eta_l}(t)\right)^2 - 1\right]. \tag{5.4b}$$

In these expressions, $\Pi^{\eta_l}(t)$ equals the right hand side of Eq. (5.1) without the exponential decay factor, and k put equal to η_l, the l's are positive integers running from 1 to ∞, and the dimensionless wavevectors η_l are given by

$$\eta_l = 2\pi(a/x_0)(2l - 1). \tag{5.5}$$

The ruling period is x_0, and the coefficients A_l are

$$A_l = \left[\frac{8}{\pi^2(2l-1)^2}\right]\left[\frac{\widetilde{\Pi}^{\eta_l}(1/\tau)}{\tau}\right]^2. \tag{5.6}$$

The last two expressions show the infinite number of spatial Fourier modes that the rectangular ruling (square wave) select.

Because the expressions are simpler in the decay stage of the experiment, the corresponding result in Eq. (5.4b) is plotted in Fig. 5.1 for purely coherent motion ($\alpha = 0$) along with attempts to fit it with results derived for purely incoherent motion (α infinite).

It is convenient to introduce a "transport length" l_T as a generalization of the well-known diffusion length. It is the average distance travelled by the exciton during its lifetime, whatever its degree of coherence. For our system it is given by[4]

$$l_T = 2a(V/\alpha)\sqrt{\alpha\tau - 1 + e^{-\alpha\tau}}. \tag{5.7}$$

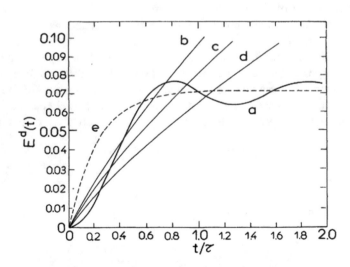

Fig. 5.1 Time dependence of the calculated decay signal $E^d(t)$ in a Ronchi ruling experiment plotted as a function of the dimensionless time t/τ for purely coherent motion of the triplet exciton (solid line marked a showing oscillations) along with fits from the incoherent expressions. The fits are based on the evolution in the region $t < \tau$ for b, c, d and in the $t > \tau$ for e. They are all unsatisfactory and thus illustrate how extreme coherence effects would be manifested in Ronchi ruling experiments. Adapted with permission from fig. 4 of Ref. Kenkre et al. (1983a); copyright (1983) by the American Physical Society

---

[4] As an exercise, provide precise reasons to motivate the expression in Eq. (5.7).

That it is a generalization of the diffusion length is clear by taking the limit $\alpha \to \infty$, $V \to \infty$, calling the limit of $2V^2/\alpha$ as the incoherent rate F, and Fa^2 as the diffusion constant D in the standard manner. In the coherent limit $\alpha \to 0$, it reduces to the length a particle would move in its lifetime if it were to move with the average group velocity in the tight-binding band. The ratio l_T/x_0 is obviously of crucial importance to the Ronchi ruling experiment as it matches the relevant length of the exciton to the measuring length, the Ronchi ruling period.

Many different ways of using this theoretical analysis in the interpretation of Ronchi ruling experiments exist and have been presented in detail in the three 1983 publications referred to at the beginning of this section. Here we point out Fig. 5.2 in the upper panel of which the ratio l_T/x_0 is kept constant and the degree of coherence varied: the values of $\alpha\tau$ vary as shown, from the coherent value 0 to the value 100 which is already the fully incoherent extreme. Oscillations are clearly representative of coherence. In the lower panel, the bandwidth of the exciton is kept constant, and the degree of coherence varied, with a similar qualitative consequence. In a realistic situation, which panel to use depends on what is known, the l_T from an independent experiment (such as sensitized luminescence perhaps) or the bandwidth V.

Fig. 5.2 Effect of coherence on the time dependence of the calculated decay signal $E^d(t)$ in a Ronchi ruling experiment plotted as a function of the dimensionless time t/τ for various degrees of coherence depicted by values of $\alpha\tau$ as shown. Held constant in (**a**) is the ratio l_T/x_0 of the transport length to the ruling period at the value 0.2 and in (**b**) the ratio $V\alpha\tau/x_0$ at the value 0.25. Reprinted with permission from fig. 7 of Ref. Kenkre et al. (1983a); copyright (1983) by the American Physical Society

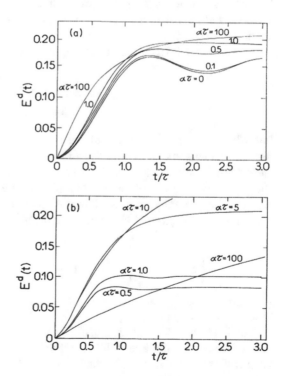

5.1.2 Steady State Accumulated Signal and Anisotropy Investigations

Of the variety of ways in which Ronchi ruling observations can be carried out, a simple one consists of focusing on the *steady-state* signal (SSS) accumulated during the build-up stage. This is a time-independent quantity in contrast to $\Delta\Phi^d(t)$ or $\Delta\Phi^b(t)$ analyzed above. Let us study it in this final part of the chapter on Ronchi rulings first, following Kenkre et al. (1983b), and then comment on anisotropy investigations as well.

Accumulated Steady State Signal
Under illumination, which is made to remain constant in time after once being switched on, one gets from the analysis shown,

$$\lim_{t\to\infty} \sum_m P_m^2(t) = \text{const.} \times \left(1 + \sum_{l=1}^{\infty} A_l\right). \tag{5.8}$$

The constant includes the square of the intensity of illumination and A_l is known from Eq. (5.6) which contains information about the geometry of the ruling and the motion of the exciton.

Let us then write the accumulated signal as

$$sss = \frac{1}{2}\left(1 + \sum_{l=1}^{\infty} A_l\right), \tag{5.9}$$

and, although it is not necessary to make the simplification, use the fact that η_l turns out to be much smaller than 1 for all cases of practical interest and approximate the sine by its argument inside the expression (5.6) for A_l. This works well because the lattice constant a is of the order of 10 A whereas the grating period is typically of the order of a micron. This allows us to approximate $(4V\tau)^2 \sin^2(\eta_l/2)$ as the square of $(4\pi V\tau a/x_0)(2l - 1)$ and write the accumulated steady state signal as

$$sss = \frac{1}{2}\left(1 + \frac{8}{\pi^2}\sum_{l=1}^{\infty}\frac{1}{(2l-1)^2}\left(\frac{1}{\sqrt{(1+\alpha\tau)^2 + c^2\tau^2(2l-1)^2} - \alpha\tau}\right)^2\right), \tag{5.10}$$

where $c = 4\pi Va/x_0$.

The important quantities in Eq. (5.10) happen to be $c\tau$ which is a measure of the ratio of the exciton lifetime to the time taken by the exciton to travel a distance equal to the ruling period x_0, and $\alpha\tau$ which is a measure of the ratio of the exciton lifetime to the mean time between scattering events. In the extreme limits of pure coherence (vanishing α) and of total incoherence (infinite α) respectively, the infinite sums in

✱ (5.10) can be evaluated analytically as shown in Kenkre et al. (1983b).[5] The limiting results for the signal are found to be

$$sss = 1 - \frac{1}{2}\left(2^{5/2}\frac{l_T}{x_0}\right)\tanh\left(\frac{x_0}{l_T 2^{5/2}}\right),$$ (5.11a)

$$sss = 1 - \frac{3}{4}\left(2^{3/2}\frac{l_T}{x_0}\right)\tanh\left(\frac{x_0}{l_T 2^{3/2}}\right) + \frac{1}{4}\operatorname{sech}^2\left(\frac{x_0}{l_T 2^{3/2}}\right).$$ (5.11b)

These extreme limits of the calculated steady state signal are plotted as a function of the dimensionless ratio l_T/x_0 in Fig. 5.3. The hope that the dependence might be different in the two cases allowing one to use this form of the experiment to gauge the degree of coherence is not fulfilled. We see that there is little possibility of graphical distinction. It is for this reason that analysis of the more detailed time-dependent signal given above earlier becomes important.

Anisotropy Investigations

The time-dependent signal equations, e.g., (5.3b) with (5.4b), are applicable to a system of arbitrary number of dimensions and can thus be used in investigations of anisotropy. A thorough analysis, as well as actual experimental demonstration where the complete anisotropic diffusion tensor of triplet excitons was mapped out in the incoherent regime, may be found in Ern (1969). This work was based on the convenient product property of k-space propagators $\Pi^m(t)$ in the time domain. The property arises from the fact that the propagators are exponential. A review may be found in Ern and Schott (1976). What is remarkable is that, although the exponential property does not hold for propagators for arbitrary degree of coherence, Fort et

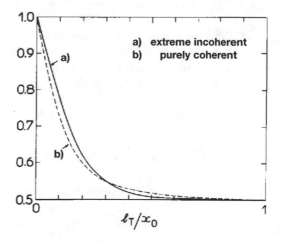

Fig. 5.3 The steady-state accumulated delayed fluorescence signal SSS plotted against the ratio l_T/x_0 of the transport length to the ruling period showing very little difference for fully incoherent and purely coherent cases. This necessitates the study of the time-dependent signal presented earlier. Reprinted with permission from fig. 2 of Ref. Kenkre et al. (1983b); copyright (1983) by Elsevier Publishing

a) extreme incoherent
b) purely coherent

---

[5]For the reader interested in honing her skills at practical calculus, this would be a fine exercise to undertake. The solution is outlined in the published paper mentioned.

al. were able to present an approximate procedure to map out the velocity tensor in the coherent limit (Fort et al. 1983). They based it on an unexpectedly good approximation whereby one can represent the product of three Bessel functions by a single Bessel function for small arguments.[6] We refer the reader to Fort et al. (1983) for details.

5.1.3 Initial Condition Subtlety

In the application of the grating *or* ruling experiments, there remains, however, an open question. It stems from the nature of the initial condition. In principle, it could influence the evolution of the exciton distribution in a profound manner. Realizing this, as we worked on coherence effects in Ronchi ruling experiments in 1983, I persuaded my coauthors to include the following footnote (no. 34) in our article Kenkre et al. (1983a) which is reproduced here with a slight change.[7]

> An interesting illustration of initial condition wherein the GME may not be used in the form given is provided by the amplitude distribution $C_m(0) = \text{const}\cos(\eta m/2)$. It corresponds to populating two exciton k states with equal and opposite quasimomenta, $\eta/2$ and $-\eta/2$, respectively, and to an initial probability distribution $P_m(0) = \text{const}[1 + \cos(\eta m)]$. If the latter form of $P_m(0)$ is used..., one would conclude that the exciton distribution evolves in time essentially as explained... However, if the above $C_m(0)$ is substituted in the Schrödinger equation..., one concludes that $P_m(t) = P_m(0)$ for all time. The reason for the contradiction is that, in this particular example, the initial driving term which is dropped in the form...of the GME, makes a contribution to the evolution of $P_m(t)$ which is exactly equal and *opposite*...In a realistic situation the above amplitudes $C_m(0)$ are expected to have phase factors which vary slightly but randomly from ensemble member to ensemble member with the result that, while $\rho_{mn}(0)$ has a zero ensemble average, $\rho_{mm}(0) = P_m(0)$ is still given by $\text{const}[1 + \cos(\eta m)]$. The initial condition then belongs to what has been called the third class in the text and the GME...is again valid.

Two years later, through what must be an independent realization of this subtlety we had pointed out in the Ronchi ruling context, Garrity and Skinner (1985) wrote about this initial condition problem in the context of Fayer's transient grating experiment. As a continuation of our investigation of the issue, working in collaboration with Schmid and Tsironis, I presented a somewhat thorough study of the issue in Kenkre and Tsironis (1985), Kenkre and Schmid (1985), Kenkre et al.

---

[6]This surprising approximation, whose validity is examined graphically in the quoted reference, is

$$J_0(u_1)J_0(u_2)J_0(u_3) \approx J_0\left(\sqrt{u_1^2 + u_2^2 + u_3^2}\right)$$

for small u_1, u_2, and u_3.

[7]In the first sentence of the quoted footnote I have replaced here "form of (2.1)" by "form given" to avoid distraction of the equation number from the original paper.

Table 5.1 Triplet exciton parameters given by Ronchi ruling experiments. Reprinted with permission from Ref. Kenkre and Schmid (1987); copyright (1987) by Elsevier Publishing

Crystal	$T(K)$	$D(10^{-4}cm^2/s)$	$a(nm)$	$V(10^{11}s^{-1})$	$\alpha(10^{11}s^{-1})$	Λ/a
Anthracene	371	1.6	0.524	4.24	63	0.1
	298	1.5			67	0.1
	160	2.5			40	0.16
	118	4.0			25	0.24
Naphthalene	300	0.3	0.510	2.36	95	0.03
1,4-DBN	300	3.5	0.409	13.8	186	0.11

(1985b), Kenkre and Schmid (1987), and Tsironis and Kenkre (1988).[8] As will be shown in detail in the second part of the present chapter when discussing the motion of singlets and the transient grating method of observing them, the initial condition problems tend to disappear as a result of superposition of states. Values of coherence parameters can be extracted, therefore, from experiments both on triplet and singlet excitons. For triplets they take the form shown in Table 5.1.

5.2 Transient Gratings for Singlet Exciton Coherence

We have seen how conceptually and practically convenient the Ronchi ruling experiment is, being direct in scope, and devoid of the shortcomings of sensitized luminescence that will become clear to us in Chap. 11. The Ronchi ruling procedure has resulted in accepted values for the triplet exciton diffusion constant, provided a clear picture of anisotropy in motion in the crystals studied, and is awaiting being put to use at low temperatures.

In the singlet realm, a conceptually identical technique was pioneered by Fayer (Salcedo et al. 1978; Rose et al. 1984) and announced by him in Agranovich and Hochstrasser (1983). It was developed as a result of dramatic progress in picosecond methodology, and goes under the name of the transient grating experiment.

A picosecond laser pulse is split into two parts and the two parts made to arrive simultaneously at a variable angle in the crystal. Optical absorption creates a singlet excitation population varying sinusoidally in space as a result of the *interference* of the two pulses. This interference plays the role here that the mechanical etchings on the ruling placed under the crystal in the Ronchi ruling method performs: it creates the inhomogeneity. The transient exciton grating then evolves in time. The evolution

---

[8]Respectively, the fundamental study of the initial condition problem was published in the first, an application to experiments reported on anthracene crystals at various temperatures was initiated in the second, an announcement of the results at a conference was made in the third, an updated extraction of the degree of coherence and of the magnitude of the diffusion constant in anthracene was made in the fourth from new experiments reported, and a theoretical analysis of temperature effects was supplied in the fifth of these publications.

is monitored through the diffraction of a third laser pulse delayed appropriately. The diffraction process plays the role of the signal produced by mutual annihilation of the triplet excitons in the Ronchi ruling experiments.

These experiments have been successfully conducted also in inorganic solids by Powell, Tyminski, Yen, and others (Lawson et al. 1982; Tyminski et al. 1984; Morgan et al. 1986) who appeared to have used in part our theory (Kenkre 1978d, 1981a; Kenkre and Reineker 1982; Wong and Kenkre 1980) in the interpretation of their observations. Although the basic considerations are essentially the same as we have explained for Ronchi ruling observations in the first part of the present chapter, the transient grating analysis takes even a simpler form because of the conceptual simplicity of both the excitation and the detection processes. Little more than a single Fourier mode is excited.[9] Similarly, the detection process is performed by diffraction for transient grating observations whereas delayed fluorescence caused by mutual annihilation of triplets, a nonlinear process, is used in the ruling experiment. In the next section, the transient grating theory is developed step by step as we gradually make more complex the motion description.

5.2.1   Conceptual Simplicity of the Transient Grating Experiment

Consider a sinusoidal inhomogeneity (the transient grating) of excitons produced in the crystal by the optical absorption of two laser beams crossed at an angle θ, λ being the wavelength. The fringe spacing of the grating is $d = \lambda/2 \sin(\theta/2)$ and the initial condition on the probability density in the crystal, considered continuous for simplicity, is $P(x,0) = (1/2)(1 + \cos \Delta x)$ where $\Delta = 2\pi/d$. If the exciton motion proceeds via the diffusion equation while the exciton is simultaneously undergoing decay with lifetime τ,

$$\frac{\partial P(x,t)}{\partial t} = D \frac{\partial^2 P(x,t)}{\partial x^2} - \frac{P(x,t)}{\tau},$$

the solution for the probability is

$$P(x,t) = \frac{e^{-t/\tau}}{2} \left[1 + e^{-\Delta^2 Dt} \cos \Delta x \right].$$

The observed signal is essentially proportional to the square of the grating depth, i.e., to $[P(0,t) - P(d/2,t)]^2$. The diffusion equation predicts the signal to be exponential, e^{-Kt}, the exponent being

---

[9]This terminology refers to the fact that just three dimensionless wavevectors of the *probability*, 0 and two equal and opposite ones, need to be excited to produce an inhomogeneity that is a single sinusoid (what happens in the transient grating) whereas an infinite number of modes are excited in the ruling experiment because of the rectangular geometry of the ruling.

$$K = 2\left(\Delta^2 D + \frac{1}{\tau}\right). \tag{5.12}$$

It is obvious that this is a direct experiment that would allow us to measure the exciton diffusion constant D. When the experiment was announced by Fayer along with the above diffusion analysis (Salcedo et al. 1978), it was easy to generalize the argument immediately, first to one appropriate to a discrete lattice and then to one appropriate to coherent motion (Kenkre 1978d). The first generalization used Bessel function propagators instead of Gaussian ones appropriate to the continuum diffusion equation, and arrived at an exponential signal (as in the case of the diffusion equation) but with the slightly different exponent

$$K = 2\left(4F\sin^2(\Delta a/2) + \frac{1}{\tau}\right). \tag{5.13}$$

Surely, when the lattice constant $a \to 0$ as in the passage back to the diffusion, Eq. (5.13) yields (5.12).

The next level of generalization is to a situation that involves an arbitrary degree of coherence in the transport of the exciton as described by the SLE (2.25). The SLE is for quantum evolution and consequently requires for its solution specification of the initial off-diagonal as well as diagonal elements of the density matrix. Continuity with the preceding classical calculations (from the diffusion or the Master equation) makes it natural to take the initial density matrix to be random-phase in the site-diagonal representation. We use thus the corresponding GME propagator (5.1).

The degree of (in)coherence is determined in the experimental context by the value of the scattering rate α to the intersite interaction V or, more appropriately, to the quantity b which contains both V and the fringe spacing. It is defined as

$$b = 4V\sin(\eta/2) = 4V\sin(\Delta a/2). \tag{5.14}$$

It can be varied in principle either by choosing different materials with differing exciton bandwidths, i.e., by varying V, or, more practically by varying the dimensionless ratio $\eta = \Delta a$. The latter can be controlled very simply by changing the angle of crossing of the exciting laser beams.

The signal $S(t)$ is generally non-exponential. There are experimental conditions that can mask this nature and make the signal look exponential. In such a case, one can always refer to an effective exponent, using the approximation $S(t) \approx S(0)\exp(-Kt)$. Equating the time integrals of both the signal and the approximating exponential, we get the value of the exponent:

$$K = \frac{S(0)}{\int_0^\infty dt\, S(t)}.$$

Fig. 5.4 Representative
transient grating exponent
$R = (K/2) - (1/\tau)$ plotted
on a log-log scale with
respect to b, see text, showing
the transition from the
incoherent asymptote b^2/α to
the coherent asymptote b as b
is increased. All quantities are
taken relative to α. Modified
with permission from fig. 3 of
Ref. Kenkre (1978d);
copyright (1978) by the
American Physical Society

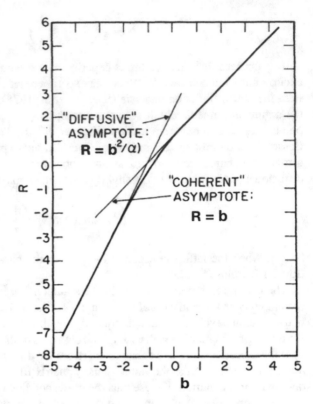

This was done in Kenkre (1978d) and the exponent calculated to be

$$K = 2\left(\sqrt{\alpha^2 + 16V^2 \sin^2(\Delta a/2)} - \alpha + \frac{1}{\tau}\right). \tag{5.15}$$

The precise quantity appropriate to the discussion in Kenkre (1978d) is $R = \frac{K}{2} - \frac{1}{\tau}$.
A modified version of the relevant log-log plot from that publication is displayed
in Fig. 5.4 as b, defined in Eq. (5.14), is varied. A continuous transition of $R =$
$\sqrt{\alpha^2 + b^2} - \alpha$ from linear to quadratic dependence on b is seen. Interestingly, the
variation is similar to that in the plot of the transfer rate used in Fig. 3.6 for the
resolution of the R^3 versus R^6 puzzle, and arises for similar reasons, but from quite
different starting expressions.

5.2.2 Oscillations in the Transient Grating Signal

The full non-exponential time-dependent signal calculated from the GME or the Avakian SLE[10] was displayed soon after Fayer's announcement of the grating experiment in a variety of publications including (Kenkre 1981a; Kenkre and Reineker 1982; Wong and Kenkre 1980). A pedagogical chart of system features and predicted signals was displayed in the first of these publications and it was pointed out in that paper and the next (book) publication how direct the grating observation was. It was argued that the grating experiment was like neutron scattering observations in that there was a direct and very easily demonstrable connection between system features and experimental signals.

The signal we calculate on the basis of our GME propagator (5.1) is (from here on we will omit displaying the lifetime τ in the signal for simplicity, taking it to be infinite):

$$S(t) = \left[J_0(bt)e^{-\alpha t} + \alpha \int_0^t du\, e^{-\alpha(t-u)} J_0\left(b\sqrt{t^2 - u^2}\right) \right]^2. \tag{5.16}$$

In Wong and Kenkre (1980), the signal has been plotted as well as presented in the mathematically equivalent form

$$S(t) = \left[1 - e^{-\alpha t} b \int_0^t du\, e^{\alpha\sqrt{t^2 - u^2}} J_1(bu) \right]^2. \tag{5.17}$$

✵ The interested reader is invited to discover the simple derivative relation that the Bessel functions J_0 and J_1 bear to each other and, on its basis, establish the equivalence between Eqs. (5.16) and (5.17).

Part of the plot of the signal is reproduced here in Fig. 5.5 where t and α are both scaled to V. We see that the signal decreases in time as in the case of the diffusion or the Master equation but exhibits *oscillations* representative of transport coherence. The extreme coherent limit ($\alpha = 0$) is reflected in the signal as the square of the zero-order J-Bessel function in Eq. (5.16). Figure 5.5 shows the closeness of the exact intermediate signal for (A) $\alpha/V = 0.02$ to the coherent limit (square of the Bessel function) and of the exact intermediate signal for (B) $\alpha/V = 2$ to the incoherent limit (exponential).

The result of our theory, Eq. (5.16) or (5.17), applies whatever the actual values of the lattice constant, the intersite matrix element V, and the scattering rate α.[11] For large scattering and the continuum limit of vanishing a, it reduces to the diffusion equation expression (5.12) given by Salcedo et al. (1978).

---

[10]We also calculated the signal as predicted by the full SLE that contained the additional hopping terms as in Eq. (4.3).

[11]Provided the assumptions of the nearest-neighbor nature of the interaction and of a single decay rate α for all off-diagonal density matrix elements represent the transport properly.

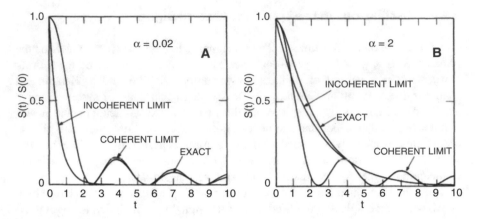

Fig. 5.5 Time-dependent transient grating signal plotted against time t in units of $1/V$, curve labeled as 'exact' along with the pure coherent limit for which $\alpha/V \to 0$ and the extreme incoherent limit for which $\alpha/V \to \infty$. Left panel (A) has $\alpha/V = 0.02$ and right panel (B) has $\alpha/V = 2$. The high coherence in the former case is seen in the near identity of the coherent limit and the actual example. The low coherence in the latter case is seen in the exponential nature of the signal and near identity with the incoherent extreme. See text. Adapted with permission from fig. 1 of Ref. Wong and Kenkre (1980); copyright (1980) by the American Physical Society

Numerical estimates might be helpful. The values $\alpha = 10^{12} \text{sec}^{-1}$, $V = 10^{12} \text{sec}^{-1}$, $\tau = 10^{-8} \text{sec}$, appear reasonable. They allow the neglect of the lifetime in the expressions given that $\alpha\tau \gg 1$. Because $(4V/\alpha)$ is a small quantity, the validity of further reduction depends on the value of the lattice constant a and the wavelength λ. For a concentration of about 1.6×10^{-3} mol/mol mentioned in the experiment reported by Salcedo et al., $a \approx 50A$ and $\lambda \approx 5 \times 10^4 A$, surely we have here $b \ll \alpha$ despite the fact that the bandwidth and the scattering rate are of the order of each other. The diffusion treatment used by Salcedo et al. (1978) in the interpretation of their reported experiment is reasonable. Yet, lowering the temperature and changing the crossing angle θ could certainly necessitate coherence considerations and introduce signal oscillations into the picture, at least in principle.

Kenkre and Schmid (1985) applied this theory to later observations of Fayer and collaborators (Rose et al. 1984) and determined the degree of coherence in pure anthracene crystals from observations reported at various temperatures. Several auxiliary points of theory were also clarified by us there. A table of the conclusions drawn about the degree of coherence initiated in Kenkre and Schmid (1985) was updated in Kenkre and Schmid (1987) and is presented at the end of this section.

5.3   Initial Condition Issue

As pointed out by Kenkre et al. (1983a) in their footnote no. 34, and discussed in Sect. 5.1.3 above, a pair of Bloch states of the Frenkel exciton, if initially excited, would produce the same sinusoidal inhomogeneity as the random-phase initial condition we have used in the above analysis. Yet, in a perfectly coherent situation, it would result in a signal that remains constant in time. There would be no decay and no oscillations. A careful analysis of the initial condition issue is clearly warranted for the interpretation of experiments. Let us set it out in this section, following the arguments in Kenkre and Tsironis (1985).

We begin with the simple (Avakian) SLE in Eq. (2.25) characterized by the nearest-neighbor interaction matrix element V and the scattering or dephasing rate α and seek its solution for arbitrary initial conditions. While its solution can be obtained in many ways, it is most appropriate to follow the one outlined in Chap. 15 of the present book, which was carried out by Kenkre and Brown (1985) for their evaluation of the van Hove scattering function. The method is particularly interesting and simple, and results directly in the Fourier transform of the probability density that we seek for the transient grating experiment.

The starting point is Eq. (15.1) and the solution we require for the Fourier-Laplace transform of the diagonal elements of the density matrix is Eq. (15.4). We anticipate that solution, and rewrite it here explicitly, employing the symbol $\mathcal{H}$ in place of η to avoid confusion with the dimensionless grating wavevector η used here for the transient grating analysis,

$$\widetilde{\rho}^k(\epsilon) = \frac{\widetilde{\mathcal{H}}^k(\epsilon)}{1 - \alpha \widetilde{\Pi}^k(\epsilon)}. \tag{5.18}$$

We already know the propagator $\Pi^k(t)$ in the time domain (to be calculated in the absence of scattering, see Chap. 15) to be simply $J_0[4Vt \sin(k/2)]$. As shown in Kenkre and Tsironis (1985) transparently, this solution leads to the grating signal as the square of $\rho^\eta(t)$. All that is necessary is to evaluate, from the given initial condition on the density matrix, the quantity $\widetilde{\mathcal{H}}^k(\epsilon)$ at $k = \eta$, from the exciton dynamics in the absence of scattering.

Specifically, we have

$$\mathcal{H}^k(t) = \sum_{m'n'} J_{n'-m'}[4Vt \sin(k/2)]\, \rho_{m'n'}(0) \exp[i(k/2)(m' + n')], \tag{5.19}$$

an expression that can be evaluated for whatever initial condition on the density matrix one desires.

5.3.1 Calculations for the Pair State

The random-phase initial condition that produces a grating of the dimensionless wavevector η is

$$\rho_{mn}(0) = (2/N)[\cos(m\eta/2)][\cos(n\eta/2)].$$

Fig. 5.6 Brillouin zone and
energy band of the exciton
showing the pair-state initial
condition that produces the
same sinusoidal initial
inhomogeneity probability as
does the standard
random-phase initial
condition but very different
time evolution of the grating
signal. The dimensionless
wavevector whose 2π span
forms the zone is depicted by
k. The two states forming the
pair are separated by η but
centered at κ in the Brillouin
zone. Reprinted with
permission from fig. 1 of Ref.
Kenkre and Tsironis (1985);
copyright (1985) by Elsevier
Publishing

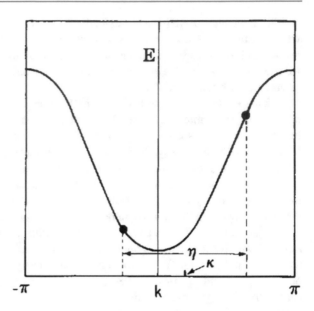

The pair state we will consider is a slight generalization of the one we commented
on in footnote 34 of Kenkre et al. (1983a) in that it is not placed symmetrically in
the Brillouin zone (see Fig. 5.6) but consists of the occupation of two Bloch states
at $k = \kappa \pm (\eta/2)$. The probability inhomogeneity has the dimensionless wavevector
η but the center of the pair is at κ, an arbitrary location in the Brillouin zone. The
initial condition on ρ elements involves simply a factor multiplying the random-
phase case

$$\rho_{mn}(0) = (2/N)[\cos(m\eta/2)][\cos(n\eta/2)]\exp[i\kappa(m-n)],$$

but with profound consequences for the signal evolution.

Combining these respective initial conditions with (5.19) to calculate $\mathcal{H}^k(t)$, it is
straightforward to obtain a very simple relation between the random-phase and the
pair-state signals. The former leads, in the Laplace domain, to

$$\tilde{\rho}^{\eta}{}_{rand}(\epsilon) = \frac{1}{\sqrt{(\epsilon + \alpha)^2 + b^2} - \alpha}, \tag{5.20}$$

where the reader is reminded of the dependence of b on η, the dimensionless
wavevector: $b = 4V\sin(\eta/2)$. The pair-state result is simply given by multiplying
this random-phase expression by a factor $\tilde{f}_\kappa$

$$\tilde{f}_\kappa(\epsilon) = \frac{(\epsilon + \alpha)\sqrt{(\epsilon + \alpha)^2 + b^2}}{(\epsilon + \alpha)^2 + b^2 \sin^2 \kappa}. \tag{5.21}$$

In other words,

$$\tilde{\rho}^{\eta}{}_{pair}(\epsilon) = [\tilde{f}_{\kappa}(\epsilon)][\tilde{\rho}^{\eta}{}_{rand}(\epsilon)]. \tag{5.22}$$

The signal in the time domain is, in either case,

$$S(t) = \left[\rho^{\eta}(t)\right]^2. \tag{5.23}$$

The close relationship of the pair-state and the random-phase signals has many interesting consequences that are essential to the understanding of this initial condition issue. First, let us examine only the pair state. One discovers that changes in the location κ in the Brillouin zone can lead to widely changing signals if the motion is coherent. Figure 5.7 depicts these differing signals as κ takes on different values in the zone. The value of the coherence parameter $\zeta = b/\alpha$, which measures the mean free path of the exciton relative to the fringe spacing is 10 in (A) but a factor of a 100 less in (B). In the latter case, all the different κ signals collapse into a single line as shown.

In a typical experimental situation, it is obviously relevant to consider *superpositions* of pair states forming a wave-packet. A simple choice of the weight function $g(\kappa)$ is a Gaussian,

$$g(\kappa) = \frac{e^{-\kappa^2/2\sigma^2}}{\sqrt{2\pi}\sigma}.$$

In Fig. 5.8 we show the effect of varying the width σ of such a wavepacket for rather coherent conditions. In Fig. 5.9 we show the same for incoherent conditions. See the captions for explanation.

5.3.2   How Superposition of Pair-State Signals Yields the Random-Phase Signal

In understanding the initial condition issue, it is crucially important to notice, by inspection of Fig. 5.8, that, for coherent conditions, as the width σ of the packet of pair states increases, the signal from their superposition tends in appearance to the signal from the random-phase initial condition given in Wong and Kenkre (1980).

The transition is even more clear through the following analytic exercise. If you integrate the multiplicative factor f_{κ} (see Eq. (5.21)) with equal weight throughout the Brillouin zone, you obtain 2π. The equal-weight superposition of pair states yields, therefore, the random-phase signal *precisely*!

$$\frac{1}{2\pi} \int_{-\pi}^{\pi} d\kappa \, \tilde{\rho}^{\eta}{}_{pair}(\epsilon) = \left[\frac{1}{2\pi} \int_{-\pi}^{\tilde{\pi}} d\kappa f_{\kappa}(\epsilon)\right] \tilde{\rho}^{\eta}{}_{rand}(\epsilon) = \tilde{\rho}^{\eta}{}_{rand}(\epsilon). \tag{5.24}$$

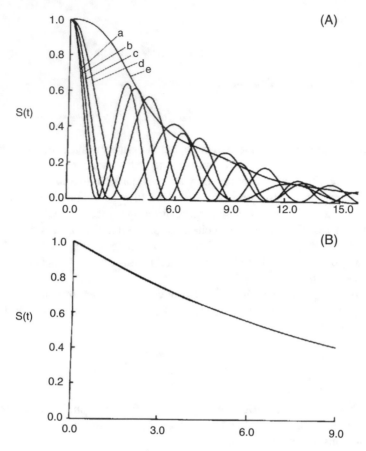

Fig. 5.7 Large differences in the pair-state signal for different locations (values of κ) in the Brillouin zone for coherent transport shown in (**A**) which disappear for incoherent transport shown in (**B**). The coherence parameter $\zeta = b/\alpha$, which measures the mean free path relative to the fringe spacing, equals 10 in (**A**) but only 0.1 in (**B**). Each line corresponds to equal initial occupation of four states, i.e., a pair state centered at κ and its symmetrical counterpart at $-\kappa$, each pair as in Fig. 5.6. Values of κ in both (**A**) and (**B**) are as shown: (a) 1, (b) 0.75, (c) 0.5, (d) 0.25, and (e) 0. All collapse onto a single line in the incoherent case (**B**). Reprinted with permission from fig. 2 of Ref. Kenkre and Tsironis (1985); copyright (1985) by Elsevier Publishing

Thus, one understands the connection between the signal for the single pair straddling the Brillouin zone origin ($\kappa = 0$),

$$S(t) = e^{-2\alpha t}\left[1 + \alpha \int_0^t du\, J_0(bu) + \alpha^2 \int_0^t du \int_0^u ds\, e^{\alpha s} J_0\left(b\sqrt{u-s}\right)\right]^2,$$

and the random-phase signal for a uniform combination of pair states throughout the zone,

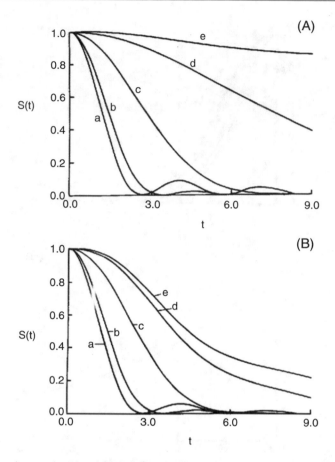

Fig. 5.8 Transient grating signal for largely coherent motion and for a realistic initial condition consisting of a Gaussian superposition of pair states with a width σ of κ values in the Brillouin zone. Lines correspond to values of the width σ as follows: (a) 1, (b) 0.7, (c) 0.3, (d) 0.1, (e) 0. The degree of the coherence is very high in (**A**), $\zeta = b/\alpha = 100$ and high in (**B**), $\zeta = 10$. Note that the signal looks like the random-phase signal for (a) (large width). Reprinted with permission from fig. 3 of Ref. Kenkre and Tsironis (1985); copyright (1985) by Elsevier Publishing

$$S(t) = \left[J_0(bt)e^{-\alpha t} + \alpha \int_0^t du\, e^{-\alpha(t-u)} J_0\left(b\sqrt{t^2 - u^2}\right) \right]^2.$$

Kenkre and Tsironis (1985) also showed that if, instead of a uniform weighting that arrives at the random-phase signal or a Gaussian weighting that we have used in the plots above, one employs the pseudo-thermal distribution[12] within the band, viz.,

---

[12]Observe that the right hand side of Eq. (5.25) is simply the ratio of the exponential Boltzmann factor for a thermalized state to the Bessel function normalization constant.

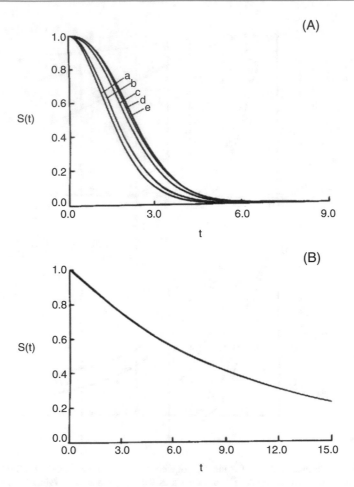

Fig. 5.9 Transient grating signal for largely incoherent motion and for a realistic initial condition consisting of a Gaussian superposition of pair states with a width σ of κ values in the Brillouin zone. Lines correspond to values of the width σ precisely as in Fig. 5.8: (a) 1, (b) 0.7, (c) 0.3, (d) 0.1, (e) 0. The motion is incoherent in (**A**), $\zeta = b/\alpha = 1$ and very incoherent in (**B**), $\zeta = 0.1$. Oscillations in the signal disappear, the curves are all similar in (**A**) and identical to one another in (**B**). Reprinted with permission from fig. 4 of Ref. Kenkre and Tsironis (1985); copyright (1985) by Elsevier Publishing

$$g(\kappa) = \frac{e^{-2V \cos \kappa / k_B T}}{2\pi I_0(2V/k_B T)}, \tag{5.25}$$

one obtains a grating signal that reduces, in the fully coherent limit, to an expression announced by Fayer in Agranovich and Hochstrasser (1983). It is perhaps remarkable that this expression, which was argued on the basis of an argument that associates a grating wavevector spatial variation with each k-state exciton happens to contain, although only in the fully coherent limit, both the pair-state result and the random condition result for the special cases of 0 and infinite temperature.[13] The interested reader will find a careful discussion of these pseudo-thermal distributions in the analysis given in Tsironis and Kenkre (1988).

Thus, we have shown here that all the theoretical predictions for the transient grating signal that have appeared in the literature are intimately related to one another. What scattering and dephasing processes happen to occur immediately after excitation is of considerable significance. Various opinions have been presented about these processes in the inorganic realm (Tyminski et al. 1984; Lawson et al. 1982; Morgan et al. 1986) as well as in the area of aromatic hydrocarbon crystals but no quantitative measurements of those initial processes are available. In the absence of that information, the natural investigation to undertake was to estimate the magnitude of possible departures from the random-phase signal that could occur as a result of incomplete dephasing processes. Let us briefly inspect that study.

5.3.3 Extraction of Coherence and Diffusion Parameters from Grating Experiments

Let us focus only on experiments performed in the Fayer group on anthracene crystals at various temperatures and argue from the theory set out in Kenkre and Tsironis (1985). Let us, however, restrict our considerations strictly to the exponential representation of the time dependence of the signal. This we will do because the observations reported by this experimental group exhibit no oscillations.[14] The procedure explained in Sect. 5.2.1 allows us to extract the relevant exponent from a time integration of the signal predicted by our theory. Whatever may be the sources of the experimental masking of the oscillations, we can use that exponent to compare theory to observations.

To make it easier for the reader to study the original investigation that the description in this section follows, (Kenkre and Schmid 1987), I will follow here the notation in that publication and use the word 'signal' to mean $s(t)$ rather than $S(t) = s^2(t)$. Furthermore, let us denote by k the apparent exponent of the signal from

$$k = \frac{1}{\int_0^\infty dt\, s(t)} = \frac{1}{\lim_{\epsilon \to 0} \tilde{s}(\epsilon)}, \tag{5.26}$$

---

[13]We have called the distribution here pseudo-thermal to eliminate any misunderstanding that might occur that the initial condition assumed in Eq. (5.25) is an actual thermalized one. There would be no grating in that case.

[14]Tyminski et al. (1984), on the other hand, do report oscillations in the inorganic doped crystals they study.

and write the random-phase and pair-state versions of the signal that we have derived earlier respectively:[15]

$$k_{rand} = \sqrt{\alpha^2 + b^2} - \alpha, \tag{5.27a}$$

$$k_{pair} = \frac{\sqrt{\alpha^2 + b^2} - \alpha}{\sqrt{1 + (b/\alpha)^2}}. \tag{5.27b}$$

Much can be learned from a comparative inspection of the two k's in Eqs. (5.27). It is clear that k_{rand} and k_{pair} are identical to each other for small enough b/α, i.e., for large enough incoherence. However they can differ *sharply* from each other for systems in which the transport is highly coherent on the length scale of the interference fringes. The two k's coincide with each other for large α but k_{rand} increases monotonically to the value b as $\alpha \to 0$ whereas k_{pair} increases to a maximum and then decreases with decreasing α, eventually vanishing at $\alpha = 0$.

This is a fundamental difference in the pair-state signal and the random-phase signal. The difference is depicted graphically in Fig. 5.10.

In the random-phase case, the signal $s(t)$ is the Bessel function of zero order for completely coherent transport. Its decay becomes slower as the scattering rate increases. This behavior arises very simply from the fact that exciton motion, which is responsible for destroying the initial inhomogeneity, becomes slower and

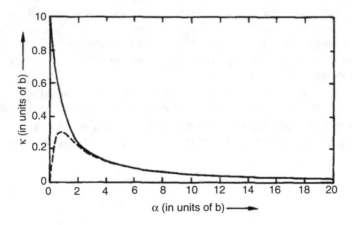

Fig. 5.10 Exponent k of the grating signal plotted as a function of the scattering or dephasing rate α, both quantities being expressed in units of b. The pair-state initial condition is represented by the dashed curve, the random-phase initial condition by the solid curve. The two exponents differ significantly for large coherence but coincide for small coherence. Reprinted with permission from fig. 1 of Ref. Kenkre and Schmid (1987); copyright (1987) by Elsevier Publishing

---

[15]The pair-state considered here is the one with $\kappa = 0$ shown in Kenkre and Tsironis (1985) to describe the other extreme. See the discussion in Sect. 5.3.2.

consequently less efficient as the scattering of the exciton increases. In the limit of infinite α, the decay rate vanishes.

Complexities arise for the pair-state case. Although, similarly to the random-phase case, the exponent decreases as the scattering increases, that is true for the pair-state case only in the high scattering region. In the limit of low scattering, the grating starts out tending to be permanent in the pair-state case. As indicated in footnote (no. 34) in our article on the Ronchi ruling analysis (Kenkre et al. 1983a), this happens because of the coherent construction of a perfect standing wave of the two Bloch states comprising the initial grating. As α increases, the scattering first causes mixing of Bloch states, destroys the perfection of the standing wave, and k increases from zero. With sufficient scattering, the effect mentioned for the large α limit in the random-phase case occurs and the exponent decreases. The behavior of k for the pair-state case is, thus, *not monotonic*. This is transparent in Fig. 5.10.

Here is a simple way of understanding in simple physical terms what might be happening in the experiment on a real system of such enormous complexity. If the initial illumination produces a random-phase excitation, our theory of the signal applies without modification. If it produces, instead, a pair state, the countless processes of dephasing and scattering that operate in the complex crystal could bring about randomizing of the location of the pair state in the Brillouin zone. At ultra short times, the pair-state grating may have the tendency to become a "permanent" grating but the dephasing and scattering processes would mix the state with other pair states. The width σ of the values of κ in the wavepacket of pair states would increase and, as the dotted line in Fig. 5.10 shows, the signal exponent k would *increase* initially along the dotted line. At the same time, the character of the packet would become more and more random; more of the solid line depicting the random-phase k would mix in. As time progresses, randomization would result in the solid line becoming more descriptive of the signal exponent. At long enough times, the signal that started out, even with an initially non-random character, would be described well by the random-phase theory constructed through our GME method.

In the absence of detailed experimentation, which I believe is out of reach at the moment in the laboratory, we have no idea whatsoever about the time scale of these various processes. There is no guarantee that, in a given system, they would follow the particular sequence I have outlined here. But if, you assign a time dependence to the width σ which brings it from its initial value to one that spreads the exciton over the full Brillouin zone, in the simple manner I have described, you could have the signal start out in the pair-state form but *evolve* into the random-phase form as explained.

An important question to ask is, therefore, the following. How much quantitative inaccuracy would enter our conclusions regarding the values of transport parameters extracted from experiments such as Rose et al. (1984) if we make the assumption that the initial state is random-phase when it is actually a pair-state? A detailed analysis, directed at specifically the observations reported by Rose et al. (1984) at various temperatures in pure crystals of anthracene, is available for the interested reader in Kenkre and Schmid (1987). Its conclusion is that, with the exception of the 1.8 K (lowest temperature) observation, there is no discernible effect of the initial

Table 5.2 Singlet exciton parameters given by transient grating experiments on anthracene at various temperatures. Reprinted with permission from Ref. Kenkre and Schmid (1987); copyright (1987) by Elsevier Publishing

$T(K)$	τ (ns)	$d\ (\mu m)$	$K(10^8 s^{-1})$	$\alpha(10^{11} s^{-1})$	Λ/a	$\Lambda (nm)$
20	10	3.2	7.6	6.7	20	10
	4.6	4.2	8.3		67	0.1
10	10	4.1	10	2.9	46	24
		9.6	3.1	3.4	39	20
1.8				≤ 0.47	≥ 280	≥ 150

condition issue. Furthermore, in this "worst-case scenario", if the nature of the initial state were a pair-state rather than the assumed random-phase, the error deduced in α at 1.8 K would involve an *underestimation* of the coherence, specifically of the mean free path, by merely a factor of 1.2. The conclusion that singlet motion in anthracene is highly coherent at that 1.8 K is thus reinforced. The updated table of coherence and diffusion parameters for singlet excitons in anthracene at various values of T is thus given here as Table 5.2.

The symbols T, τ, d, K, α, Λ, a, D, and V, represent in Table 5.2, respectively, the temperature, the exciton lifetime, the fringe spacing, the grating signal exponent, the scattering rate, the mean free path of the exciton (sometimes called the coherence length), the intersite distance, the exciton diffusion constant, and the intersite matrix element proportional to the bandwidth.

5.4   Chapter 5 in Summary

Applications of the memory function formalism were discussed here from the viewpoint of experiments of the *gentle* kind, in which the system being studied is not modified strongly by the probe used to make the measurements. A simple discussion of exciton coherence was given at the beginning of the chapter with emphasis on theoretical concepts underlying its quantitative measurement. It focused on the ratio of the exciton mean free path to the lattice constant, and avoided philosophical distractions. A theory of coherence in the context of experimental studies using Ronchi rulings pioneered by Ern and others in the Dupont laboratories was developed next. The analysis addressed the time evolution of an initially periodic distribution of triplet excitons. It was followed by a similar analysis of transient grating investigations for singlet excitons introduced in the field by Fayer. This latter method uses the same concept as the method of Ronchi rulings but is conceptually cleaner as it excites initially just one, or perhaps more accurately two, spatial modes of the exciton density by crossing laser beams into the crystal. These two classes of experiment, Ronchi ruling and transient grating, are particularly suitable to probe exciton coherence. In the last part of this chapter, an interesting analysis of possible effects of initial phases on grating and ruling phenomena was provided. It took as its starting point, an early study (Kenkre 1978b), of the initial condition term in the Zwanzig derivation of the GME, carried out long before transient gratings entered the molecular crystals field. The connection between initial pair states and initially random states was elucidated. Explicit tables of coherence parameters were made available through the analysis of a series of reported experiments, in several crystals at room temperature on triplets and in anthracene crystals at several temperatures on singlets.

6

Application to Charges Moving in Crystals: Resolution of the Mobility Puzzle in Naphthalene and Related Results

As a physical observable, temperature exhibits schizophrenic properties. It is typically simple to heat or cool a system and, if one is not dealing with extremes, to measure the temperature of a system with a suitable thermometer. Plots of the temperature dependence of physical observables are, therefore, met with in great abundance in physics research. Yet, attempt thinking about what the meaning of the underlying processes is, and it is not simple at all. This happens because temperature is not a mechanical variable. When we graduated from our kindergarten understanding of that physical quantity as the average kinetic energy, we all found that coming to terms with temperature as having to do with statistical ensembles and probability distributions was by no means straightforward. In contrast to distances, times and frequencies, temperature requires quasi-philosophical constructs. It is associated with phase-space considerations on the one hand and, on the other, with interactions with (dirty) baths and reservoirs that introduce broadening into lineshapes and decays into motion. As a consequence, it is much easier to build theories capable of analyzing experiments of the kind treated in Chap. 5 than it is to treat experiments to be described in the present Chap. 6.

Our subject for this chapter is another important series of experiments, also in organic materials, but the system under investigation is now not Frenkel excitons. Instead, it is photo-injected charge carriers. It has ultimate technological relevance to the industrial processes of photo-copying and printing. The primary purpose of the chapter is to briefly recount the story of Naphthalene that played out over a couple of decades in the 1970s and 80s and dealt with understanding the peculiar temperature T-dependence of the mobility of photo-injected holes and electrons in the aromatic hydrocarbon crystals.

© The Author(s), under exclusive license to Springer Nature Switzerland AG 2021   115
V. M. (Nitant) Kenkre, *Memory Functions, Projection Operators, and the Defect Technique*, Lecture Notes in Physics 982,
https://doi.org/10.1007/978-3-030-68667-3_6

6.1   The Mobility Puzzle in Naphthalene

We begin by describing that puzzle regarding the T-dependence of photo-injected charge carrier mobilities in hydrocarbon crystals and showing how its resolution came to pass.

6.1.1   Experimental Background

The puzzle existed for many years and consisted of the fact that the mobility of photoinjected electrons in naphthalene in the c' crystallographic direction over the temperature range $100 < T < 300$ K was T-independent and decreased rapidly with an increase in T for $30 < T < 100$ K. The behavior of the mobility in different directions was also considerably different.

Investigators were trained to think of any decreasing mobility of electrons in crystals with increasing T to correspond to "band motion", i.e., motion described in terms of velocities of carriers in the band scattered by impurities and phonons, often acoustic. The theory was well known for electrons in metals for many decades. Polaron mobilities that increase with an increase in temperature because the relevant carriers are self-trapped within lattice distortions as a result of strong interactions (and are thus helped by an increase in T to escape the self-generated wells) were also well-known since the time of Landau, Pekar, Fröhlich, and especially Holstein, but flat mobilities presented a puzzle in themselves. Additionally, there was the point of anisotropy which meant that different mechanisms seemed to be operating in different directions.

Anisotropy was not difficult to address, given the fact that, unlike in inorganic substances (e.g., ordinary metals and inorganic semiconductors), what lies at a lattice point of an organic crystal is a complex object not impartial to directions. Thus in the case of naphthalene, one has to contend with two molecules oriented in a specific way, each with 10 carbon atoms forming a planar object, with other atoms linked to the carbon atoms. This is nothing like a spherical atom that one may take as sitting at a lattice point in a copper lattice. The motion of all those entities, and the interactions of their electron clouds, made for a situation full of complex dynamic disorder which inorganic theorists were not accustomed to analyze.

Among the experimentalists, there was a healthy balance between the enthusiastic and creative approach of Larry Schein (Schein et al. 1978) in Webster, and the measured and steady methodology of Norbert Karl (Warta and Karl 1985) in Stuttgart. With Charlie Duke as coauthor, Schein wrote an article (Duke and Schein 1980) in the journal *Physics Today* inviting theorists to come up with an understanding. It was an important undertaking and several of us who thought we could make our correlation functions and master equations knock sense into the exciting reported observations were drawn into the activity. The whole subject is represented, competently and thoroughly, in the comprehensive book by Pope and Swenberg (1999).

6.1.2 Theoretical Situation

Various specific aspects of the measurements had been interpreted qualitatively (Madhukar and Post 1977; Efrima and Metiu 1979; Sumi 1979b,a; Roberts et al. 1980; Silbey and Munn 1980; Reineker et al. 1981; Andersen et al. 1983) but no single theoretical model had succeeded in providing a simultaneous description of all of them.[1] Several theorists felt they had each solved the essential problem.[2] To make frame-independent progress (if such was possible), in the context of an nsf-funded university-industry collaboration (between the University of Rochester and the Webster Research Labs of Xerox), Charlie Duke and I undertook a joint investigation. John Andersen, a student of mine, led the charge. He began by examining each proposed theory, not cursorily but in the detailed light of experiment, applying to each of them, stringent requirements of observations. Surprisingly, as will be seen below, all the theories tested fell by the wayside as a result of the Andersen-led testing. Finally, extending somewhat the GME theory (explained in Chap. 4 for Frenkel excitons) we were able to achieve (Kenkre et al. 1989) a successful description of the naphthalene data for all T and all directions.

6.1.3 Putting Existing Theories to Quantitative Test

In order to understand the excellent qualitative (and even quantitative) explanation coupled with what must be called eventual failure of the tested theoretical attempts, let us start by examining two theories that we were involved in ourselves. In the first, let us inspect the impressive fit presented from the SLE based essentially on the Haken-Strobl model as used for this purpose by Reineker et al. (1981), and as the second the partial, but also impressive, fits given from acoustic phonon scattering in band theory by Andersen et al. (1983). Invoking the SLE (4.1) for the density matrix elements as describing the coupled coherent and incoherent motion of the photo-injected carriers, where m, n represent the sites of the naphthalene crystal in the c' direction, Reineker et al. (1981) showed that the mobility was proportional to

$$\frac{1}{k_B T}\left[2\gamma_1 + \frac{V^2}{\Gamma + \bar{\gamma}_1}\right],$$

---

[1]The last two of the publications mentioned provided visually impressive quantitative fits of the theory to the data as well.

[2]Throughout my scientific life I have wondered what precisely "essential" means in this context. When do I know I have captured the essence of a problem through my theory? Did Einstein capture the essence in his theory of specific heats of insulators based on dispersionless oscillators? Does Debye's innovation in this subject address an essential ingredient or is it a "mere matter of detail"? Is knowing whether one's theory has captured the essence of an experiment an inner phenomenon only? Are there no criteria more objective than for knowing if one has fallen in love?

where $\Gamma = \sum_i \gamma_i$, (and α in Eq. (4.1) equals 2Γ), and that it reduced to various expressions in appropriate limits (such as those of Madhukar and Post (1977)). The parameters γ describe fluctuations depending on phonon occupation numbers[3]

$$n = \frac{1}{e^{\hbar\omega/k_B T} - 1}$$

where ω is a characteristic frequency.

Figure 6.1 shows a fit of these simple arguments to the mobility data as reported by two different laboratories on the temperature dependence of photo-injected charge carrier mobility in the c' direction of naphthalene. How did such an economical theory providing the impressive fit to data fail as a contender when Andersen applied his test? It turns out that the frequency required for the fits is lower than the lowest known libration in naphthalene. Precisely the same occurred when the test was applied to the theory of Sumi (1979a,b).

We turned to the Boltzmann equation treatment of Andersen et al. (1983) next. The underlying mechanism is scattering by acoustic phonons and the theory is

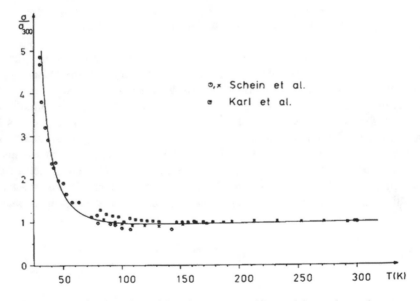

Fig. 6.1 A fine-looking fit to the mobility data as reported by two laboratories on the temperature dependence in the c' direction of naphthalene, given on the basis of the SLE in Reineker et al. (1981). While optically excellent, it requires for the fit a librational phonon branch at a frequency that is lower than that of the lowest known libration in naphthalene. Reprinted with permission from fig. 1 of Ref. Reineker et al. (1981); copyright (1981) by Elsevier Publishing

---

[3]In this chapter we show all occurrences of $\hbar$ explicitly to avoid any confusion because of the plentiful discussion of experimental data here.

Fig. 6.2 Good fits to partial
mobility data in the three
directions of naphthalene, on
the basis of acoustic phonon
scattering as given in
Andersen et al. (1983).While
the fitting certainly appears
reasonable qualitatively, the
mean free paths that
correspond to these fits are
smaller than a lattice
constant, making the very
basis of the application of
band transport here invalid.
Reprinted with permission
from fig. 1 of Ref. Andersen
et al. (1983); copyright
(1983) by the American
Physical Society

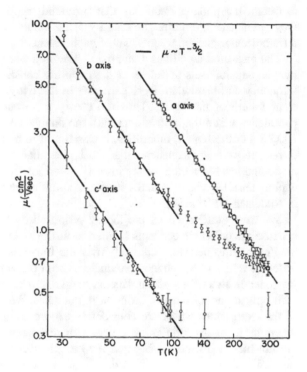

the standard one borrowed from the inorganic tradition but modified in detail for
naphthalene. Once again, the fits are optically excellent as clear from Fig. 6.2 when
we restrict ourselves to data below 100 K in all crystallographic directions. Yet,
when mean free paths were calculated, they were found to be less than a lattice
constant for most of the data in the c' direction. The theory had to be therefore
discarded for lack of internal consistency.

We had high hopes for the polaronic theory of Silbey and Munn (1980) because
the ideas introduced by Silbey and his collaborators into the subject appeared
physical and justified. They offered as their mobility expression,

$$\mu = \left(\frac{ea^2}{k_B T}\right) \left(\frac{1}{Q(y)} \frac{\tilde{B}^2 + \Gamma^2}{\sqrt{\tilde{B}^2 + 2\Gamma^2}} + \frac{Q(y)\tilde{B}^2}{4\sqrt{2\tilde{B}^2 + \Gamma^2}} \exp\left[\frac{-\tilde{B}^4}{4(k_B T)^2 \left(2\tilde{B}^2 + \Gamma^2\right)}\right]\right)$$

where a is the lattice constant, $\tilde{B} = B \exp\left[-g^2(2n+1)\right]$ is the polaron bandwidth,
B being the bare bandwidth given by $4V$ times a constant of the order of unity, n is
the phonon occupation number from Bose statistics, $y = 4g^2\sqrt{n(n+1)}$, $Q(y) = \sqrt{\pi}\left[I_0(y) - 1\right]$, I_0 is the modified Bessel function of order 0, Δ is the phonon
bandwidth, and Γ is a parameter defined to be Δ if $y \leq 1$ but $\Delta\sqrt{y}$ if $y \geq 1$.

Andersen et al. (1984) proceeded to test the Silbey-Munn theory vis-a-vis the
data in two steps. First, they ignored the specific identification of the phonons
involved with any of the known modes in naphthalene and varied the parameters

to obtain the phonon frequency that provided the best fit to the data. Second, they attempted to describe the data by using values of the known librational and intramolecular phonon frequencies of naphthalene.

The measured mobilities along the c' direction were divided into two distinct T ranges. Separate sets of values of the coupling constant, bare bandwidth, phonon frequency and bandwidth were extracted in the two ranges. Figure 6.3 shows the clear failure of the fitting. Thus the theory was found to result in perturbation parameters that are not consistent with the analysis of the low-T mobilities ($T <$ 100 K). For the high-T mobilities, it was found to be consistent with the data but to require phonon frequencies that are higher than the highest known librational mode and lower than that of the lowest known totally symmetric internal mode in naphthalene. Figures 6.3 and 6.4 show this situation rather dramatically, as analyzed in Andersen et al. (1984).

We were quite taken by the theory proposed by Silbey and his collaborators because the polaronic concepts seemed to fit the system adequately. Because good fits appeared to be given separately over the low-T and the high-T regimes using different values of the phonon parameters (both frequency and coupling constant), we generalized the Silbey-Munn theory to take into account interaction of the carrier with optical phonons of two different frequencies. We do not show here the details of the expressions which are considerably more complex than the original ones. Attempts to fit over the entire range with the extended expression we obtained did not meet with success because the best fit occurred when one of the phonon frequencies was zero. For further details of our work with the Silbey-Munn theory we refer the reader to Andersen et al. (1984).

Fig. 6.3 Best fit of the Silbey-Munn expression denoted by lines (see text) to data denoted by open circles with error bars, in low, high, and entire temperature ranges: (**a**) $30 \leq T \leq 100$ K, (**b**) $100 < T < 300$ K, and (**c**) $30 < T < 300K$. Reprinted with permission from fig. 1 of Ref. Andersen et al. (1984); copyright (1984) by Elsevier Publishing

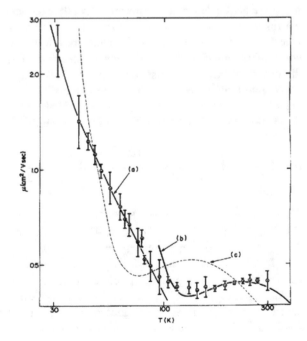

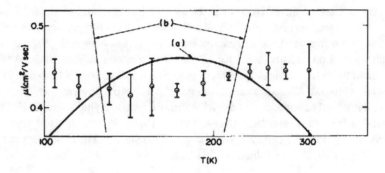

Fig. 6.4 Best fit of the Silbey-Munn expression (see text) to high-T $100 < T < 300$ K data for mobilities in the c' direction in naphthalene using (**a**) the highest frequency librational mode (16 meV) and (**b**) the lowest frequency totally symmetric intramolecular mode (60 meV). While the best fit procedure produces quite good fits over separate temperature ranges, attempting the fit over the entire range of T is quite unsatisfactory. Reprinted with permission from fig. 3 of Ref. Andersen et al. (1984); copyright (1984) by Elsevier Publishing

As a result of similar careful studies of all theories we could subject to such scrutiny, we had to draw the inevitable conclusion that there was no satisfactory explanation of the mobility puzzle among the existing theories in the literature. This was particularly striking, and instructive, given that they all seemed to predict qualitatively similar behavior of the temperature dependence.

6.2 A Successful Theory

Early discussions in Chap. 4 have referred to a theory for the transport of Frenkel excitons (Kenkre 1975c) in which I had combined the polaronic concepts introduced by Silbey and collaborators (Grover and Silbey 1971; Silbey 1976) with a perturbation scheme that differed from (indeed, competed with) the one used by that group. A natural question to ask at this point was whether a blend of the dressing transformation used in situations involving strong coupling with phonons (large g) and the GME method along with the perturbation scheme developed for its use in the exciton field would describe better the transport of photo-injected charge carriers in naphthalene. The theoretical exercise we undertook for that purpose produced the expression for the ith diagonal element of the mobility tensor

$$\mu_{i,i} = 2|V_i/\hbar|^2 \left(ea_i^2/k_BT\right) e^{-2g_i^2 \coth(\hbar\omega_0/2k_BT)} \int_0^\infty dt\, e^{-\alpha t}$$

$$I_0\left(\frac{2g_i^2}{\sinh\left(\frac{\hbar\omega_0}{2k_BT}\right)}|J_0(\Delta_1 t)J_0(\Delta_2 t)J_0(\Delta_3 t)| \left[1 + J_1^2(\Delta_i t)/J_0^2(\Delta_i t)\right]^{1/2}\right),$$

$$(6.1)$$

where the J's are ordinary Bessel functions and I is a modified Bessel function. The a_i and g_i are, respectively, lattice constants and coupling constants for each direction, and α measures the site energy fluctuations. On comparing the expression to mobility data on naphthalene,[4] for all temperatures and all directions, Andersen found them to be in excellent agreement (see Fig. 6.5) and was able to extract very reasonable values of all parameters concerned as Table 6.1 shows.

This theory was announced at the Molecular Crystals Symposium in Lugano in 1985 and published a little later (Kenkre et al. 1989). It was received well as the detailed description in Pope and Swenberg (1999) shows. Those authors analyzed our work vis-a-vis that of Silinsh and Capek (1994) on the carrier effective mass

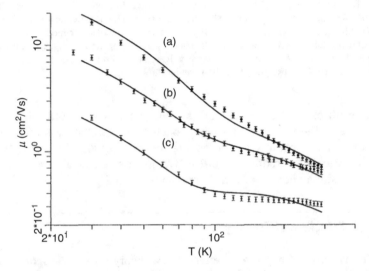

Fig. 6.5 Good fits to complete mobility data in the three directions, 1, b and 3, of naphthalene, represented by (a), (b), and (c) respectively, on the basis of the memory function computation of the mobility given in Kenkre et al. (1989). Filled circles with error bars represent the data. Reprinted with permission from fig. 2 of Ref. Kenkre et al. (1989); copyright (1989) by the American Physical Society

---

[4]David Dunlap who collaborated on this fitting did not work further on crystal mobilities. He went on to produce, in collaboration with Paul Parris, a remarkable theory of the field-dependence of mobilities in the presence of dipolar disorder that will be mentioned passingly in Chap. 13 (Dunlap et al. 1996, 1999; Novikov et al. 1998; Kenkre et al. 1998a; Parris et al. 2001b). It has not been described in this book because it does not fit in with the flow of the book's topics. However, I heartily recommend to the interested reader that fine work by Dunlap and Parris on the description of carrier transport in statically disordered systems.

Table 6.1 Values of parameters extracted via Eq. (6.1) from observations of the electronic drift mobility data in naphthalene. Reprinted with permission from Ref. Kenkre et al. (1989); copyright (1989) by the American Physical Society

Parameters extracted from fitting procedure	1	Principal axis direction	
	1	b	3
g	1.615 ± 0.014	1.830 ± 0.013	1.879 ± 0.013
B (meV)	8.41 ± 0.15	10.47 ± 0.14	7.88 ± 0.14
$\hbar\alpha$ (meV)	0.1694 ± 0.0062	0.1694 ± 0.0062	0.1694 ± 0.0062
Phonon parameters			
$\hbar\omega_0$ (meV)	16.0 ± 0.05	16.0 ± 0.05	16.0 ± 0.05
$\hbar\Delta$ (meV)	0.70 ± 0.05	1.40 ± 0.05	0.24 ± 0.05
Figure of merit			
$g^2\hbar\omega_0/B$	5.0 ± 0.01	4.6 ± 0.01	7.2 ± 0.01

in naphthalene and found a theoretical explanation in the former for an empirical constant introduced in the latter.[5] Our theory for naphthalene was also appreciated by other theorists such as Wu and Conwell who commented on how it eliminates a divergence problem in Holstein's theory. They went on to apply it successfully to transport in a different material, α-sexithiophene (Wu and Conwell 1997).

Although it does not make too much sense to use data fits to select one perturbation scheme over another (given the many assumptions and approximations that are always associated in the fit procedures), we show Fig. 6.6 that compares the fit of naphthalene c' direction mobility data to our theory (dotted line) on the one hand and to the Silbey-Munn theory (solid line) on the other. Our procedure fares better.

Values for the fitting parameters for naphthalene as obtained in Kenkre et al. (1989) are displayed in Table 6.1. Those of the bandwidths Δ_i and frequency ω_0 used in the fitting procedure were obtained from neutron diffraction measurements on deuterated naphthalene. Those of V_i, g_i and α were extracted from the observations through a least-squares-fitting procedure. It is important to examine the resulting values of the ratio of the polaron binding energy $g^2\hbar\omega_0$ to the electronic bandwidth B_i. These ratios (see Table 6.1) lie between 4.6 and 7.2. If the value were less than 1, polaron theory would be inapplicable. There are thus no problems of internal consistency, or wished-for but non-existent librations. Indeed, a subtlety is worth noting. While the observed mobilities obey $\mu_1 > \mu_b > \mu_3 >$, the deduced B's, equivalently V's (see Table 6.1) do not follow this sequence. In fact, the key parameter controlling the value of the mobility is the combination $F = 2(V^2/\alpha)\exp(-g^2)$ of V with the coupling constant g and the inhomogeneity parameter α. Normalized to its value in the 3 direction, F equals 2.9, 2.2 and 1 in

---

[5]They went on to remark (Swenberg and Pope 1998) that "This surprisingly close agreement between the two approaches strengthens the validity of each...".

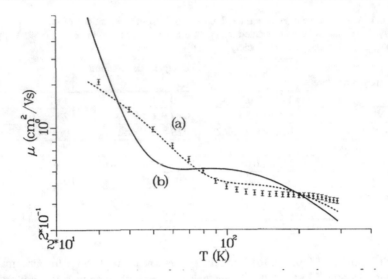

Fig. 6.6 Comparison of the result of two different calculational schemes applied to a polaronic starting point for the description of photo-injected carrier mobility in the c' direction in naphthalene. Solid line corresponds to Silbey and Munn (1980) and dotted line to Kenkre et al. (1989). Filled circles with error bars represent the data. Reprinted with permission from fig. 1 of Ref. Kenkre et al. (1989); copyright (1989) by the American Physical Society

the respective directions 1, b, and 3. Thus, the sequence $F_1 > F_b > F_3 >$ does apply precisely in keeping with the observations.

The observed T-independence of the mobility for the high temperature region is not associated in this theory with any qualitative feature; rather only with the quantitative combination of parameters that happens to make the mobility effectively T-independent. Nevertheless, the theory demands that we think of the carriers as polaronic in all directions, there being no band-hopping transition as was originally thought. The decrease in the mobility with increasing T in the low temperature region is not an indication of carriers in band states getting scattering by acoustic phonons. Instead, it stems from the increase in polaronic dressing, i.e., the bandwidth reduction that strong interaction with phonons imposes on the carriers. This is an inescapable feature of the theory. The static disorder, called "diagonal disorder" by Pope and Swenberg (1999) in their excellent summary of our theory, and the strong interaction with the librational phonon branch at energy about $16\,meV$ with small dispersions in each crystallographic direction are other important ingredients. See also Dunlap and Kenkre (1993) where these matters are discussed in summary form.

The picture that finally emerges from the successful theory is thus that, starting from very low temperatures, increase of T first decreases the velocity of the carriers as more and more virtual phonons dress the carrier, with a consequent decrease in the mobility. Further increase in T causes the phonon-assisted effect to increase in magnitude as the carrier is helped by temperature increase to escape the self-trapping well. The combination appears as a T-independent mobility as a result of the specific values of the parameters. An increase in mobility with an increase in T could be expected beyond that interval but phase change makes a measurement impossible.[6]

The Hamiltonian underlying the theory in Kenkre et al. (1989) is

$$H = \sum_m \epsilon_m a_m^\dagger a_m + \sum_{m,n} V_{m-n} a_m^\dagger a_n + \sum_q \hbar\omega_q \left(b_q^\dagger b_q + \frac{1}{2} \right)$$

$$+ \frac{1}{\sqrt{N}} \sum_{m,q} \hbar\omega_q g_q e^{iq \cdot R_m} (b_q + b_{-q}^\dagger) a_m^\dagger a_m. \tag{6.2}$$

Here, with m and q as vectors, ϵ_m is the site energy of the carrier at site located at R_m (labelled by m), a_m is the destruction operator for the carrier at that site and b_q is the destruction operator for a phonon of wavevector q with frequency ω_q and coupling constant g_q, and N measures the number of sites. The formula (6.1) for the mobility tensor is obtained from the velocity auto-correlation function which, in turn, is calculated as explained in Chap. 4 from the memory function of the carrier appearing in the GME.

6.3 Further Transport Theory Work on Related Problems

Around the turn of the millenium there was a burst of activity in photo-injected carrier transport theory. The activity had been stimulated by reports of exciting observations in other aromatic hydrocarbons, pentacene in particular. Unfortunately, what appeared as falsification of data, a rare occurrence in any field of science, caused a temporary cooling of enthusiasm on the part of many workers in the field who steered clear of the field for some time. Fueled by the experimental reports, a number of interesting theoretical pieces of research were performed during this period and I think it intriguing to make brief mention of a few of those that I know of from close, even though they lie on the borders of the theme of this book. The ones I will mention dealt with (1) variational calculations for partial polaronic dressing, (2) mobility saturation at high electric fields, and (3) effects in intermediate (rather than small) bandwidth carrier transport. Let us describe them in turn.

6.3.1 Variational Calculations for Partial Polaronic Dressing

A bare charged particle interacting weakly with phonons is taken to be scattered by them and the zero-order particle states we consider in the scattering have zero dressing. If the charged particle interacts strongly with the phonons, we take the

---

[6]The late Larry Schein said to me when I asked him to push the measurements to that region, "I would if I could but the *&%*%& crystal melts at that point."

carriers to be dressed *fully*. What this means is that we carry out the dressing transformation with the coupling constants as stated in the Hamiltonian and regard these fully dressed states as the zero-order states to use in perturbation calculations. This is what we have done in the theory described thus far. While such an approach works if the strength of the interaction lies either in the weak or strong limit, the annoying question remains of how we should treat the case of intermediate coupling.

This fascinating issue has been treated by a variety of authors (Merrifield 1964; Yarkony and Silbey 1977; Nasu and Toyozawa 1981; Silbey and Harris 1984; Harris and Silbey 1985; Parris and Silbey 1985; Brown et al. 1997). The basic approach is variational. The idea is to assume that the dressing is to be done with a value of the coupling constant lying in between 0 and the one stated in the Hamiltonian, and to determine the value from a variational principle applied to the free energy of the system. I first learned about the approach through the writings of Yarkony and Silbey.[7] Just as we had done for the naphthalene case in the previous section in attempting to amplify the work of Silbey and Munn (1980), I thought it might be important to consider variational analysis of simultaneous interaction with two different phonon branches. Such a study was undertaken at my request by Parris who, during his stay at MIT had worked with Silbey on variational investigations (Parris and Silbey 1985).

Thus, Parris and Kenkre (2004) investigated properties of a carrier moving through a 3-dimensional isotropic band in interaction with two separate branches of optical phonons. They generalized the variational approach set forth by Yarkony and Silbey originally for excitons in molecular crystals and subsequently extended by Silbey and his collaborators. The generalization produced phase diagrams showing regions of parameter space where coupling to both phonon branches leads to significant dressing of the carrier. The absence of a sharp transition was noted between a singly dressed and doubly dressed polaronic phase. The investigations indicated that, when one high frequency branch is coupled to the carrier with an intermediate to strong coupling strength g_1, the effective bandwidth, already reduced by this coupling, was stable with respect to small to intermediate values of the coupling g_2 associated with additional low frequency optical phonon branches. The authors concluded that their study lent support to the basic picture of a carrier moving through a narrowed polaron band being scattered by its interaction with additional weakly coupled phonon branches.

Further work in this area (Cheng and Silbey 2008) performed several years later has come to our attention and appears to touch upon a number of interesting issues and offer a comprehensive picture of dressing.

---

[7]I find this topic powerful but personally rather mystifying and beset with logic that, to my uneducated eye, appears a tad confusing.

6.3.2   Should Mobility Saturate with Increasing Electric Field?

During the feverish activity that occurred on the experimental front in this field at the turn of the millenium, there were reports of observed velocity saturation with increased field and attempts to use those observations to claim coherence in the motion and to extract values of the bandwidth.

Motivated in part by these reports and in part by the more general wish to understand nonlinear field dependence of the mobility in narrow-band materials, we did three calculations that we published around that time: (Parris et al. 2001a; Kenkre and Parris 2002a,b). Let us start with a brief description of the Fokker-Planck analysis that we undertook in collaboration with Marek Kuś.

The system analyzed is a set of mutually noninteracting carriers of charge q moving in the presence of electric field E, not necessarily weak so we must step outside linear response, in a one-dimensional tight binding band arising from nearest-neighbor matrix elements V. The carriers interact with acoustic or optical phonons of energy much smaller than the electronic bandwidth so that continuum equations in k-space apply for $f(k, t)$, the probability distribution of the carriers. A constant E causes a continuous linear change of k of a carrier in a Bloch state of that quasimomentum, this ballistic motion in the band resulting in Bloch oscillations (Anderson 1997) of period h/qEa where a is the lattice constant. In the absence of scattering, one has zero average group velocity and an absence of dc current. However, scattering events disrupt this cycle to produce a finite dc mobility.

The evolution equation is

$$\frac{\partial f_k}{\partial t} = \frac{\alpha}{\hbar^2} \frac{\partial}{\partial k} \left[\left(\frac{d\epsilon_k}{dk} - \frac{\hbar q E}{\alpha} \right) f + k_B T \frac{\partial f}{\partial k} \right] \tag{6.3}$$

where ϵ_k is the band state energy of a k-state and α is the carrier-phonon scattering rate. From its solution, along with the prescription

$$< v >= \frac{1}{\hbar} \int_{-\pi/a}^{\pi/a} f_k \frac{d\epsilon_k}{dk} dk \tag{6.4}$$

to calculate the average velocity of the carrier, we can find a remarkably interesting nonlinear dependence of the drift velocity on the electric field:

$$< v >= \frac{qE}{\alpha} - \frac{1}{\pi} \frac{a \sinh(\pi \hbar \beta q E / \alpha a)}{\beta \hbar I_{-i\hbar \beta q E / \alpha a}(2\beta V) I_{i\hbar \beta q E / \alpha a}(2\beta V)}. \tag{6.5}$$

where $\beta = 1/k_B T$. The I_ν are modified Bessel functions of order ν. Note that in Eq. (6.5), the orders of the Bessel function are strictly imaginary! They are proportional to the field E and inversely proportional to the scattering rate and the temperature. The ratio of the carrier drift velocity in the steady state to the maximum attainable band velocity $v_0 = 2Va/\hbar$ is plotted in Fig. 6.7 as a function of the ratio of the applied field E to $E_0 = 2V\alpha a/\hbar q$ for various temperatures as shown.

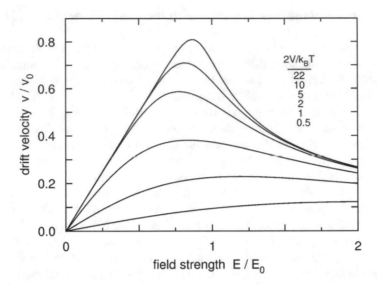

Fig. 6.7 Nonlinear effects in the field dependence of the current as obtained by an analytical solution (6.5) of the Fokker-Planck equation. Plotted is the steady state carrier drift velocity v/v_0 as a function of applied electric field E/E_0 for different temperatures as shown. There is an increase with an increase in the field, followed by a decrease but no saturation of any kind. Reprinted with permission from fig. 1 of Ref. Parris et al. (2001a); copyright (2001) by Elsevier Publishing

One can conclude from this result of Parris et al. (2001a) that there is an Ohmic, linear increase of drift velocity at low fields and a decrease at very large fields. For low temperatures ($k_B T \ll 2V$), the peak separating the two regimes of field dependence is very pronounced, i.e., easily visible. At higher T it is broadened and the field dependence is flat over a larger region. The achievements of this work by Parris et al. (2001a) thus are two: bringing this fascinating nonlinear field dependence of the carrier velocity to light with the explicit solution of a nonlinear Fokker-Planck equation in k-space, and establishing that the simple model does *not* result in the reported field saturation.

There are three more papers that we wrote on this subject of field saturation. One showed, as in the above calculation, that other standard theoretical arguments did not result in saturation contrary to what had been claimed in the literature.[8] Another gave an interesting mechanism that could produce saturation. I resist a strong temptation to explain those interesting calculations here because the subject is devoid of memory functions, projection operators or the defect technique, the stated subject of this book. The interested reader is directed to those two papers (Kenkre and Parris 2002b,a) as also to an expository lecture delivered at the PanAmerican

---

[8]Cautionary conclusions of this kind were also drawn by Conwell and Basko (2003) on independent grounds.

Study Institute (PASI) (Kenkre and Lindenberg 2003). In the latter publication you will find a combined narration–even a doggerel you might enjoy if you like the poetry of Lord Tennyson. To cater to the rare reader whose inclinations might be as weird as mine, and to whet the appetite of the rest of the readers, I give here the doggerel I felt compelled to compose at the conclusion of that work:

PLIGHT OF THE CHARGE (brigade?)

"I am the carrier," said the hole, "I will not stop."
"Whatever the barrier, I will cross, by flow or hop."
"You forget, you puny charge," replied the wall.
"How steep I rise, above your fields, large or small."
"From this height, I pity your plight, you have much gall."
"Fear not, hole, you ar'n't the sole, wisher of currents."
Said a friendly field, "We will not yield, to scattering tyrants."
Much she tried, the friendly field, to push the hole.
But whatever the p or hbar k, the scattering stole.
Every time, the hole was forced, to the bottom of the bowl.

6.3.3 Materials with Intermediate Bandwidths

Let us leave doggerels behind us, shift sides of our brain, and examine the subject of phenomena in materials with intermediate bandwidths. One typically avoids the use of standard band-theoretic approaches in narrow-band materials. One also has no hesitation in using them in materials with broad bands as in inorganic crystals. What about systems in which the bandwidth is of intermediate value?

Around the year 2000, this was another question raised by experimentalists, particularly with pentacene in mind. The large low-T mobility reported in such systems, sharp *power-law* dependence on the temperature in a wide range, and nonlinear saturation phenomena were some of the observations that needed analysis. The task before the theorist was to develop a theory for situations which combined, somehow, the lessons learned in both extremes. It might be interesting to inspect the first of two calculations (Kenkre 2002; Giuggioli et al. 2003) we published then and then briefly comment on the second. What I attempted in Kenkre (2002) was exploring finite-bandwidths on the T-dependence of the mobility of injected carriers for the simplified case of impurity scattering and then combining those elementary band-theoretic procedures with polaronic considerations. In its first part, the calculation was successful in describing power-law mobilities. It was also able to produce, in its second part, as consequences of a single theory, flat mobilities (no T-dependence at high temperatures) as had been observed for decades in naphthalene and mobilities that decreased and then increased as expected from polarons, simply by changing parameter values. See Fig. 6.8.

The calculation begins (Kenkre 2002) with the widely known textbook formula used in inorganic materials which assumes the carrier bandwidth B to be infinite:

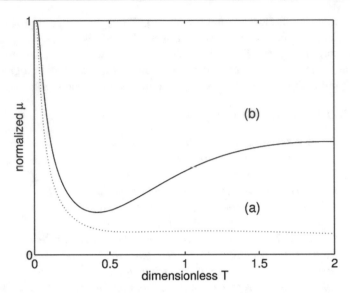

Fig. 6.8 Mobility dependence on temperature, $\mu(T)$, plotted from Eq. (6.12) below for two (arbitrary) parameter combinations. The coupling constant g is 1.5 in (**a**) but 1.8 in (**b**). In units of $\hbar\Omega$, the value of the bare bandwidth B_0 is 0.5 in (**a**) but 2 in (**b**). The temperature T is plotted in units of $\hbar\Omega/2k_B$ and $\mu(T)$ is normalized to its value at $T = 0$. Qualitatively only, prediction (**a**) (dotted line) is similar to reported data in the c' direction of (narrow-band) naphthalene, and prediction (**b**) (solid line) to data claimed to have been seen in (intermediate-band) pentacene. See Kenkre (2002). Reprinted with permission from fig. 1 of Ref. Kenkre (2002); copyright (2002) by Elsevier Publishing

$$\mu = \left(\frac{q}{k_BT}\right)\frac{\int_0^\infty d\epsilon\, v^2(\epsilon)\tau(\epsilon)\rho(\epsilon)e^{-\epsilon/k_BT}}{\int_0^\infty d\epsilon\, \rho(\epsilon)e^{-\epsilon/k_BT}}. \tag{6.6}$$

The symbols have their usual meaning. In a one-band model (band gap large enough to make interband transitions unimportant) with large enough bandwidth, the upper limit of the energy integrations is ∞ to facilitate analytic integrations. Let us take the density of states to be given by the simple free-electronic form within the band and to vanish outside,

$$\rho(\epsilon) = N\left(\frac{3\sqrt{\epsilon}}{2B\sqrt{B}}\right)[\theta(\epsilon) - \theta(\epsilon - B)] \tag{6.7}$$

where θ is the Heaviside step function. Given that the group velocity for a carrier having the energy ϵ can be taken to be proportional to the square root of ϵB, and taking the relaxation time τ to be independent of the band energy for simplicity (as in an early analysis by Erginsoy (1950)), we can write down the finite-band generalization of Eq. (6.6) as

$$\mu' = \left(\frac{B}{S}\right)\frac{\int_0^{B/k_BT} dx\, x\sqrt{x}e^{-x}}{\int_0^{B/k_BT} dx\, \sqrt{x}e^{-x}} = \left(\frac{B}{S}\right)\left(\frac{B}{k_BT}\right)\frac{\int_0^1 dx\, x\sqrt{x}e^{-xB/k_BT}}{\int_0^1 dx\, \sqrt{x}e^{-xB/k_BT}}. \tag{6.8}$$

Here $\mu' = \mu/\mu_0$ is the mobility expressed in dimensionless terms, with $\mu_0 = [(4a^2q/3\hbar)(3\pi^2)^{-2/3}]$ and S equals the "scattering energy" $\hbar/\tau$. The first of these expressions shows that, as $B/k_BT \to \infty$, μ' tends to $(3/2)(B/S)$ but as $B/k_BT \to 0$, μ' tends to $(3/5)(B/S)(B/k_BT)$.[9]

These considerations lead to a compact and useful formula that expresses μ' in terms of incomplete gamma functions,

$$\mu' = \left(\frac{B}{S}\right)\frac{\gamma(5/2, B/k_BT)}{\gamma(3/2, B/k_BT)}, \tag{6.9}$$

and allow one to address (see Kenkre 2002) observations of power-law T-dependence as were claimed for pentacene.

To blend these simple band-theoretic considerations with polaronic considerations, let us we introduce a quantity α^* to recast our final formula (6.1) for narrow band polaron mobility in the ith principal axis direction as

$$\frac{\alpha}{\alpha^*} = \frac{\int_0^\infty dt\,\alpha e^{-\alpha t} I_0\left(\dfrac{2g_i^2}{\sinh\left(\frac{\hbar\omega_0}{2k_BT}\right)}|J_0(\Delta_1 t)J_0(\Delta_2 t)J_0(\Delta_3 t)|\left[1+J_1^2(\Delta_i t)/J_0^2(\Delta_i t)\right]^{1/2}\right)}{I_0\left(\dfrac{2g_i^2}{\sinh\left(\frac{\hbar\omega_0}{2k_BT}\right)}\right)}. \tag{6.10}$$

In terms of this new quantity, our Eq. (6.1) takes the simple form

$$\mu_{i,i} = \frac{2e}{k_BT}\left(\frac{\widetilde{B}_i a}{\hbar}\right)^2\left(\frac{1}{\alpha^*}\right)I_0\left(\frac{2g_i^2}{\sinh\left(\frac{\hbar\omega_0}{2k_BT}\right)}\right) \tag{6.11}$$

where $\widetilde{B}_i = V_i \exp\left[-2g_i^2 \coth\left(\frac{\hbar\omega_0}{2k_BT}\right)\right]$.

We know that α arises from site energy fluctuations as a source of scattering. If it happens to be so much larger than other quantities such as the widths of the phonon bands, Δ_i, the term $\alpha e^{-\alpha t}$ in the integrand will behave like a δ-function and pick out the value of the rest of the integrand at $t = 0$. Under such conditions, α^* becomes identical to α. If the condition does not apply, α^* will be affected by the magnitudes of the phonon bandwidths and will have its own temperature dependence arising from the I_0. Let us assume the simplified situation to apply as was done by Kenkre (2002) and, thus, take $\alpha^* = \alpha$. One can then obtain the following simple normalized mobility formula for *intermediate* bandwidths:

---

[9]This B to B^2 transition as temperature increases is reminiscent of the V to V^2 behavior we discussed in the Perrin-Förster problem of excitation transfer in Chap. 3 but arises here from a completely different source!

$$\frac{\mu(T)}{\mu(0)} = \frac{e^{-G^2\left[\coth\left(\frac{\hbar\Omega}{2k_BT}\right)-1\right]}}{3/2} \left[\frac{\gamma\left(\frac{5}{2},\frac{\tilde{B}}{k_BT}\right)}{\gamma\left(\frac{3}{2},\frac{\tilde{B}}{k_BT}\right)}\right] I_0\left(\frac{2G^2}{\sinh\left(\frac{\hbar\Omega_0}{2k_BT}\right)}\right). \qquad (6.12)$$

Here $\tilde{B} = B_0 \exp\left[-2G^2 \coth\left(\frac{\hbar\Omega}{2k_BT}\right)\right]$, the coupling constant is G, and the bare bandwidth of the carrier is B_0.[10] It is this result that is plotted in Fig. 6.8 to show that it is able to describe mobilities which are respectively flat at high temperatures and exhibit a turn-over depending on the values of the parameters used. In all the extreme limits such as those of T and B, the expression yields results known earlier.

The arguments in Kenkre (2002) do not form an exact theory but possess an interpolation flavor. Further work was started in Giuggioli et al. (2003).[11] Briefly, it led to the following. Three specific scattering mechanisms were considered: impurities, dispersionless optical phonons, and acoustic phonons. The finite width of the carrier band gives rise to expressions which have multiple forms depending on the magnitude of $\hbar\omega_0$ for optical phonons and of $\hbar\omega_D$ for acoustical phonons relative to the carrier bandwidth B. Here ω_D is the Debye frequency. The expressions in Giuggioli et al. (2003) reduce to well-known counterparts in the respective limits of narrow and wide bands. They also reduce to the ones in Kenkre (2002) when only impurity scattering, or narrow bands (i.e. high temperatures) are considered with a single phonon scattering mechanism. A particular feature when considering intermediate carrier bands is that, for intermediate T, carriers in a middle region of the band cannot scatter to arbitrary k values because the optical phonon energy is too large. This suppression of scattering gives rise to an increasing mobility with increased T. For higher T, the electrons which now occupy states above the middle region of the band, can scatter so the mobility begins to decrease again. Thus, the dip in the mobility that had been claimed to be observed in pentacene, can be explained. Another conclusion of that work was that the announced T-dependence of the hole mobility in pentacene could be explained with reasonable parameters for optical and impurity scattering if one optical band is weakly coupled to the carriers and gives rise to scattering, while a second band is strongly coupled and gives rise to polaron formation.

---

[10] We have kept to the definition of B here used in Kenkre (2002) rather than in Kenkre et al. (1989) although there is unfortunately a discrepancy of a factor of 4 in the two. B is simply the assumed matrix element V in the tightbinding model in the former but the actual bandwidth that results from it in the latter. Quantitative precision is not being sought here and I hope my pointing this out will help rather than harm the explanations.

[11] It was unfortunately not pursued further because of the data falsification situation in the field alluded to earlier. I believe that serious theoretical analysis should return with renewed intensity now that the field appears to have bounced back again under the leadership of inspired scientists such as Biscarini, Dodabalapur, and their collaborators. See, e.g., Cramer et al. (2009), Shehu et al. (2010), Wang and Dodabalapur (2018), Wang et al. (2019), and Wang and Dodabalapur (2020).

6.4   Chapter 6 in Summary

The nature of the motion of charge carriers is reflected in the temperature dependence of their mobility. Photo-injected carriers in aromatic hydrocarbon crystals had shown puzzling behavior particularly in naphthalene in which the mobility drops with an increase in temperature from 30 to 100 K and is then unaffected by it as temperature varies further from 100 to 300 K. Many theories had been developed to explain the phenomenon. The chapter showed how they all fail when put to quantitative test. These qualitatively (apparently) successful explanations included standard acoustic phonon scattering formalisms, SLE considerations, and polaron considerations as well. It was shown how the GME theory with its specific calculational scheme applied to a system of polaronic transfer finally succeeded for the entire temperature interval and in all crystallographic considerations. Further calculations, some of them about the saturation of mobility with increasing field, and others designed to address materials with carrier bands broader than in narrow-band polaronic materials were developed. The latter showed how a mobility dependence could arise with the appearance of a band-hopping transition for intermediate width bands.

Projections and Memories for Microscopic Treatment of Vibrational Relaxation

Unlike the last three chapters that have dealt with the construction of theories to explain specific reported observations, the present chapter is addressed at how to do certain calculations suggested by queries that are primarily theoretical.[1] So far in the book, we have described the evolution of a system among its energetically equivalent (degenerate) quantum states exemplified by the site-localized states of a perfect crystal. We now address such evolution, instead, among energetically *inequivalent* states. We will see that the calculations we perform target not only a practical issue in chemical and condensed matter physics but also help with questions one might have regarding some conceptual matters in statistical mechanics. The conceptual issues are weighty, our own contributions that I will describe are somewhat light in spirit. Let us inspect some analysis that Tiwari, Chase, and Ierides, performed with me on the borders of the subject in parts of their dissertations, and published in four recent papers. In order, the first analyzed the inequivalent dimer, the second concentrated on three systems including the vibrationally relaxing molecule (the system of interest in this chapter) with detailed focus on various interactions with a phenomenologically specified reservoir, and the third on microscopic realizations with a boson bath, with both strong and weak coupling to the system (Tiwari and Kenkre 2014; Kenkre and Chase 2017; Kenkre and Ierides 2018; Ierides and Kenkre 2018).

The conceptual issue is concerned with the approach to thermal equilibrium of any ordinary system in contact with a canonical bath (reservoir). It will be commented on only at the end of the chapter. We will start with the practical matter. It has to do with the relaxation, typically vibrational, of a molecule embedded in a larger system that we look upon as the reservoir.

---

[1]The calculated quantities are, nevertheless, all observable in principle.

V. M. (Nitant) Kenkre, *Memory Functions, Projection Operators, and the Defect Technique*, Lecture Notes in Physics 982,
https://doi.org/10.1007/978-3-030-68667-3_7

7.1   Vibrational Relaxation of a Molecule Embedded in a Reservoir

Often, light made to shine on a material, or a different similar agency, raises a molecule to an excited state, and its relaxation back into normal equilibrium needs to be described and manipulated (Laubereau and Kaiser 1978; Lin and Eyring 1974; Oxtoby 1981; Dlott and Fayer 1990; Tokmakoff et al. 1994; Nitzan and Jortner 1973; Metiu et al. 1977; Diestler 1976; Adelman et al. 1991; Kenkre et al. 1994). An understanding of the process of relaxation into equilibrium can be of interest in a variety of processes including unimolecular dissociation (Buff and Wilson 1960) and other chemical reactions (Wan et al. 2000), gentle or explosive (Shuler 1955; Tokmakoff et al. 1993). The search for practical insights into the relaxation of specific quantum systems and a detailed understanding of quantum mechanical evolution at very short times are of obvious importance in the light of femtoscale experiments and the advent of super fast spectroscopy.

Although a number of theoretical approaches have been developed for such investigations more recently, the Master equation that has been a workhorse in chemical physics for the study of vibrational relaxation for almost 70 years is the Montroll-Shuler equation (Montroll and Shuler 1957). Many details of the properties of that equation, the system evolution it describes, and extensions for various purposes have been worked out in the literature. See, for example, the investigation of the canonical nature of the Master equation (Andersen et al. 1964; van Kampen 1971; Thiele et al. 1981), the interaction of relaxation with luminescence and related decay processes (Seshadri and Kenkre 1976, 1978, 1979; Kenkre and Seshadri 1977) and such interaction with the motion of excitation (Kenkre 1977b).

What is not known well in this context is the tool for analysis we should employ if our interest is also in the short-time evolution. Let us assume, as in earlier chapters, that the initial conditions of our system justify the assumption of random phases, and that the observables that our interest is focused on, all are determined by the probabilities rather than the off-diagonal density matrix elements in a given representation. Then, in that representation, for describing short-time as well as long-time evolution, we might adopt the GME (see, e.g., Eq. (3.2)):

$$\frac{dP_M(t)}{dt} = \int_0^t dt' \sum_N \left[\mathcal{W}_{MN}(t-t') P_N(t') - \mathcal{W}_{NM}(t-t') P_M(t') \right]$$

that we have worked with in the last few chapters. Notice that, for the purposes of the present chapter, we have reverted to the notation of using capital letters M, N to represent the states of the system (and m, n to represent reservoir states as we did earlier.)

7.2 The Montroll-Shuler Equation and Its Generalization to the Coherent Domain

The equation introduced into the theory of vibrational relaxation by Shuler with his collaborators (Rubin and Shuler 1956, 1957) and solved completely in its general, discrete, form by Montroll and Shuler (1957), represents the relaxing molecule as a simple harmonic oscillator, its energy levels labeled by integers M, N, etc., and is given by,

$$\frac{dP_M}{dt} = \kappa \left[(M+1)P_{M+1} + Me^{-\beta\Omega}P_{M-1} - \left(M + (M+1)e^{-\beta\Omega} \right) P_M \right].$$

$$(7.1)$$

Here $P_M(t)$ is the probability that the oscillator can be found in the Mth energy level of the harmonic oscillator, Ω is the difference in energy between levels, equivalently the frequency of the oscillator since we have put $\hbar = 1$ throughout most of the book, κ is the relaxation rate and $\beta = 1/k_B T$ as mentioned above. All frequencies are given in units of κ. The well-known basis of the Master equation Eq. (7.1) is the so-called Landau-Teller transitions that arise from an interaction of the harmonic oscillator with the bath, taken to be linear in the oscillator coordinate.

On the basis of the Zwanzig procedure of diagonalizing projection operators generalized via coarse-graining as explained in Chap. 3, we find the GME for vibrational relaxation under the assumption that everything else remains the same in the theory (Kenkre and Chase 2017). The GME obtained is capable of describing the combined process of decoherence and population for conditions in which the oscillator density matrix is initially diagonal in its Hamiltonian eigenstates. In order to understand the formal connection of our GME to the Montroll-Shuler equation Eq. (7.1), let us first observe that the latter has $(M+1)P_{M+1} - MP_M$ as the terms describing energetically upward transitions and $e^{-\beta\Omega}[MP_{M-1} - (M+1)P_M]$ as those describing downward transitions; the detailed balance factor $e^{-\beta\Omega}$ makes the difference particularly transparent. Let us, accordingly, rearrange the terms in Eq. (7.1):

$$\frac{1}{\kappa}\frac{dP_M}{dt} = [(M+1)P_{M+1} \quad MP_M] - e^{-\beta\Omega}[(M+1)P_M - MP_{M-1}]. \quad (7.2)$$

Straightforward calculations indicate that the generalization for the description of decoherence results simply in the respective terms being multiplied by memories $\phi_-(t)$ and $\phi_+(t)$, respectively. The generalization we present is, thus, the GME

$$\frac{1}{\kappa}\frac{dP_M(t)}{dt} = \int_0^t dt' \phi_-(t-t') \left[(M+1)P_{M+1}(t') - MP_M(t') \right]$$

$$-\phi_+(t-t') \left[(M+1)P_M(t') - MP_{M-1}(t') \right]. \quad (7.3)$$

The memory functions $\phi_\pm(t)$ do not bear to each other a detailed balance ratio at every t but their time integrals, as t goes from 0 to ∞, do. This ensures accurate short-time behavior on the one hand and correct long-time evolution on the other, the latter showing detailed balance between energetically upward and downward transition *rates*. Typically, the memories are rapidly decaying functions and the Markoffian approximation

$$\phi_-(t) \approx \delta(t) \left[\int_0^\infty dt' \phi_-(t') \right], \qquad \phi_+(t) \approx \delta(t) e^{-\beta\Omega} \left[\int_0^\infty dt' \phi_-(t') \right],$$

allows the Montroll-Shuler equation to be recovered from our GME, Eq. (7.3). For times short with respect to the decay time of the memories, Eq. (7.3) describes coherent phenomena such as oscillations not present in predictions of Eq. (7.2).

We have focused on the evolution of only probabilities. The effects of the evolution of the off-diagonal elements of the oscillator density matrix are by no means neglected, however. Taking them, as well as the reservoir dynamics, into account leads to the introduction of the ingredient in Eq. (7.3) that makes it an *extension* of the Montroll-Shuler equation capable of describing very short time behavior.[2] This ingredient is the memory function pair $\phi_\pm(t)$.

What is the precise form of these memory functions? They arise from the interactions of the bath with the relaxing molecule. Our explicit calculations use the general, physically transparent, Fourier-transform prescription elaborated on in Kenkre (1977a) and explained in the present book in Chap. 3. It can be said to express the memories in terms of the spectral function $Y(z)$ of the bath with which the relaxing molecule interacts:

$$\kappa \phi_\pm(t) = \int_{-\infty}^\infty dz \, Y(z) \cos((z \pm \Omega)t). \tag{7.4}$$

The Markoffian approximation of this equation makes clear the relation,

$$\kappa = \pi Y(\Omega) = \pi Y(-\Omega) e^{\beta\Omega},$$

---

[2] A student at an oral exam in Rochester in the early 80s had kindly referred to my own work with the Montroll-Shuler equation in his writeup. I was one of the examiners. Another examiner on the committee attacked that equation and the rest of us with it because it was, after all, a Master equation and consequently incapable of analyzing very short-time behavior. On hearing that objection, I had quietly resolved to generalize the Montroll-Shuler equation as soon as the opportunity arose. It only came to pass, almost four decades later, during conversations with two of my last students, Matt Chase and Anastasia Ierides, and is now presented to you in this chapter. I must confess it certainly is not yet capable of addressing questions that depend on off-diagonal density matrix elements *directly*, or on initial conditions that are not random phase. Incidentally, the student was Wayne Knox, the son of my collaborator Bob Knox, and an illustrious scientist in his own right who went on to direct the Institute of Optics at Rochester as also the Advanced Photonics Research at Bell Labs.

of the relaxation rate κ and the bath spectral function, as well as the thermal property of the bath spectral function required by detailed balance. The nature of the spectral function is determined by details of the interaction of the system with the bath and is analyzed in detail in Kenkre and Chase (2017). It has been shown there that, if the bath is truly deserving of the name, it would bring any system put in interaction with it to equilibrium at the same temperature $1/k_B\beta$ characteristic of itself and that this requires the property

$$Y(-z) = Y(z)e^{-\beta z}.$$

An alternative way of expressing this property is that, in terms of an arbitrary symmetric function $Y_s(z) = Y_s(-z)$,

$$Y(z) = \frac{Y_s(z)}{1 + e^{-\beta z}}.$$

We see that we may define, for later use, a sum-memory $\phi_S(t)$

$$\kappa\phi_S(t) = \kappa\left[\phi_-(t) + \phi_+(t)\right], \tag{7.5}$$

which is the product of $2\cos\Omega t$ and the cosine transform of $Y(z)$, and a difference-memory $\phi_\Delta(t)$,

$$\kappa\phi_\Delta(t) = \kappa\left[\phi_-(t) - \phi_+(t)\right], \tag{7.6}$$

which is the product of $2\sin\Omega t$ and the sine transform of $Y(z)$.

Using the factorial moment

$$f_M(t) = \sum_N N(N-1)\ldots(N-M+1)P_N(t), \tag{7.7}$$

Montrol and Shuler obtained the simple equation

$$\frac{1}{\kappa}\frac{df_M}{dt} + M(1 - e^{-\theta})f_M = M^2 e^{-\theta} f_{M-1}. \tag{7.8}$$

The generalization from the GME given in Kenkre and Chase (2017) turns out to be also simple:

$$\frac{1}{\kappa}\frac{df_M(t)}{dt} + M\int_0^t dt'\phi_\Delta(t-t')f_M(t') = M^2\int_0^t dt'\phi_+(t-t')f_{M-1}(t'). \tag{7.9}$$

Here we see the natural appearance of the difference memory function $\phi_\Delta(t)$ defined in Eq. (7.6) and note that its time integral from 0 to ∞ equals $1 - e^{-\theta}$. The evolution equations for the direct moments have a slightly more complicated form

$$\frac{1}{\kappa}\frac{d\langle M^n\rangle}{dt}+n\phi_\Delta*\langle M^n\rangle = \sum_{p=1}^{n-1}\binom{n}{p-1}\left[\frac{(n+1)}{p}\phi_+-(-1)^{n-p}\phi_-\right]*\langle M^p\rangle+\phi_+*1,$$

$$(7.10)$$

where we use the notation $g*f = \int_0^t dt'\, g(t-t')f(t')$. The first and second moments obey:

$$\frac{1}{\kappa}\frac{d\langle M\rangle}{dt}+\int_0^t dt'\phi_\wedge(t-t')\langle M\rangle(t') = \int_0^t dt'\phi_+(t'),\qquad(7.11a)$$

$$\frac{1}{\kappa}\frac{d\langle M^2\rangle}{dt}+2\int_0^t dt'\phi_\Delta(t-t')\langle M^2\rangle(t')\qquad\qquad(7.11b)$$

$$= \int_0^t dt'\left[\phi_\Delta(t-t')+4\phi_+(t-t')\right]\langle M\rangle(t')+\int_0^t dt'\phi_+(t').$$

It is thus seen that the memory combinations that occur prominently are $\phi_\Delta(t)$ and $\phi_+(t)$. Under the Markoffian approximation, they are the products of $\delta(t)$ and, respectively, $1-e^{-\theta}$ and $e^{-\theta}$. We remind the reader that, as in the analysis of
⊛ Montroll and Shuler,[3] we use the symbol $\theta = \Omega/k_B T$. The explicit derivation of the above-given moment equations from Eq. (7.3) is recommended as an exercise to the reader.

7.2.1 Extension of the Bethe-Teller Result for the Average Energy

In an important investigation of deviations from thermal equilibrium in shock waves, Bethe and Teller found that the average energy $E(t)$ of a molecule in a reservoir would relax exponentially from its initial value $E(0)$ to the thermal value $E_{th} = (\Omega/2)\coth(\theta/2)$ according to

$$E(t) = E(0)e^{-\kappa(1-e^{-\theta})t}+E_{th}\left(1-e^{-\kappa(1-e^{-\theta})t}\right).\qquad(7.12)$$

---

[3]Montroll and Shuler had been successful in achieving great simplifications in the analysis of their equation on the basis of a property known as canonical invariance. For instance, it allowed one to show that an initial probability distribution of the Boltzmann form among the molecular energy levels at a temperature T other than that of the environment, maintains that Boltzmann form throughout its evolution from the initial T to the environment T. This permits the introduction of the concept of a *time-dependent T* in the evolution of the probabilities. This basic property, also explored further in Andersen et al. (1964), van Kampen (1971), and Thiele et al. (1981), is unfortunately lost in going from the Master equation to the GME. Given the interesting phenomena that Seshadri and I had been able to describe by exploiting canonical invariance in our extensions of the equation to include additional processes (Seshadri and Kenkre 1976; Kenkre and Seshadri 1977; Kenkre 1977b), we wanted to be able to restore it to the GME in some modified form. Chase and I made multiple attempts in this direction. So far they have not met with success.

This result can also be seen in the later work of Rubin, Shuler, and Montroll (Rubin and Shuler 1956, 1957; Montroll and Shuler 1957). There are two important questions here. The first is whether the time dependence of the relaxation of the energy is different for different initial values of the energy; the second is whether it depends on the particular form of the initial distribution of the energy among the energy levels of the molecule for a given value of the initial energy. The answers in the work of Bethe and Teller, which are the same as those emerging from the Montroll-Shuler equation, are that the time dependence is exponential throughout the evolution and that this exponential nature is independent of the initial details of the distribution. The latter result is in contrast to the evolution of the probabilities themselves which do depend on their initial values.

In the context of our present investigation which addresses an arbitrary degree of coherence in the vibrational relaxation, we also find, as in the Bethe-Teller analysis, that the initial details of the probability distribution do not influence the time dependence of the average energy; and that the initial value of the average energy does so only in a trivial (multiplicative) way. However, we do expect that if the time evolution shows an exponential nature, it would do so only in the long time limit when the Markoffian approximation becomes valid. We study now the departures that occur at short times.

The answers in the presence of coherence are in the moment equations that we have derived above. We find in the Laplace domain our new result for the dynamics of the energy valid for an arbitrary degree of coherence:

$$\widetilde{E}(\epsilon) = \frac{E(0)}{\epsilon + \kappa \widetilde{\phi}_\Delta(\epsilon)} + E_{th} \left[\frac{\widetilde{\phi}_S(\epsilon)}{\widetilde{\phi}_\Delta(\epsilon)} \tanh\left(\frac{\theta}{2}\right) \right] \left[\frac{1}{\epsilon} - \frac{1}{\epsilon + \kappa \widetilde{\phi}_\Delta(\epsilon)} \right]. \quad (7.13)$$

The Markoffian limit of the energy relaxation dynamics we have obtained yields the Bethe-Teller or Montroll-Shuler result in an interesting way. Abelian theorems can be invoked to obtain the time-domain limiting values at large times of the memories $\phi(t)$ as the limiting values of $\epsilon\widetilde{\phi}(\epsilon)$ as ϵ tends to zero. We recall that the ratio of the time-integrals from 0 to ∞ of the sum and difference memories, $\phi_S(t)$, $\phi_\Delta(t)$, equals $\coth(\theta/2)$. This renders the contents of the first square bracket in Eq. (7.13) equal to 1 as ϵ tends to zero. If we write

$$\widetilde{\eta}(\epsilon) = \frac{1}{\epsilon + \kappa \widetilde{\phi}_\Delta(\epsilon)},$$

in other words denote the first term on the right hand side of Eq. (7.13) as the Laplace transform of $E(0)\eta(t)$, we see that the last square bracket in Eq. (7.13) is the Laplace transform of $1 - \eta(t)$. Thus, the generalization of Eq. (7.12) that arises from our GME analysis,

$$E(t) = E(0)\eta(t) + E_{th} \int_0^t dt' \xi(t - t') \left[1 - \eta(t') \right], \quad (7.14)$$

maintains a form that is almost identical to the Bethe-Teller evolution. The only differences are the non-exponential nature of $\eta(t)$ in the presence of coherence and the existence of the convolution with $\xi(t)$, a function that is defined through its Laplace transform via

$$\widetilde{\xi}(\epsilon) = \frac{\widetilde{\phi}_S(\epsilon)}{\widetilde{\phi}_\Delta(\epsilon)} \tanh\left(\frac{\theta}{2}\right).$$

We display in Fig. 7.1 the oscillatory relaxation of the energy for coherent conditions using parameter values that are by no means expected for normal systems but exaggerated to make clear the oscillation effects. The Lorentzian bath memory is considered for different assumed spectral widths. The resulting oscillations in the relaxation, our generalization of the Bethe-Teller result, are shown along with the incoherent limit. The memories are responsible for the coherent behavior at short times. At long times they reproduce the Montroll-Shuler or Bethe-Teller behavior. This happens because in the Markoffian limit $\eta(t)$ becomes $e^{-\kappa(1-e^{-\theta})t}$, and the

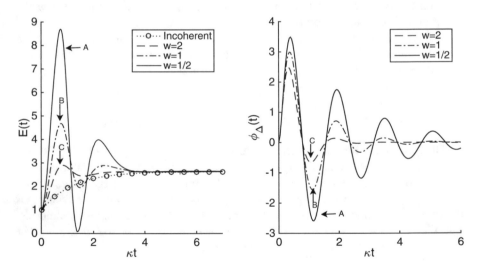

Fig. 7.1 Generalization of the Bethe-Teller result for coherent conditions. Shown in the left panel is the time evolution of the average energy of the relaxing molecule, $E(t)$, normalized to its initial value $E(0)$, as a function of the dimensionless time κt. The energy rises from its initial value and saturates to the thermal value E_{th} (see Eq. (7.14)), the incoherent limit given by the Bethe-Teller result, Eq. (7.12). This incoherent result is a simple exponential rise and is labeled by the symbol o. Coherent cases of the bath spectral function, assumed Lorentzian, with respective (illustrative) scaled values of an internal parameter w as 2, 1, 1/2 lead to oscillations in the energy with larger and larger amplitudes of oscillation. The three difference memory functions $\phi_\Delta(t)$ corresponding to the three examples are shown in the right panel. Assumed scaled values of other parameters, temperature and molecular frequency, are illustrative, $\beta = 1/2$ and $\Omega = 4$. Reprinted with permission from fig. 1 of Ref. Kenkre and Chase (2017); copyright (2017) by World Scientific publishing

ratio of the time integrals from 0 to ∞ of the sum and difference memories, $\int_0^\infty dt \phi_S(t) / \int_0^\infty dt \phi_\Delta(t)$, exactly equals $\coth(\theta/2)$.

7.2.2 Explicit Memory Functions for Specified Baths

Chase has performed a comprehensive analysis of various bath-system interactions and compiled a catalog as part of his dissertation. We refer the reader to Kenkre and Chase (2017) for the prominent features of the catalog but show here just one item relevant to the temperature dependence of the spectral function $Y(z)$ whose Fourier transform determines the memory functions $\phi(t)$.

In order to understand the effect of temperature on the shape of the spectral function, we show Fig. 7.2, in which the Gaussian spectral function,

$$Y(z) = \left[\frac{1}{1 + e^{-\beta z}} \right] e^{-\frac{z^2}{w^2}},$$

is displayed for four values of the dimensionless energy ratio $\Omega/k_B T = 0, 1, 4, \infty$. The interval between the solid vertical lines indicates the width of the spectral function and the horizontal axis is normalized by Ω. For very high temperatures, the spectral function is symmetric (solid line). An increase in the parameter $\Omega/k_B T$ leads to a loss of this symmetry in keeping with detailed balance between the energetically upward and downward transitions, respectively. While for

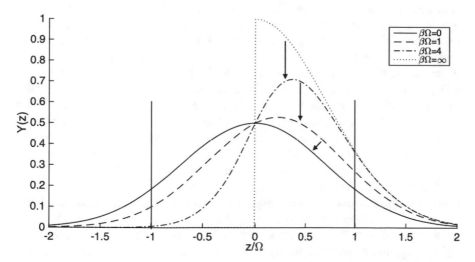

Fig. 7.2 Gaussian bath spectral function $Y(z)$ (see text) is displayed for four values of the parameter $\beta\Omega = 0, 1, 4, \infty$. The vertical lines indicate the width of the spectral function and Ω normalizes z on the horizontal axis. The arrows show the effect of increasing temperature from 0 to ∞. Reprinted with permission from fig. 4 of Ref. Kenkre and Chase (2017); copyright (2017) by World Scientific publishing

small values of $\Omega/k_B T$ (high temperatures) the spectral function is approximately a Gaussian, for high values (low temperatures) the function tends to a half-Gaussian, being non-negligible only for positive values of z (dotted line) at zero temperature.

The behavior of the memories for particular bath spectral functions is intimately related to the relative values of the three (dimensionless) energies that characterize the system and the reservoir: the system energy Ω, the thermal energy of the bath $k_B T$, and w, the energy that characterizes the spectral resolution of the bath. Physically, the effects caused by the variation of these parameters are related to the change in the ratios of the energy scales, $w/k_B T$, $\Omega/k_B T$ and Ω/w. The first ratio compares the thermal energy of the reservoir to its average spectral energy. The second and third ratios compare the respective reservoir energies to the system energy. Small values of either lead to incoherent motion. The exchange of energy is suppressed when the ratios are large, which extends the time-scale of coherent system evolution.

7.3 Reservoir Effects on Vibrational Relaxation for Specific Microscopic Interactions

Let us now turn to the examination of the effects of specific microscopic interactions postulated via Hamiltonian expressions. The interaction of the vibrating molecule with the bath that led to the well-known Montroll-Shuler relaxation equation was based on the so-called Landau-Teller rates. For the sake of simplicity, these assumed the interaction to be linear in the molecule displacement. Little more had to be specified in such a derivation because rates in the incoherent Master equation were all that was required. Our present interest, however, lies in a more detailed description that is capable of describing stages of the approach to equilibrium that are earlier in time as well and can give rise to effects like oscillations that occur during decoherence of the initial state. The memory functions required for this purpose demand more information about the basic system-bath interaction. Therefore, while we maintain the linear dependence feature of the Landau-Teller rates, we have to specify additional details of the bath observables in the interaction.

Two types of interaction appear of interest and relevance. Both are compatible with the earlier analysis in the literature. One of them is associated with weak coupling and single-phonon transitions while the other is best described as arising from strong coupling involving transitions of single phonons leading, after a transformation, to multiphonon transitions. Our primary focus is in the latter because of its interesting consequences. We specify the two interactions in the Hamiltonian below. We comment on the noteworthy features of the approach to equilibrium that arise in the strong, multiphonon interaction case and show how they do not arise in the other, weak, single-phonon case.

7.3.1   The Interaction Hamiltonians

Although the formalism can be developed for a completely general Hamiltonian, we consider here separable cases of the form (Oxtoby 1981; Kenkre et al. 1994)

$$\mathcal{H} = \mathcal{H}_S + \mathcal{H}_B + \sum_j \mathcal{V}_{S,j} \mathcal{V}_{B,j}, \tag{7.15}$$

where $\mathcal{H}_S$ is the system Hamiltonian and $\mathcal{V}_{S,j}$ is the j-th system contribution to the system-bath interaction, respectively.

Let us define the operators

$$H_B = \sum_q \omega_q \left(b_q^\dagger b_q + \frac{1}{2} \right), \tag{7.16a}$$

$$V_B = \sum_q g_q \omega_q \left(b_{-q} + b_q^\dagger \right), \tag{7.16b}$$

where the b_q's describe bath bosons, for instance phonons, of frequency ω_q and wavevector q and the g_q's are the dimensionless coupling constants determining the interaction strength. In the first instance, we take the unperturbed bath Hamiltonian $\mathcal{H}_B$ to equal H_B and assume a single bath contribution j in the interaction, $\mathcal{V}_{B,j} = V_B$, weak enough so that the calculational procedure can use a perturbation in orders of the third term of Eq. (7.15). The system operators for this first system, the vibrating molecule, are

$$\mathcal{H}_S = \Omega \left(a^\dagger a + \frac{1}{2} \right), \tag{7.17a}$$

$$\mathcal{V}_S = a + a^\dagger, \tag{7.17b}$$

where Ω is the oscillator frequency and $a + a^\dagger$ is essentially the displacement of the molecular oscillator.

In order to analyze strong coupling, we begin by considering, as our second case, the same $\mathcal{H}_S$ as in Eq. (7.17a) and the same $\mathcal{H}_B = H_B$ as in Eq. (7.16a), but $V_B a^\dagger a + u(a + a^\dagger)$ as the system-bath interaction. The c-number u which describes the constant force exerted by the bath on the molecular oscillator is weak but V_B in this case is not because the coupling constants g_q are taken to be large. The term proportional to V_B cannot be treated in this second case perturbatively. In order to take as much of its effects into account exactly, we perform the standard polaronic transformation taking the Hamiltonian from a representation in terms of the bare operators a, b, to a representation in terms of their dressed versions A, B. The transformation, as is well known, corresponds to a shift of the equilibrium position of the bath oscillators and allows the Hamiltonian to be written as

$$\mathcal{H} = \left(\Omega - \sum_q g_q^2 \omega_q \right) A^\dagger A + \sum_q \omega_q \left(B_q^\dagger B_q + \frac{1}{2} \right)$$

$$+ u \left[A e^{\sum_q g_q \left(B_{-q} - B_q^\dagger \right)} + A^\dagger e^{-\sum_q g_q \left(B_{-q} - B_q^\dagger \right)} \right]. \tag{7.18}$$

Here, the dressed operators A and B are given by $A = a e^{-\sum_q g_q \left(B_{-q} - B_q^\dagger \right)}$ and $B_q = b_q + g_q a^\dagger a$. We see that Eq. (7.18) is precisely of the form of Eq. (7.15) with $\mathcal{H}_S = \left(\Omega - \sum_q g_q^2 \omega_q \right) A^\dagger A$ and $\mathcal{H}_B = \sum_q \omega_q \left(B_q^\dagger B_q + 1/2 \right)$. The interaction term in the second line of Eq. (7.18) is amenable to a perturbative treatment because of the assumed smallness of the c-number u. The presence of the so-called phonon clouds (the exponentials of the dressed bath operators B_q, $B_q^\dagger$) forces the transformed Hamiltonian of Eq. (7.18) to have a *multiphonon* nature although the original Hamiltonian before the transformation has only single-phonon interactions.[4]

7.3.2   Bath Correlation Functions

As shown in Sect. 3.2, the memory functions appearing in the GME, exemplified by Eq. (3.3), can be written down explicitly in terms of the bath partition function $Z = \sum_n e^{-\beta \epsilon_n}$, the bath energy differences $\omega_{mn} = (\epsilon_m - \epsilon_n)$, and the system energy differences $\Omega_{MN} = (E_M - E_N)$.

While $\mathcal{V}$ may be generally of arbitrary form, we have taken it here, as explained in the context of Eq. (7.15), as a sum of products of independently contributing bath and system components. For a single j for simplicity, we see that

$$\mathcal{W}_{MN}(t) = |\langle M|\mathcal{V}_S|N\rangle|^2 \sum_{m,n} \frac{e^{-\beta \epsilon_n}}{Z} |\langle m|\mathcal{V}_B|n\rangle|^2 \left[e^{i(\omega_{mn} + \Omega_{MN})t} + e^{-i(\omega_{mn} + \Omega_{MN})t} \right],$$
$$\tag{7.19a}$$

$$\mathcal{W}_{NM}(t) = |\langle M|\mathcal{V}_S|N\rangle|^2 \sum_{m,n} \frac{e^{-\beta \epsilon_n}}{Z} |\langle m|\mathcal{V}_B|n\rangle|^2 \left[e^{i(\omega_{mn} - \Omega_{MN})t} + e^{-i(\omega_{mn} - \Omega_{MN})t} \right].$$
$$\tag{7.19b}$$

Distributing the exponents in Eqs. (7.19) so that bath quantities are absorbed into factors separated from the system quantities, we obtain

---

[4]It should be evident that we are not offering any new Hamiltonians and methods to deal with them in what follows. The techniques of solution are borrowed from earlier work of many authors in the excitonic polaron field, most notably Silbey (we have given examples and explanations in previous chapters). What is new is the consequences for the present vibrational relaxation problem, in particular a comparison of results for strong versus weak coupling, as we shall see below.

$$\mathcal{W}_{MN}(t) = |\langle M|\mathcal{V}_S|N\rangle|^2 \left[e^{i\Omega_{MN}t}\mathcal{B}(t) + e^{-i\Omega_{MN}t}\mathcal{B}^*(t) \right], \tag{7.20a}$$

$$\mathcal{W}_{NM}(t) = |\langle M|\mathcal{V}_S|N\rangle|^2 \left[e^{-i\Omega_{MN}t}\mathcal{B}(t) + e^{i\Omega_{MN}t}\mathcal{B}^*(t) \right]. \tag{7.20b}$$

Equation (7.20) presents the memories as a direct combination of, on one hand the system detail in the matrix elements of $\mathcal{V}_S$ and the difference Ω_{MN}, and on the other the bath correlation function $\mathcal{B}(t)$, a property entirely of the bath given by

$$\mathcal{B}(t) = \sum_{m,n} \frac{e^{-\beta\epsilon_n}}{Z} \langle m|\mathcal{V}_B|n\rangle\langle n|\mathcal{V}_B^\dagger(t)|m\rangle = \frac{\mathrm{Tr}\left(e^{-\beta\mathcal{H}_B}\mathcal{V}_B e^{i\mathcal{H}_B t}\mathcal{V}_B^\dagger e^{-i\mathcal{H}_B t}\right)}{\mathrm{Tr}\left(e^{-\beta\mathcal{H}_B}\right)}. \tag{7.21}$$

This fact is calculationally significant. It stems from the reasonable, normally made, supposition that the bath always remains in its thermal equilibrium even as it influences the time evolution of the system. Since the system analyzed is the harmonic oscillator, M and N differ only by 1 such that only transitions between nearest neighbor states in energy space can occur (this is a consequence of the assumed Landau-Teller interaction). Their energy difference is $\pm\Omega$, depending on the respective transition direction.

Explicit calculation from Eq. (7.16) gives, for the weak-coupling case,

$$\mathcal{B}(t) = \sum_q g_q^2\omega_q^2\left[\coth\left(\frac{\beta\omega_q}{2}\right)\cos\omega_q t + i\sin\omega_q t\right]. \tag{7.22}$$

The evaluation is more involved for the strong-coupling case of Eq. (7.18), but follows standard polaronic manipulations that yield

$$\mathcal{B}(t) = u^2 e^{-\sum_q g_q^2\coth\left(\frac{\beta\omega_q}{2}\right)}e^{\frac{1}{2}\sum_q g_q^2\mathrm{csch}\left(\frac{\beta\omega_q}{2}\right)\left(e^{\frac{\beta\omega_q}{2}}e^{i\omega_q t}+e^{-\frac{\beta\omega_q}{2}}e^{-i\omega_q t}\right)}. \tag{7.23}$$

In order to perform the summation over the reservoir modes q implicit in $\mathcal{B}(t)$ in Eqs. (7.22) and (7.23), let us use the simplification of optical phonons with a peak frequency ω_0 and small dispersion $\sigma \ll \omega_0$, with an average coupling constant g. If the density of states considered is that of a Gaussian peaked at ω_0 and width σ, Eqs. (7.22) and (7.23) (in the continuum limit) have the forms

$$\mathcal{B}(t) = g^2\omega_0^2 e^{-\frac{\sigma^2 t^2}{2}}$$

$$\times \left\{ \coth\left(\frac{\beta\omega_0}{2}\right)\left[\left(1 + \frac{\sigma^2}{\omega_0^2} - \frac{\sigma^4 t^2}{\omega_0^2}\right)\cos(\omega_0 t) - 2\frac{\sigma^2}{\omega_0}t\sin(\omega_0 t)\right]\right.$$

$$+ i \left[\left(1 + \frac{\sigma^2}{\omega_0^2} - \frac{\sigma^4 t^2}{\omega_0^2} \right) \sin(\omega_0 t) + 2\frac{\sigma^2}{\omega_0} t \cos(\omega_0 t) \right] \Bigg\},$$

(7.24)

in the weak-coupling case, and

$$\mathcal{B}(t) = u^2 e^{-g^2 \coth\left(\frac{\beta\omega_0}{2}\right)\left(1 - e^{-\frac{\sigma^2 t^2}{2}}\cos\omega_0 t\right)} e^{ig^2 e^{-\frac{\sigma^2 t^2}{2}} \sin\omega_0 t},$$

(7.25)

in the strong-coupling case respectively. The latter result is widely known in the literature and can be found discussed, for instance, in Kenkre (2003). The former

⊛ has been derived in detail and used by Ierides in her dissertation work. The exercise of the explicit derivation of these expressions for the bath correlation function from Eq. (7.21) would serve the reader well in familiarizing herself with the calculational technique.

It is expected that the memory functions resulting from the use of Eqs. (7.24) and (7.25) would decay to zero at long times due to the interaction of the system with the thermal reservoir. However, the linear interactions considered in this study cannot broaden the zero-phonon lines present in optical spectra (Kenkre 1975c; Fitchen 1968) (see also Chap. 4 of the present book), and the memories produced never quite decay to zero, oscillating about a nonzero value determined by the microscopically specified parameters. The solution to this problem, discussed in the literature earlier (Kenkre 1975c; Kenkre et al. 1989), lies in the unavoidable introduction of sources of disorder not taken explicitly into account in the microscopic description of the Hamiltonian. This is done through the incorporation of an extra decay factor introduced into the memories (Kenkre et al. 1989) as we have learnt in Chaps. 4 and 6. In all three systems, the bath correlation functions given in Eqs. (7.22) and (7.23) may be thus considered as being calculated for a single mode of vibrational mode and multiplied by a time-dependent factor $f(t)$ which takes into account both the phonon spectral line shape and sources of disorder mentioned above. This allows Eqs. (7.24) and (7.25) to be represented as

$$\mathcal{B}(t) = g^2 \omega_0^2 \left[\coth\left(\frac{\beta\omega_0}{2}\right) \cos(\omega_0 t) + i \sin(\omega_0 t) \right] f(t),$$

(7.26)

and

$$\mathcal{B}(t) = u^2 e^{-g^2 \coth\left(\frac{\beta\hbar\omega_0}{2}\right)(1 - \cos\omega_0 t)} e^{ig^2 \sin\omega_0 t} f(t),$$

(7.27)

respectively. Equations (7.26), (7.27) are analytically more manageable versions of (7.22), (7.23) which use the full sums over the wavevectors.

In terms of these bath correlations that have been calculated, the memory functions $\phi_\pm$ can be written down explicitly. For the strong-coupling case, the sum memory $\phi_S = \phi_- + \phi_+$ is

$$\kappa\phi_S(t) = 4u^2 e^{-g^2 \coth(\beta\omega_0/2)(1-\cos\omega_0 t)} \cos\left(g^2 \sin\omega_0 t\right) \cos(\Omega t) f(t), \qquad (7.28)$$

whereas the difference memory $\phi_\Delta = \phi_- - \phi_+$ is

$$\kappa\phi_\Delta(t) = 4u^2 e^{-g^2 \coth(\beta\omega_0/2)(1-\cos\omega_0 t)} \sin\left(g^2 \sin\omega_0 t\right) \sin(\Omega t) f(t). \qquad (7.29)$$

For the weak-coupling case, the sum memory ϕ_S is

$$\kappa\phi_S(t) = 4g^2\omega_0^2 \coth(\beta\omega_0/2) \cos(\omega_0 t) \cos(\Omega t) f(t), \qquad (7.30)$$

while the difference memory ϕ_Δ is

$$\kappa\phi_\Delta(t) = 4g^2\omega_0^2 \sin(\omega_0 t) \sin(\Omega t) f(t). \qquad (7.31)$$

Ierides and Kenkre (2018) have also provided explicit formulae to obtain the relaxation rate κ from these expressions. The most convenient representation of $f(t)$ is the exponential $e^{-\alpha t}$ which is compatible with early derivations of the GME in the presence of disorder (Kenkre and Reineker 1982; Kenkre et al. 1989). See also the discussion in Chap. 6 in the present book. The κ expressions are found through integration of the respective memories and are, in terms of modified Bessel functions I_ℓ,

$$\kappa = 2u^2 e^{-g^2 \coth\left(\frac{\beta\omega_0}{2}\right)} \sum_{\ell=-\infty}^{\infty} I_\ell\left[2g^2 n(\omega_0)\right] \sum_{k=0}^{\infty} \frac{g^{2k}}{k!} \frac{\alpha}{\alpha^2 + [(k+\ell)\omega_0 - \Omega]^2} \qquad (7.32)$$

in the multiphonon (strong-coupling) interaction case. They are given by

$$\kappa = 2g^2\omega_0^2 \left\{ [n(\omega_0) + 1] \frac{\alpha}{\alpha^2 + (\omega_0 - \Omega)^2} + n(\omega_0) \frac{\alpha}{\alpha^2 + (\omega_0 + \Omega)^2} \right\} \qquad (7.33)$$

in the single-phonon (weak-coupling) interaction case. While the case of weak interaction with the bath (single-phonon interaction) results in an Arrhenius behavior wherein the rate is activated and increases with increasing T, in the case of a strong-coupling (multiphonon) interaction, a non-monotonic behavior of the rate is observed. Thus, whether the relaxation rate increases or decreases with T is determined by what T-interval is used for the measurement. The interested reader should consult (Ierides and Kenkre 2018; Kenkre and Ierides 2018) for plots and discussion. As noted in the second of the quoted publications, our analysis might have relevance to the so-called inverted temperature dependence of relaxation rates (e.g., in $W(CO)_6$) mentioned in Tokmakoff et al. (1994).

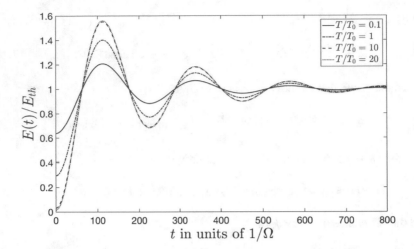

Fig. 7.3 Straightforward temperature dependence of the time evolution of the average energy of the relaxing molecule of characteristic frequency Ω for weak-coupling (single-phonon) interactions ($g = 0.02$). The interaction is with optical phonons of peak frequency $\omega_0 = \Omega + \alpha/2$ and narrow dispersion $\alpha/\Omega = 0.01$. As the temperature T is increased, the average energy initially rises above the incoherent limit value, with oscillations decaying faster to E_{th}. The frequency of oscillation remains the same for all T, while the amplitude shows a direct T dependence, increasing with increasing T. Reprinted with permission from fig. 2 of Kenkre and Ierides (2018); copyright (2018) by Elsevier Publishing

7.3.3 Energy Time Evolution: Interesting Results in the Temperature Dependence for Strong Coupling

We have seen the evolution of the average energy of the relaxing molecule from its initial value to its thermalized value obeys Eq. (7.12) and is exponential as given by the original Montroll-Shuler equation. We have also seen that, in the generalized form of Kenkre and Chase, it is nonexponential, and given as in Eq. (7.14) in terms of the initial energy of the system $E(0)$, the thermal energy $E_{th} = (\Omega/2) \coth(\beta\Omega/2)$, and the functions $\eta(t)$ and $\xi(t)$. These are the respective Laplace inverses of

$$\tilde{\eta}(\epsilon) = \frac{1}{\epsilon + \kappa\widetilde{\phi}_\Delta(\epsilon)} \qquad \text{and} \qquad \tilde{\xi}(\epsilon) = \frac{\widetilde{\phi}_S(\epsilon)}{\widetilde{\phi}_\Delta(\epsilon)} \tanh\left(\frac{\beta\Omega}{2}\right). \tag{7.34}$$

We now use the expressions for microscopic memories we have derived to investigate whether any interesting effects of T and other parameters emerge.

Figure 7.3 shows the unremarkable temperature dependence of the average energy of the relaxing molecule in the case of the weak-coupling (single-phonon) interaction. The figure caption is self-explanatory. By contrast, striking results are visible in Fig. 7.4 where the strong-coupling (multiple-phonon) case is analyzed: the dependence on temperature is non-monotonic. The inset in the plot shows a

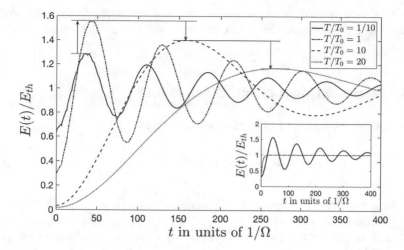

Fig. 7.4 Non-monotonic temperature (T) dependence of the energy of the relaxing molecule of characteristic frequency Ω in strong interaction ($g = 2$) with optical phonons. Other parameters are as in Fig. 7.3. Plots are of $E(t)$ scaled to its thermal equilibrium value of E_{th} for varying temperatures, i.e., for various values of $\beta\Omega$ as shown in the legend. A striking dependence is observed as T is increased. The values of T in units of T_0 for the various curves are as shown and the corresponding peaks are represented by arrows. As T is increased, the peak value increases from the first curve to the second but decreases from the second to the third and from the third to the fourth. The inset at the lower right is a mere reminder of the difference between the incoherent limit (Bethe-Teller) (dashed line) and our GME predictions (solid line) for one representative case, $\Omega/k_BT = 1$. Reprinted with permission from fig. 1 of Kenkre and Ierides (2018); copyright (2018) by Elsevier Publishing

comparison of the coherent (oscillatory) behavior of the energy predicted by the GME to the incoherent (non-oscillatory) behavior predicted by the original Master equation for one temperature. For each value of T shown in the main part of Fig. 7.4, the approach to equilibrium of the energy is dominated by initial T-dependent oscillations about the steady-state value E_{th}, and then dissipating to it. The non-monotonic T dependence is seen clearly in the rise and fall of the peak amplitude of the energy as T is increased. As the figure shows, the peak value of the energy rises initially (from a to b) with increasing T and then falls (from c to d) as the period of oscillation increases monotonically. This non-monotonic behavior is also visible in the T dependence of the relaxation rate in Eq. (7.32).

7.4 Two Simplified Descriptions

We have seen how the vibrational relaxation rate κ can be obtained through the Fourier transform of the microscopically derived bath correlation function $\mathcal{B}(t)$ evaluated at the oscillator frequency. In this section, we develop two simplified approaches, one based on a relaxation *memory* and an *effective* rate, and another on the so-called half-Markoffian approximation which produces an interpolation formula.

7.4.1　Relaxation Memory and Effective Rate

The incoherent result of Eq. (7.12), may be rewritten as

$$\frac{E(t) - E_{th}}{E(0) - E_{th}} = e^{-\kappa\left(1 - e^{-\beta\Omega}\right)t} \tag{7.35}$$

to emphasize that the effective rate of relaxation is $\kappa\left(1 - e^{-\beta\Omega}\right)$. Although our GME generalization reflected in Eq. (7.13) results in a non-exponential time-dependence of the left hand side of Eq. (7.35), its Laplace transform can always be expressed as

$$\frac{\tilde{E}(\epsilon) - E_{th}/\epsilon}{E(0) - E_{th}} = \frac{1}{\epsilon + \tilde{\zeta}(\epsilon)}, \tag{7.36}$$

through the introduction of the Laplace transform of a *relaxation memory* $\zeta(t)$. In the incoherent Master equation approximation, $\zeta(t)$ is obviously the product of the Dirac delta function $\delta(t)$ and $\kappa\left(1 - e^{-\beta\Omega}\right)$, while in the general case, $\zeta(t)$ describes coherent oscillations present in the exact result of $E(t)$ obtained from our generalization. The memory is found in the Laplace domain as

$$\tilde{\zeta}(\epsilon) = \frac{\epsilon\left[E(0) - E_{th}\tilde{\xi}(\epsilon)\right]\left[1 - \epsilon\tilde{\eta}(\epsilon)\right]}{E(0) - E_{th} - \left[E(0) - E_{th}\tilde{\xi}(\epsilon)\right]\left[1 - \epsilon\tilde{\eta}(\epsilon)\right]}. \tag{7.37}$$

On the basis of the memory, we can provide a formula for an *effective* rate of relaxation in the presence of non-exponential time evolution of the energy $E(t)$ that can take into account the consequences of all non-exponential features including oscillations. We define it from Eq. (7.37) by applying the Markoffian approximation to $\zeta(t)$. This means putting the Laplace variable equal to zero in the right hand side of Eq. (7.37), and yields

$$\tilde{\zeta}(0) = \frac{1}{\int_0^\infty \frac{E(t) - E_{th}}{E(0) - E_{th}} dt}. \tag{7.38}$$

If the evolution were totally incoherent (exponential time-dependence), there would be no oscillations and this effective rate would equal the incoherent rate $\kappa\left(1 - e^{-\beta\hbar\Omega}\right)$. But if there are oscillations due to coherence, Eq. (7.38) would take them into account as they would enter the effective rate through the integration of the oscillations in the denominator of its right hand side. Eq. (7.38) is a prescription that can be used in this fashion on experimental observations of the energy time dependence.

An interesting result is seen to emerge (Ierides and Kenkre 2018) when one uses the prescription on the numerical solutions obtained above. When the Markoffian rate $\tilde{\zeta}(0)$ obtained in this manner is plotted with respect to temperature for the strong interaction case using the derived microscopic memories, *multiple oscillations* with increasing T are observed. On the other hand, only a simple and expected monotonic decay with T is seen in the weak-coupling (single-phonon) case. An overall non-monotonic T dependence exists in the strong interaction regime, where the increasing or decreasing behavior is determined by the T interval in which observations are made.

7.4.2 Half-Markoffian Approximation and an Interpolation Formula

Let us now develop a useful interpolation formula to describe the time evolution of the energy on the basis of an approximation that has been appropriately dubbed (Kenkre and Reineker 1982; Kenkre and Sevilla 2007) the *half-Markoffian* approximation. It is found to be exact at very short and very long times, and to provide an approximate description at intermediate times. The half-Markoffian approximation is defined by taking an expression which is a convolution of two functions and approximating it by a convolutionless expression as

$$\int_0^t ds\, f(t-s)g(s) \approx g(t) \int_0^t ds\, f(s). \tag{7.39}$$

To this end, let us begin by converting Eq. (7.14) to the integro-differential equation

$$\frac{dE(t)}{dt} + \kappa \int_0^t dt'\, \phi_\Delta(t-t')E(t') = \frac{\kappa\Omega}{2} \int_0^t dt'\, \phi_S(t'). \tag{7.40}$$

Let us now apply the half-Markoffian approximation (7.39) to its second term, replacing it,

$$\kappa \int_0^t dt'\, \phi_\Delta(t-t')E(t') \approx \mathcal{I}_\Delta(t)E(t),$$

where $\mathcal{I}_\Delta(t) = \kappa \int_0^t dt'\phi_\Delta(t')$. In this simplified form, the differential equation for $E(t)$ can be solved immediately as a driven linear equation to give

$$E(t) = \frac{\Omega}{2} + \left[E(0) - \frac{\Omega}{2}\right] e^{-\int_0^t dt'\mathcal{I}_\Delta(t')} + \Omega \int_0^t dt'\mathcal{I}_+(t')e^{-\int_{t'}^t ds\mathcal{I}_\Delta(s)}, \tag{7.41}$$

where $\mathcal{I}_+(t) = \int_0^t dt'\kappa\phi_+(t')$.

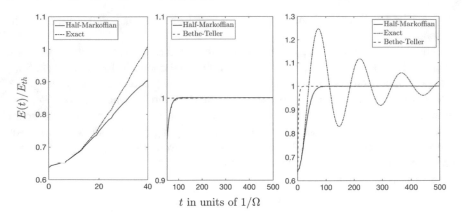

Fig. 7.5 Interpolation formula developed as Eq. (7.41) on the basis of the half-Markoffian approximation (see text) showing excellent agreement with the exact result at extreme (both short and long) times but considerable departures at intermediate times. The energy, scaled to its thermal value, is plotted for very low T for $g = 2$, $\omega_0 = \Omega + \alpha/2$, and $\alpha/\Omega = 0.02$. The kinks visible at short times (left panel) and equilibration to the steady-state result of the energy (middle panel) are faithfully reproduced by the interpolation formula of Ierides and Kenkre (2018). The full evolution over the three regimes of time are depicted in the right panel. Reprinted with permission from fig. 8 of Ref. Ierides and Kenkre (2018); copyright (2018) by Elsevier Publishing

The first two terms in Eq. (7.41) show a tendency of the molecule to relax to its zero point energy which would be appropriate in the absence of the bath. However, the forcing term brings the average energy in line with the value dictated by the bath, given enough time. Then the system comes into coincidence with the Bethe-Teller evolution

$$E(t) = E(0)e^{-\kappa\left(1 - e^{-\beta\Omega}\right)t} + E_{th}\left[1 - e^{-\kappa\left(1 - e^{-\beta\Omega}\right)t}\right].$$

Figure 7.5 shows the half-Markoffian approximation in comparison to the exact and Bethe-Teller results. An excellent agreement of the interpolation formula is seen at both short and long times with the exact result. As certainly expected, it is unable to predict the intermediate behavior accurately.

7.5   Conceptual Discussion

Given the fact that the theory of excitation transfer given in Kenkre and Knox (1974a) already contains a formula for energy transfer analysis in an inequivalent molecular dimer, it is appropriate to ask why it was necessary to devote an entire chapter in this book to analyze energetically inequivalent systems. A part of the answer lies in the need to generalize the Montroll-Shuler equation for vibrational relaxation. The usefulness of the new results generated by Chase and by Ierides that we have displayed in that quarter surely justifies the detailed explanations. But there is another reason which has to do with conceptual questions connected with approach to equilibrium.

Consider an isolated system with Hamiltonian H and therefore a spectrum of energies corresponding to the various eigenstates of H. If we put the system initially in one of those eigenstates, nothing will obviously happen, in the sense that the system will stay in that state, the phase merely evolving periodically in time. When we encounter the system without any given preparation into a specific state, we typically invoke the postulate of equilibrium statistical mechanics (see any common text such as Huang (1987)), that the system density matrix is diagonal in the representation of the eigenstates and that its diagonal elements have Boltzmann weights in that representation. May I draw attention here to the split nature of the postulate, made particularly clear by the text mentioned: one part referring to the probabilities and the other to the phases. Invoking the postulate means that we tacitly (and naturally) assume that the system is not really isolated but that there is an interaction of the system with the rest of the universe, or, simply put, with a bath. The system-bath interaction must have certain properties for the postulate to be valid. The interaction must be *strong* enough, and of such a nature, that, whatever the initial state, the evolution will drive the system to a state with random phases and Boltzmann weights characteristic of the temperature of the bath. Yet the interaction must be *weak* enough so that, in the representation of the eigenstates of the given isolated system, the state reached at equilibrium is diagonal and Boltzmann-weighted, *no vestige* of the bath, other than the value of the temperature, being left in the state that the system arrives at.

Maintaining this weak-interaction qualification, let us ask for an evolution equation which can describe the approach to equilibrium accurately, *including at times short* with respect to the equilibration. Clearly, the GME for density matrix elements diagonal in the representation of the *eigenstates of the system* is an appropriate candidate for such an evolution equation.[5]

I have always wanted to understand, at a level as simple as possible, how a bath (the rest of the universe) in weak interaction with a system (any system) does this combined job of thermalizing the populations in the system and randomizing the phases.[6] This is the problem that Tiwari and I examined for a simple two-state system. What has often intrigued me is how the non-resonant oscillations at short times in a dimer evolve into the thermalized decay with Boltzmann weights at long times. Notice Fig. 7.6 which is over an enormously stretched scale on the time axis. At short times the non-degeneracy produces oscillations that are non-resonant. At

---

[5]How this situation is connected to some recent discussions of coherence among photosynthesis scientists will be remarked on in Chap. 16.

[6]For instance, do the two happen sequentially? Simultaneously? What in the system or bath decides the order? The non-resonant oscillations at short times in an inequivalent dimer are independent of the sign of the energy difference. The populations at long times are crucially dependent on the sign. How does the sensitivity to sign come about as time proceeds?

long times the same non-degeneracy produces weights that are Boltzmann. How and when do these happen?

Indeed, the manner in which some of the questions about approach to equilibrium of a two-state inequivalent dimer can be asked very simply are as follows. We know that for a *degenerate* dimer the two memories $\mathcal{W}_{12}(t)$ and $\mathcal{W}_{21}(t)$ are equal to each other and if we call that equal quantity $\mathcal{W}(t)$, we can write the governing equation for the probability difference $p(t) = P_1(t) - P_2(t)$ as

$$\frac{dp(t)}{dt} + 2\int_0^t \mathcal{W}(t-s)p(s)ds = 0. \tag{7.42}$$

The time dependence of $\mathcal{W}(t)$ determines, in appropriate fashion, the transition from coherent to incoherent behavior. The amplitude of oscillations of $p(t)$ simply die to a zero value. Bath interactions decide the decay behavior of $\mathcal{W}(t)$ which, in turn, dictates the loss of quantum coherence. However, in the nondegenerate system, additional questions arise from the fact that two unequal memories exist, one for each direction of transfer. Is detailed balance obeyed by these two memories the way it is by their time integrals, i.e., by the rates $F_{12} = \int_0^\infty \mathcal{W}_{12}(t)dt$ and $F_{21} = \int_0^\infty \mathcal{W}_{21}(t)dt$? Does this mean that $\mathcal{W}_{12}(t)$ and $\mathcal{W}_{21}(t)$ are in Boltzmann ratio at every instant of time t? Does such a time-independent ratio provide an accurate description of transfer at all times? In other words, is it possible to have a separation situation of the memory in the form $\mathcal{W}_{12}(t) = F_{12}\phi(t)$? If so, how could

$$\frac{dp(t)}{dt} + (F_{12} + F_{21})\int_0^t \phi(t-s)p(s)ds = (F_{12} - F_{21})\int_0^t \phi(s)ds. \tag{7.43}$$

provide an accurate description of both the randomization process (which would be taken care of by the time dependence of $\phi(t)$ as in the degenerate case) and the detailed balance process which would depend on the energy state difference of the system which are independent of the reservoir?

The analysis in Tiwari and Kenkre (2014) did provide some of the answers, one of which was the stretched plot in Fig. 7.6 which shows the non-resonant oscillations at short times evolving into the detailed balance population difference at long times. That analysis also showed that Eq. (7.43) was incorrect and that what described the situation was

$$\frac{dp(t)}{dt} + \int_0^t ds\, \mathcal{W}_+(t-s)p(s) = \int_0^t ds\, \mathcal{W}_-(s). \tag{7.44}$$

where

$$\mathcal{W}_\pm(t) = [\mathcal{W}_{12}(t) \pm \mathcal{W}_{21}(t)].$$

The formal similarity of Eq. (7.44) with Eq. (7.40) is worth noting as is the fact that the symmetric and difference combinations of memories appear in different terms in

⊛ the two cases. The reader is encouraged to derive Eq. (7.44) and check the derivation against the published article.

At very short times the driving term on the right hand side of Eq. (7.44) can be neglected and the probability difference $p(t)$ begins its evolution as for a degenerate dimer. When the driving term picks up, the non-resonant oscillations enter the fray, and we have

$$p(t) = \frac{\Delta^2}{V^2 + \Delta^2} + \frac{V^2}{V^2 + \Delta^2} \cos\left(2t\sqrt{V^2 + \Delta^2}\right),$$

which accurately describes the insensitivity to the sign of the energy difference 2Δ between states 1 and 2. The value around which the oscillations occur is the dashed line in Fig. 7.6. As time evolves further and the bath interactions cause the memories to decay, one has the ultimate passage into the detailed balance ratio of populations whose magnitude $\tanh(\Delta/k_B T)$ is depicted by the dashed dotted line in Fig. 7.6. The sensitivity of the equilibrium configuration to the sign of Δ arises, transparently, from the odd nature of the tanh function.

The preliminary dimer studies of Tiwari and Kenkre (2014) were all confirmed and enriched by our more recent investigations as we have described. Even revivals that occur in the short-time evolution of $p(t)$ in spin-boson models (Kenkre et al. 1996) which can be seen in unrelated work by Raghavan et al. were reproduced by the work of Ierides and Kenkre (2018).

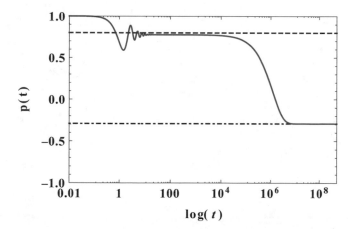

Fig. 7.6 Time evolution (solid line) of the probability difference $p(t)$ for a nondegenerate dimer showing two different regimes in which the short time behavior is dominated by the so-called dc stark value (dashed line) and the long time behavior settles to the value given by the detailed balance condition (dashed-dotted line). Here $\Delta/V = 2$, $V/k_B T = 0.15$, other values are arbitrary and $p(0) = 1$. Reprinted with permission from fig. 4 of Ref. Tiwari and Kenkre (2014); copyright (2014) by Springer Nature

7.6   Chapter 7 in Summary

The process of vibrational relaxation was studied in this chapter, specifically from the point of view of deriving generalizations of the Montroll-Shuler equation of the process into the coherent domain. The process of vibrational relaxation of molecules is interesting for various reasons including its interaction with luminescence and for its analysis as a gateway process for more drastic processes of dissociation or chemical reactions. Coarse-grained projection operators were developed to eliminate reservoir variables and focus only on the relaxation of the vibrational coordinate of the molecule under consideration. Several models of the interaction of a vibrationally relaxing molecule embedded in a reservoir were studied. Some surprising effects were described for the case of strong coupling or multiphonon processes. This chapter should be useful both to understand the (coarse-grained) projection technique in microscopically specified reservoirs and to probe the process of the approach to equilibrium of systems in contact with a bath, including its temperature dependence. The latter is of fundamental importance to the general subject of quantum non-equilibrium statistical mechanics.

Projection Operators for Various Contexts

8

The discussion in this book began by focusing attention on the fact that movements observed everywhere in nature vary from one extreme limit, which involves oscillations, to the other, which displays decays while covering the whole gamut in between; and that memory functions are a natural and powerful tool to unify these extremes. We also found that the theoretical tool of projection operators resulted effortlessly in the appearance of memory functions when applied to the problem of the passage of underlying microscopic evolution to emerging macroscopic phenomena in almost any physical system.

The projection operator thus arose as a tool for discovering memory functions in phenomena. It turns out however, that the technique goes well beyond this use and finds employment in diverse areas of statistical mechanics and condensed matter physics as long as there is need for focusing attention on one part of a whole. The present chapter will discuss several different ways of putting projection operators to use, having generally little to do with the original Nakajima-Zwanzig reason of looking at the diagonal part of the density matrix. We will start with projections that do not diagonalize but take the trace to calculate correlation functions that appear in the theory of electrical resistivity. We will revisit the dynamical localization problem we met with as an exercise in Chap. 2 and will inspect what can be done with projections for that system. We will mention, in passing, projections used for the derivation of the BBGKY hierarchy, in which the classical Liouville equation is the starting point. We will go on to the classical Torrey-Bloch equation in nuclear magnetic resonance microscopy to analyze constrained motion of spins in water, a study that is relevant to the operation of MRI's in medical science. Finally, we will return to the railway-track model and see how projections can be made to help in that little system that we discussed at the very beginning of the book, in Chap. 1.

© The Author(s), under exclusive license to Springer Nature Switzerland AG 2021
V. M. (Nitant) Kenkre, *Memory Functions, Projection Operators, and the Defect Technique*, Lecture Notes in Physics 982,
https://doi.org/10.1007/978-3-030-68667-3_8

8.1   Application to Quantum Systems

My own first contact with projection operators came when my thesis advisor, Max
Dresden, gave me a copy of Kubo's famous article on linear response (Kubo 1957),
a copy of one of the first of Zwanzing's publications on the subject of projections
(Zwanzig 1961), and asked me to combine them giving me no clue whatsoever
about what direction such a combination might take.[1] I wrote a couple of papers
based on my struggles to do my advisor's bidding (Kenkre and Dresden 1971, 1972),
and a couple of additional publications probably not worth citing. Here are some
interesting aspects of those early research attempts. They dealt with the calculation
of correlations functions done *with projection techniques*.

8.1.1   Projections for the Theory of Electrical Resistivity

Everyone with even a small familiarity with the subject knows how to write a
classical Drude equation to describe the flow of current induced by an applied dc
electric field in a metal containing n electrons per unit volume each with a charge e
and (effective) mass m, and obtain from it an expression for the conductivity σ,

$$\sigma = \frac{ne^2\tau}{m},$$

with τ as the relaxation time, or to justify it with the help of a (linear) Boltzmann-
equation derivation by considering the scattering of non-interacting electrons in a
band of the metal (Ziman 2001). Obtaining it from a fully quantum mechanical
starting point is an important task and it had been started by Kohn and Luttinger
(1957), Luttinger and Kohn (1958) and others. There was a tendency in those times
to link calculations of transport coefficients to the Kubo theory of linear response
(Kubo 1957) and this was done effectively in this context by Verboven (1960) and
by Chester and Thellung (1959).

   Let us first cast the Kubo expression for the frequency-dependent conductivity in
the form

$$\sigma(\omega) = E_\beta(\omega)C(\omega),$$

where $E_\beta = (\omega/2)\coth(\beta\omega/2)$ is the universal function introduced by Kubo,
and $C(\omega)$ is the Fourier transform $\int_0^\infty dt\, e^{i\omega t} C(t)$ of the *symmetrized* correlation
function of the velocity. The theory of linear response (Kubo 1957), touched upon
cursorily at the beginning of Sect. 4.3 (see, e.g., Eq. (4.15)), gives this correlation
function, except for unimportant constant factors, as

---

[1] It was 50 years ago almost to the dot, it was the second research problem he had suggested to me,
and these were his precise words "Go marry these two and tell me what happens."

$$C(t) = \frac{\langle v(t)v + vv(t) \rangle}{2} = Tr\left[\rho_{eq}\left(\frac{e^{itH}ve^{-itH}v + ve^{itH}ve^{-itH}}{2}\right)\right]. \quad (8.1)$$

Here the thermal density matrix ρ_{eq} has the canonical form $e^{-\beta H}/Tre^{-\beta H}$.

A careful inspection of Eq. (8.1) reveals that, as a consequence of cyclic permutation of operators within the trace, the correlation function $C(t)$ can be looked upon as the result of tracing over the product of the velocity operator v and an operator $K(t) = e^{-itH}Ke^{itH}$ whose initial value $K = K(0)$ happens to equal $(1/2)(v\rho_{eq} + \rho_{eq}v)$. Out of the blue, we have here a situation ripe for the application of Zwanzig's projection operator technique.

Given that $K(t)$ satisfies the very same Liouville-von Neumann equation that the density matrix $\rho(t)$ in Zwanzig's original problem (see Chap. 2) did, and given the possibility of our defining the projection operator $\mathcal{P}$ through

$$\mathcal{P}O = \left(\frac{A}{Tr\,vA}\right)Tr\,vO \quad (8.2)$$

for any operator O, with the stipulation that A is as yet an undefined operator, we can invoke Eq. (2.7), the equation derived by Zwanzig, in this completely unrelated context. The presence of the prefactor containing A makes the projection operator idempotent and its other properties lead us straight to Eq. (2.7) in the Zwanzig framework. But we have greater control in this problem than in the original one in that we can choose A to make the initial term vanish identically! There is no need to wait for the assumption of an initial random phase condition that was essential in the Zwanzig context. The choice $A = K(0)$ forces

$$(1 - \mathcal{P})K(0) = 0$$

which makes the initial condition term in the Zwanzig equation zero for all t and leads to a simple consequence of Eq. (2.7) in the present case:

$$\frac{dC(t)}{dt} + \int_0^t ds\,Q(t-s)C(s) = 0, \quad (8.3)$$

where the kernel $Q(t)$ equals $[1/C(0)]Tr\,vLe^{-it(1-\mathcal{P})L}(1-\mathcal{P})LK(0)$.

The usual way of computing the dc electrical conductivity is by attempting to evaluate the integral $\int_0^\infty dt\,C(t)$ of the correlation function $C(t)$. An achievement of the present development is that it provides an alternative procedure to calculate the transport coefficient. The alternative is through the calculation of its reciprocal, the resistivity, as an integral $\int_0^\infty dt\,Q(t)$ of the kernel $Q(t)$. Thus, except for constant factors,

$$\sigma(\omega = 0) \sim \int_0^\infty dt\,C(t,\lambda) = \frac{1}{\int_0^\infty dt\,Q(t,\lambda)}. \quad (8.4)$$

The fact that the conductivity or resistivity depends crucially on the scattering strength λ of the charge carriers that comprise the sample to which the electric field is applied, has been made explicit in Eq. (8.4) to draw attention to an obvious problem one encounters in the calculation of the transport coefficient. Computation of quantities like $C(t, \lambda)$ without approximation is seldom possible for any realistic system. A method that naturally suggests itself employs series expansions in terms of the scattering strength:

$$\int_0^\infty dt\, C(t, \lambda) = \sigma_0 + \lambda \sigma_1 + \lambda^2 \sigma_2 + \dots$$

However, for a system like free electrons, the first term $\sigma_0 = \int_0^\infty dt\, C(t, 0)$ would diverge and indeed, so would all the other terms in the expansion.

The purpose and result of the work in Kenkre and Dresden (1971, 1972) was to provide such alternatives to the computation of the conductivity because the methods commonly employed, such as by Verboven (1960) and by Chester and Thellung (1959), used the van Hove $\lambda^2 t$ limit. Although much venerated because of the success of van Hove in obtaining with its help the Master equation from the quantum mechanical starting point of the von Neumann equation, the precise meaning of the $\lambda^2 t$ limit has never been clear except that it involves long times and weak coupling to the scatterers in that peculiar combination.[2] When the infinite series terms that survive in that limit are collected together, van Hove's treatment is able to sum them in the form of a non-divergent combination. Similarly, in the same limit, the followers of van Hove, such as Chester and Thellung (1959) and Verboven (1960), were able to calculate a finite expression for σ. Because of dissatisfaction with the meaning of that limit, we devised this "downstairs" method (Kenkre and Dresden 1971, 1972) as an *alternative* to the "upstairs" $\lambda^2 t$ procedure that was in the literature. The reader might recall the discussion presented in Chap. 4 of blind approximation procedures and the difficulties of choosing between them. However enamored workers in the field might have been with the $\lambda^2 t$ procedure in the aftermath of its spectacular success in the problem of the derivation of the Master equation, it is certainly a blind approximation technique.[3]

There is a whole series of discussions that it could be opportune to go into at this point including the detailed approximate formulae for resistivity that were derived via the method developed in Kenkre and Dresden (1971) and Kenkre and Dresden (1972). They could include the similarity, but not identity, that we eventually found between the projection operators used by us and in the widely known publication by Mori (1965), the latter for the derivation of Langevin equations. They could

---

[2] It is worth examining an interesting treatment of the meaning of that limit given, although in passing, by Zwanzig in his excellent exposition in the book by Meijer (1966).

[3] Amusingly, a debate was initiated by Argyres and Sigel (1974) opposing our method on the basis of its lack of agreement with the other procedure, ignoring the fact that the difference was the very reason for our publishing it.

elaborate on the fact that Mori used the very initial term of the Zwanzig equation (that we forced to become zero by our specific choice of projection operator) as the distinguishing random force term that characterizes noise. Particularly they could expound on a caricature model that I developed around 1975 to examine the validity of the projection technique method of calculation versus the $\lambda^2 t$ procedure.[4] However, these matters do lie outside the margins of the subject matter of this book. Having emphasized that projection operators can be used to derive Eq. (8.3) and, with its help, surprisingly turn the upstairs problem into the downstairs problem as clear from Eq. (8.4), let us move on to other stomping grounds of projection operators.

8.1.2 Where Projections Take Us in the Quantum Control of Dynamic Localization

As an exercise in the use of projection techniques described in Sect. 2.3.2, we saw that the application of a time dependent electric field to a two-site system with the consequent Hamiltonian

$$H = \begin{pmatrix} \mathcal{E}(t)/2 & V \\ V & -\mathcal{E}(t)/2 \end{pmatrix},$$

leads to an evolution equation for the probability difference $p(t) = P_1(t) - P_2(t)$,

$$\frac{dp(t)}{dt} + 2 \int_0^t ds\, \mathcal{W}(t, s) p(s) = 0,$$

which contains a memory function that is not a function of the difference $t - s$. In fact, the memory takes the form

$$\mathcal{W}(t, s) = 2V^2 \left[\phi_c(t)\phi_c(s) + \phi_s(t)\phi_s(s) \right],$$

where the functions $\phi_c(t) = \cos\left[\int_0^t ds\, \mathcal{E}(s)\right]$ and $\phi_s = \sin\left[\int_0^t ds\, \mathcal{E}(s)\right]$ are obtained from the time variation of $E(t)$ in the Hamiltonian. A few brief remarks about what this leads to should make clear how projection operators help in varied contexts.

The problem started with the report by Agarwal and Harshawardhan (1994) who showed a plot of bursts of transfer of probability (hereafter called the AH structure)

---

[4]The model did not have a high degree of sophistication but showed that our downstairs method involving projections reproduced the exact correlation through its lowest nonvanishing term, the higher order terms being identically zero. By contrast, the upstairs method based on the $\lambda^2 t$ technique gave an approximate result. It is my hope to increase its degree of sophistication with the help of some more inventive co-worker and present it in the open literature one of these days.

that occur periodically between two sites of a dimer system when the site energy difference varies periodically, e.g., sinusoidally. The bursts can be seen in their originally published paper as well as in both curves of Fig. 8.1 reproduced here from our study (Raghavan et al. 1996) in which the field $E(t) = E_0 \cos \omega t$, the characteristic energy $\mathcal{E} = ea E_0$, and e and a are, respectively, the charge and the distance between the two sites.

It was suggested in the literature that the bursts happened when the ratio $\mathcal{E}/\omega$ equalled roots of a Bessel function. Such a condition signaled a connection to the phenomenon of dynamical localization, discovered and analyzed earlier by Dunlap and Kenkre (1986) and studied experimentally in optical lattices by Madison et al. (1998).

Raghavan et al. (1996) showed unequivocally that the phenomena were connected but by no means in the manner suggested by some. For instance, they demonstrated that the AH structure appeared both on resonance (upper curve in Fig. 8.1 with $\mathcal{E}/\omega = 30.635$, and off resonance (lower curve in Fig. 8.1 with $\mathcal{E}/\omega = 31.635$, whereas the Bessel function condition was crucial to dynamical

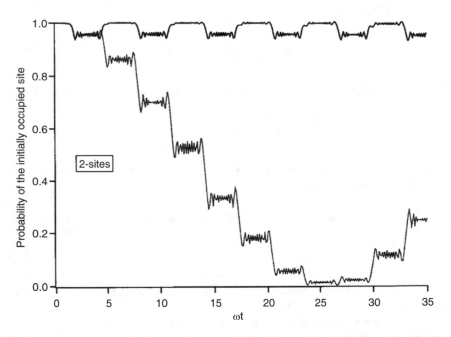

Fig. 8.1 Coexistence of the AH structure (repeated bursts of transfer) and of dynamic localization in a two-site system. The probability of the initially occupied site is plotted on resonance (upper curve) and off resonance (lower curve) as a function of the dimensionless time ωt. The fact that the probability changes drastically off-resonance but, when on resonance, remains overall localized close to 1, shows the sensitivity of dynamical localization to the Bessel root condition. The fact that the AH bursts are visible in both curves shows their insensitivity to the Bessel root condition. Reprinted with permission from fig. 4 of Ref. Raghavan et al. (1996); copyright (1996) by the American Physical Society

localization. Moreover, Raghavan et al. (1996) analyzed the importance of system size to the AH structure. They showed that, whereas the structure disappears as the size increases, dynamical localization remains unaffected.[5] The extreme sensitivity of the Bessel root condition to dynamical localization is thus clear. Yet, the AH structure (bursts of transfer) is visibly unchanged in both curves.

While fascinating, the details of the relational study, available to the reader in the quoted references, lies outside the topics of the present book. However, it is worthwhile to address the question that remains, which is where the AH bursts originate. We can provide an answer with the help of the memory $\mathcal{W}(t, s)$ that we have derived above. The following is a brief explanation.

Figure 8.2, reproduced from Kenkre (2000), shows the numerically exact variation of the probability difference in the dimer analyzed in Agarwal and Harshawardhan (1994) and an approximation I developed on the basis of the memory $\mathcal{W}(t, s)$ which shows amazing coincidence with the exact evolution at short times, $(\omega t < 8)$, with discrepancies starting to appear only for larger times. To reproduce the bursts and quiescent behavior of the exact solution all that is necessary to do is to notice that, if the parameter $\mathcal{E}/\omega$ is large enough with respect

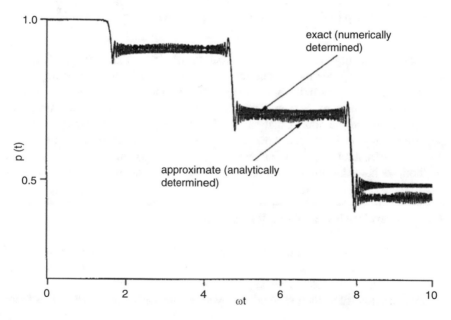

Fig. 8.2 Understanding the AH structure. The exact solution (numerical) and our analytic approximation as given in Eq. (8.7) from direct integration are in near coincidence in the initial region of time, and depart slightly for $\omega t > 8$. See text. Reprinted with permission from fig. 3 of Ref. Kenkre (2000); copyright (2000) by the American Chemical Society

---

[5]Indeed, the discovery of dynamical localization announced in Dunlap and Kenkre (1986), and extended in Dunlap and Kenkre (1988a,b), was made for an *infinite system*.

to 1, $\phi_c(t)$ and $\phi_s(t)$ oscillate rapidly. Because of the rapid oscillations, one may consider taking $p(s)$ in the right hand side of its integrodifferential equation out of the integral for short times (short with respect to the period of the field but long with respect to the time of oscillation of the memory). Under this approximation,

$$\frac{dp(t)}{dt} + 2V^2 \left[\frac{d\left(h_c^2 + h_s^2\right)}{dt} \right] p(t) = 0. \tag{8.5}$$

The characteristic functions h_c and h_s are given by

$$h_c(t) = \int_0^t ds \, \cos\left[(\mathcal{E}/\omega)\sin\omega s\right], \tag{8.6a}$$

$$h_s(t) = \int_0^t ds \, \sin\left[(\mathcal{E}/\omega)\sin\omega s\right]. \tag{8.6b}$$

Equation (8.5) can be solved analytically! It yields

$$p(t) = p(0)\exp\left[-2\int_0^t \int_0^{t'} ds \, \mathcal{W}(t',s)\right] = \exp\left[-2V^2\left(h_c^2(t) + h_s^2(t)\right)\right]. \tag{8.7}$$

The second equality uses $p(0) = 1$ and is what is plotted in Fig. 8.2. The agreement of our approximation and the exact solution is noteworthy. Both consist of rapid oscillations of small amplitude around horizontal segments with no change in the average value of $p(t)$ and then sudden step-function-like decreases. The exact solution and the approximation we have developed are in near coincidence. Departures only begin as time grows and that is understood easily from the nature of the approximation. Through a simple straightforward argument from the memory function, we have thus understood the bursts in the AH structure.[6]

8.2   Application to Classical Systems

Although projection techniques were developed for use in quantum mechanical systems, it should be clear that they can be used for classical or semiclassical systems as well. One notable such application that I am familiar with was carried out by Muriel and Dresden (1969a) and Muriel and Dresden (1969b). The focus

---

[6]Dynamic localization referred to above, or the related phenomenon that has been described by some with the phrase "coherent destruction of tunneling", is not a topic discussed in any detail in this book, only used for an exercise and an illustration. However, it is worth pointing out that it appears to have led, and is leading, to exciting new research. A comprehensive recent review is Bukov et al. (2015). Past work includes Großmann and Hänggi (1992) and Grifoni and Hänggi (1998). The reader is encouraged also to study investigations reported currently as in Moessner and Sondhi (2017).

in that study was to start from the Liouville equation for the classical Liouville density which is a function of all the momenta and coordinates in the system and derive from it by contraction equations of evolution for reduced entities involving for instance a few of the momenta and coordinates, or only the momenta and not the coordinates. One of their goals was the derivation of the BBGKY (Born-Bogolubov-Green-Kirkwood-Yvon) hierarchy (see, e.g., Reichl 2009). For their purpose, they defined projection operators that carried out integrations in classical phase-space and derived, as a consequence, equations of motion for various reduced distribution functions. They treated isolated systems as also systems subjected to external fields. The reader should consult those publications for the details of the manner it was done.

Let us focus on a more modern application to a system that may be considered classical, or perhaps better called as semiclassical, which is of relevance in the medical field. It deals with nuclear magnetic resonance (NMR) and is described next.

8.2.1 Torrey-Bloch Equation for NMR Microscopy

The diffusion of particles possessing a nuclear spin in confined geometries as studied with pulsed-gradient spin-echo nuclear magnetic resonance (NMR) is a topic of great interest and medical relevance (Callaghan 1991; Blees 1994; Mitra and Halperin 1995; Wang et al. 1995; Sheltraw and Kenkre 1996; Kenkre et al. 1997; Kenkre and Sevilla 2006; Sevilla and Kenkre 2007). Investigations in this subject are of importance to basic issues in the principles of operation of devices such as the MRI machine. A brief application of projection techniques and memory functions to this area is the subject matter of the present section.

The system under investigation consists of a large number of particles, typically protons in the water molecules, each of which possesses a nuclear spin diffusing in the presence of a strong homogeneous static magnetic field B_0 and a weak inhomogeneous static magnetic field, both of the fields taken to point along the z axis. The inhomogeneous field is due to a linear gradient $g(t) = gx\hat{x}$, where $\hat{x}$ is the unit vector in the x-direction. The spins are excited at $t = 0$ by what is called a $\pi/2$ pulse about the y-axis. The full quantum mechanical starting point in these investigations is the von Neumann equation for the density matrix ρ,

$$i\frac{d\rho(t)}{dt} = [\mathcal{H}_r + H_I + H_{xI}, \rho(t)] \tag{8.8}$$

where $H_I = \omega_0 I_z$ and $H_{xI} = -f(t)\gamma gx I_z$, and $\mathcal{H}_r$ is the Hamiltonian for the spatial coordinates in the absence of coupling to the spin variables. In the notation particular to this field, $f(t)$ is the shape function for the gradient pulse, γ is the gyromagnetic ratio of the particle, and $\omega_0 = -\gamma B_0$. The part $\mathcal{H}_r$ of the Hamiltonian contains the complexities of the kinetic energy of the particles and the collisional potential energy due to both interactions among particles and interactions between particles and the confining walls. Its effects are usually taken into account by asserting as the starting equation of motion, the mixed entity

$$\frac{\partial \rho(r,t)}{\partial t} = -i[H_I + H_{xI}, \rho(r,t)] + D\nabla^2 \rho(r,t). \tag{8.9}$$

The appelation "mixed entity" stems from the blend of the classical part that describes the diffusion (or random walk) of the spin-bearing particles with diffusion constant D with the quantum mechanical part in the commutator in the equation. This equation or its corresponding form for the magnetization density,

$$\frac{\partial M(r,t)}{\partial t} = -i\gamma g x f(t) M(r,t) + D\nabla^2 M(r,t), \tag{8.10}$$

obtained by a conversion to the interaction picture and use of $\langle I^+ \rangle = \langle I_x + iI_y \rangle = Tr\,[I^+ \rho(r,t)]$, is termed the Torrey-Bloch equation in this area of research. Why and how one might think of applying projection techniques to this semi-classical starting point represented by Eq. (8.9) or (8.10), how it was done in collaboration with Sheltraw[7] and Fukushima, (Sheltraw and Kenkre 1996; Kenkre et al. 1997) and also later with Sevilla (Kenkre and Sevilla 2006; Sevilla and Kenkre 2007), and what was achieved by these calculations, is described in the following.

8.2.2 Projections and Memory Functions in NMR

Let us define a projection operator $\mathcal{P}$ that, when applied to a position (and time) dependent operator $O(r,t)$, integrates over the spatial variables:

$$\mathcal{P}O(r,t) = \sigma(r,0) \int d^3r' O(r',t) \tag{8.11}$$

As usual in the application of projections, we have some freedom in the choice of $\sigma(r,0)$. We will take it to represent the initial distribution of the spins, i.e., to equal $\int d^3r' \rho(r',t)$. Since, initially, the gradient is off and the spatial and spin variables are uncorrelated, it follows that $\mathcal{P}$ satisfies all the required properties of a Zwanzig projection operator, including idempotency.

Cranking out the usual procedure including the weak-coupling approximation, we arrive (Sheltraw and Kenkre 1996) at the following equation of evolution for the spatially integrated time-dependent magnetization density, $M(t)$:

$$\frac{dM(t)}{dt} + f(t) \int_0^t ds\, f(s)\phi(t-s)M(s) = 0. \tag{8.12}$$

---

[7]He was an unusual Ph.D. student who, after obtaining an M.D. degree gave up pursuing a lucrative career as a medical doctor and decided to do physics, starting research by rejoining graduate school.

This simple equation is the result of our application of projection techniques to the semi-classical Torrey-Bloch equation. It has already built into it the weak-coupling approximation, $f(t)$ describes the shape of the pulse, and the consequences of the motion of the spins (typically in confined geometry) is reflected in the memory function which, except for a constant of proportionality, is nothing other than the position auto-correlation function of the random walking spins.

The auto-correlation function can be calculated in the standard manner for random walk motion confined to $-a/2 \leq x \leq a/2$, in a 1-dimensional box of length a. Sheltraw and Kenkre (1996) thus obtained the memory function for this problem explicitly as

$$\phi(t) = \gamma^2 g^2 \langle xx(t) \rangle = \frac{8(\gamma g a)^2}{\pi^4} \sum_{n=1}^{\infty} \frac{1}{(2n-1)^4} \exp\left[-\frac{(2n-1)^2 \pi^2 D t}{a^2}\right].$$

(8.13)

⊛ A suggested exercise for the reader is the calculation of the displacement autocorrelation function in Eq. (8.13).

Although the Laplace transform of this memory as an infinite series cannot be evaluated analytically, inspection shows that one can use an excellent approximation to represent the sum of exponentials in the infinite sum in Eq. (8.13) as a single exponential. Details of the argument involve Bernoulli numbers and may be found spelled out in Sheltraw and Kenkre (1996). Our approximation is *not* the same as dropping higher terms in the series indicated although even such a simple monoexponential approximation works rather well because the second and successive terms are at least a factor of 80 smaller than the first one.[8]

The result of our approximation is that the memory function is given quite accurately by

$$\phi(t) = C^2 e^{-2\lambda t},$$

(8.14a)

$$C^2 = \frac{(\gamma g a)^2}{12},$$

(8.14b)

$$\lambda = \frac{5D}{a^2}.$$

(8.14c)

For the case of a time-independent gradient, $f(t) = 1$, and an analytic solution of Eq. (8.12) is immediately possible:

$$M(t) = e^{-\lambda t}\left[\cosh\left(t\sqrt{\lambda^2 - C^2}\right) + \frac{\lambda}{\sqrt{\lambda^2 - C^2}} \sinh\left(t\sqrt{\lambda^2 - C^2}\right)\right].$$

(8.15)

This NMR signal, predicted by the memory technique introduced into the NMR microscopy theory by Sheltraw and Kenkre (1996), is formally identical to the

---

[8]This is independent of system parameter values!

displacement of a damped harmonic oscillator, the damping being controlled by the diffusion time a^2/D required by the spins to cover the extent of the confining space in their random walk.

The standard calculational technique used in the field used to be that of cumulant expansions and leads generally to

$$M(t) = \exp\left[-(\gamma^2 g^2/2) \int_0^t \int_0^t f(t_1) f(t_2) \langle xx(t_1 - t_2)\rangle \, dt_1 dt_2\right].$$ (8.16)

It is instructive to learn how simply this result, approximate but widely known in the area of NMR microscopy, was shown by Sheltraw and Kenkre (1996) to emerge as an approximation from the memory result.

Assume in our Eq. (8.12) that the memory function $\phi(t)$ decays rapidly enough to allow us to remove $M(t)$ from the integral, yielding

$$\frac{dM(t)}{dt} \approx -(\gamma g)^2 f(t) M(t) \int_0^t ds \, f(s) \langle xx(t - s)\rangle.$$

Evaluating the integral, one gets

$$M(t) = \exp\left[-(\gamma^2 g^2) \int_0^t \int_0^t f(t_1) f(t_2) \langle xx(t_1 - t_2)\rangle dt_1 dt_2\right].$$ (8.17)

For a stationary process, the quantity $f(t_1) f(t_2) \langle xx(t_1 - t_2)\rangle$ is symmetric under exchange of t_1 and t_2. Replacing the double integral of Eq. (8.17) over a triangular region in t_1 and t_2 by $\frac{1}{2}$ the double integral over the corresponding square region, leads to the result that Eq. (8.17) is identical to the cumulant result (8.16).

This demonstration shows explicitly that the (truncated) cumulant expansion expression is a further approximation to the memory result, and that it tends to the latter for those situations in which the memory function decays rapidly in time. Similar connections between the cumulant technique and memory functions had already appeared in earlier studies on vibrational relaxation (Kenkre and Seshadri 1977).[9]

How much better as an approximation do our memory calculations based on the use of projections fare? Sheltraw performed a comparison of the time dependence of the NMR signal as predicted by our result obtained with the projection-based memory method displayed in Eq. (8.15) with its exact and cumulant counterparts. By the exact is meant the result of numerical computation from the Torrey-Bloch Eq. (8.10). By the cumulant version is meant the consequence of Eq. (8.16) not

---

[9]Remarkably, we found another application of projections to NMR microscopy (Robertson 1966). However, Robertson included the partial time-local approximation, which made his results identical to those of the cumulant technique. By contrast, our calculations in Sheltraw and Kenkre (1996) refrained from making that time-local approximation so one could go beyond the cumulant results.

shown explicitly here as an expression but given in (Sheltraw and Kenkre 1996). This comparison was carried out for various values of the ratio $\zeta = C/\lambda$ ranging from 0.5 to 10. The idea was to determine in what regions of parameter space, if any, one approximation scheme was preferable to the other. The memory method was found to be always better in that it could produce oscillations in addition to being closer to the numerically found exact result.

The physical significance of the parameter ζ, which equals $\gamma g a^3/10\sqrt{3}D$, is that it measures the relative magnitudes of two characteristic times. Except for a numerical proportionality constant, ζ is the ratio of the extreme difference between the precessional frequencies in the confining space, viz., $\gamma g a$, to the reciprocal of the time it would take the spin particle to diffuse from one end of the confining segment to the other, viz., D/a^2.

Figure 8.3 displays the comparison for one of the many values of ζ considered. Others may be seen in the original publication. We see that not only is the memory function approximation closer to the exact signal but that it is qualitatively correct

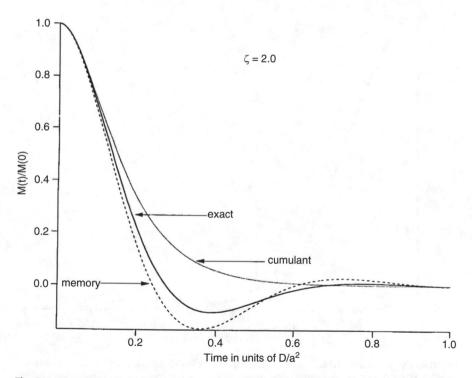

Fig. 8.3 The theoretical NMR signal $M(t)$ plotted versus time t in units of D/a^2 for the continual-gradient case showing a comparison of the numerically determined exact result to the two approximations considered. The value of $\zeta = C/\lambda$ considered is 2. The cumulant approximation is monotonic and cannot reproduce the oscillation that both the exact and the memory approximation possess. Adapted with permission from fig. 1c of Ref. Sheltraw and Kenkre (1996); copyright (1996) by Elsevier Publishing

in that it exhibits an oscillation as does the exact signal. The cumulant result is monotonic, i.e., displays no oscillations.

The situation analyzed in that plot is theoretically meaningful but seldom encountered in the laboratory. Sheltraw also applied our technique to the more realistic scenario in NMR microscopy, the two-pulse gradient experiment known commonly as the PGSE (pulsed-gradient spin echo) (Callaghan 1991). Consisting of the application of two gradient pulses of strength g and duration δ, and separated by an interval of time Δ, the second pulse is preceded by a π pulse about the x-axis. Comparison results similar to the case of time-dependence are shown in Fig. 8.4. Plotted is $M(2\Delta)/M(0)$, the normalized NMR signal, as a function of the dimensionless wavevector $\gamma g \delta a$. Of the numerous values of the parameters explored in the original publication, here the display shows only $\delta D/a^2 = 0.25$ and $\Delta D/a^2 = 1$. Again, we see that the existence of the minimum in the exact curve is reproduced by the memory approximation although at a slightly displaced

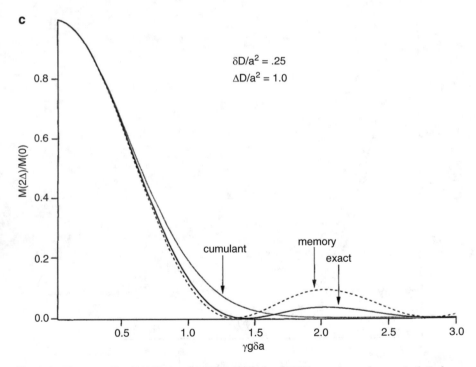

Fig. 8.4 The normalized NMR signal $M(2\Delta)/M(0)$ in a PGSE experiment (see text) plotted as a function of the dimensionless wavevector $\gamma g \delta a$, where a is the confining length. See text for the meaning of other symbols. Comparison of the numerically determined exact signal to the cumulant approximation and the memory approximation (as indicated on the plot) show clearly that the memory approximation is preferably both because it is generally closer to the exact signal but also because it displays minima which the cumulant approximation cannot reproduce. Adapted with permission from fig. 2c of Ref. Sheltraw and Kenkre (1996); copyright (1996) by Elsevier Publishing

value of the abscissa $\gamma g \delta a$, while the cumulant approximation has no minimum anywhere.

A much more complete overview is available in Sheltraw and Kenkre (1996) for the reader who is interested in further details.

8.2.3 Confinement of the Spins Analyzed via a Potential

Confinement effects are of paramount importance in this subject of NMR microscopy as they are instrumental in finding where walls, tumors, growths and such entities lie in the systems analyzed, for instance through MRI devices. We have employed thus far stated boundary conditions to investigate this aspect through the simple procedure of specifying the dimensions of the confining box. How would we analyze confinement due to a potential that provides an attractive force to a center felt by the spins? We take up this question now.

Let us return to Eq. (8.10) for the magnetization density, consider only the 1-dimensional case, do away with box boundary conditions, but augment the equation through the introduction of a potential into

$$\frac{\partial M(x,t)}{\partial t} = -i\gamma g x f(t) M(x,t) + \frac{\partial}{\partial x}\left[D\frac{\partial M(x,t)}{\partial x} + AxM(x,t)\right]. \quad (8.18)$$

Equation (8.18) is the Torrey-Bloch equation enhanced by the introduction of a potential $(A/2)x^2$, equivalently by subjecting the spins to an attractive force $-Ax$ towards the origin. Its form is that of a Smoluchowski equation and Kenkre and Sevilla (2006) have determined its exact solution not only for a constant A but for a time-dependent $A(t)$ as well.

The analytic solution in the Fourier domain for the transform of the magnetization, defined as $\hat{M}(k,t) = \int_{-\infty}^{\infty} dx\, e^{ikx} M(x,t)$, is

$$\hat{M}(k,t) = \hat{M}\left(ke^{-\int_0^t ds\, A(s)} + \gamma g \int_0^t dt'\, f(t')e^{-\int_0^{t'} ds\, A(s)}, 0\right)$$

$$\times \exp\left(-D\int_0^t dt'\left[ke^{-\int_{t'}^t ds\, A(s)} + \gamma g \int_{t'}^t ds\, f(s)e^{-\int_{t'}^s dz\, A(z)}\right]^2\right).$$

$$(8.19)$$

While inversion of the Fourier transform expression is possible, it is unnecessary if our interest is in the total NMR signal $M(t)$ which is obtained simply by taking the limit $k \to \infty$ of the Fourier transform: $M(t) = \lim_{k\to 0}\int_{-\infty}^{\infty} dx\, e^{-ikx} M(x,t) = \lim_{k\to 0} \hat{M}(k,t)$. The time-dependent NMR signal is obtained explicitly:

$$M(t) = \hat{M}\left(\gamma g \int_0^t dt'\, f(t')e^{-\int_0^{t'} ds\, A(s)}, 0\right)$$

$$\exp\left(-D\int_0^t dt' \left[\gamma g \int_{t'}^t ds\, f(s)e^{-\int_{t'}^s dz\, A(z)}\right]^2\right).\qquad(8.20)$$

As a simple example of the expression in Eq. (8.20) relevant to the study of time-dependent elasticity in colloidal gels, we display the NMR signal for initially localized spins and a sinusoidal $f(t)$ of frequency ω, the potential $A(t)$ being proportional to that dependence, i.e., $A(t) = Af(t)$. For this case,

$$\frac{-A^2 \ln M(t)}{Dg^2\gamma^2} = t\left[I_0(2a)e^{-2a\cos\omega t} - 2I_0(a)e^{-a\cos\omega t} + 1\right]$$

$$+ e^{-a\cos\omega t}\left(\frac{2}{\omega}\right)\sum_{k=1}^{\infty}\frac{\sin k\omega t}{k}\left[e^{-a\cos\omega t}I_k(2a) - 2I_k(a)\right],$$

$$(8.21)$$

where $a = A/\omega$. The logarithm of the NMR signal is shown in Fig. 8.5 plotted against ωt, for four values (respectively, 0.1, 0.5, 1 and 2) of the attractive strength of the confining force expressed as the dimensionless parameter $a = A/\omega$ as shown. The significance of a is that it is the ratio of the time period of the potential to the

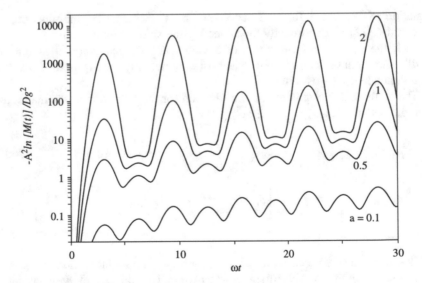

Fig. 8.5 Time dependence of the NMR signal as given by Eq. (8.21) for the case of a time-dependent gradient field and also a time-dependent attractive potential (see text). Both time dependences are sinusoidal with frequency ω. Four cases of the dimensionless ratio $a = A/\omega$ characterize the four curves. Odd-numbered and even-numbered peaks behave differently as A is varied. Reprinted with permission from fig. 1 of Kenkre and Sevilla (2006); copyright (2006) by Elsevier Publishing

time it takes the spins to get to the origin. As it increases, the behavior of the maxima of the signal change in a rather interesting manner: odd-numbered maxima increase their height while even-numbered maxima diminish in size. Equation (8.21) makes clear why this happens.

As another application of our theory, we display in Fig. 8.6 the normalized NMR signal when the attractive force does not vary in time (constant A) and the turn-on is sinusoidal from a non-zero value. Specifically, $f(t) = \cos \omega t$ and $\omega/(Dg^2\gamma^2)^{1/3} = 0.5$. Four different values of the attractive force are considered, see figure caption.

These explicit expressions we have derived, applicable to systems with the confinement produced by an attraction towards the origin through a time-dependent force, have been shown by Sevilla and Kenkre (2007) to reduce a variety of particular cases known in the NMR literature including for (i) pulsed field gradient, (ii) constant field gradient, (iii) exponential turn-on and (iv) sinusoidal turn-on cases. Case (i) is of widespread and direct observational interest in the laboratory while case (ii) is not only experimentally realizable but theoretically simple and therefore indicative of the essential effects. In actual experiments, the gradient field turn-on cannot be realized in infinitesimal time. It has been well noted in the practical NMR literature that the rapid rise and fall of gradient pulses can generate eddy currents in the surrounding conducting surfaces around gradient coils even when shielded coils are used. Those eddy currents can cause phase changes in the observed

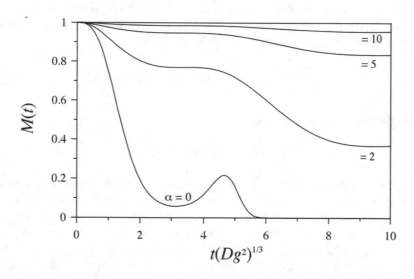

Fig. 8.6 The time dependence of the normalized NMR signal $M(t)$ as a function of the dimensionless time $t(Dg^2\gamma^2)^{1/3}$ for sinusoidal dependence of turn-on and a temporally constant attractive force towards the origin. Four values of the force, i.e., of the dimensionless ratio $\alpha = A/(Dg^2\gamma^2)^{1/3}$ are depicted: 0, 2, 5, and 10, respectively. The lowest curve is for free diffusion, i.e., no attraction. A strong reduction of the attenuation of the signal is evident as α increases. Adapted with permission from fig. 4 of Ref. Sevilla and Kenkre (2007); copyright (2007) by IOP publishing

spectrum, anomalous changes in the attenuation, gradient-induced broadening of the observed spectrum, and time-dependent but spatially invariant shift effects in the static magnetic field. Our analytic study is consequently of some value to the field.

A Simple Dimer System for the Description of the NMR Signal

A more detailed description of the spin echo signal in NMR microscopy with the help of our expressions can be found in the original publications (Kenkre and Sevilla 2006; Sevilla and Kenkre 2007). There is, however, one interesting result derived for the NMR signal that might interest the reader. If you attempt to calculate the NMR signal, not for a 1-dimensional continuum, but for a discrete linear chain on the sites of which the spins hop at nearest-neighbor rates F, you can also solve the calculational problem (Kenkre et al. 1996). Let us consider the simplest case, i.e., a chain of 2 sites, 1 and 2. The caricature Torrey-Bloch equation that the magnetization obeys on this dimer is obviously

$$\frac{dM_1}{dt} = - F(M_1 - M_2) - ibg\gamma f(t)M_1$$

$$\frac{dM_2}{dt} = - F(M_2 - M_1) - ibg\gamma f(t)M_2 \qquad (8.22a)$$

where b is the distance between the two sites with the NMR signal being the sum $M(t) = M_1(t) + M_2(t)$. Adding and subtracting the two equations, we can write immediately an evolution equation for the NMR signal:

$$\frac{dM(t)}{dt} + f(t) \int_0^t ds\, f(s)\phi(t - s)M(s) = 0 \qquad (8.23)$$

where the memory function ϕ is given by

$$\phi(t) = (bg\gamma)^2 e^{-2Ft}.$$

The remarkable observation to be made here is that this memory function is formally precisely the same as $C^2 e^{-2\lambda t}$ that was obtained with the help of projection operators and a weak-coupling approximation by Sheltraw and Kenkre (1996) that we displayed in Eq. (8.14a). Indeed, Eq. (8.12) from the Sheltraw-Kenkre analysis is identical to Eq. (8.23) here! The latter is obtained exactly for the dimer, the former required the projection operator apparatus and a weak-coupling approximation. Note that this also means that the so-called Stepisnik equation from the NMR literature that Wang et al. (1995) used emerges as an *approximation* from our result (8.23) when we take $M(t)$ outside the integral in the manner explained earlier, leading from Eq. (8.23) to

$$M(t) = \exp\left[-(bg\gamma)^2 \int_0^t ds \int_0^s dt'\, f(s)f(t')\phi(s - t')\right].$$

What this means is that all the results of Wang et al. (1995) can be recovered from our exact simple 2-state diffusion result through the above slow-$M(t)$ approximation. The necessary correspondence of parameters as given by Sheltraw and Kenkre (1996) is that the 2-state system b here equals the actual confinement length a except for a proportionality constant ($b = a/2\sqrt{3}$), and the 2-state hopping time $1/F$ here equals the diffusion time also except for a proportionality constant ($1/F = a^2/5D$), where D is the diffusion constant of the spins in the actual system. In a sense, the two states of the 2-state system represent the two halves of the confined region under consideration.

8.2.4   Projections for the Railway-Track Model

How would we develop projection operators capable of use in conjunction with the railway-track model introduced in Sect. 1.2? The idea is to invent projection operators that would sum the probability densities on the two tracks and, starting from Eqs. (1.14) for the individual densities, obtain the telegrapher's equation for their sum $P(x, t) = P_L(x, t) + P_R(x, t)$.

(✳)   Let me tempt the enthusiastic readers to undertake this exercise on their own with the following detailed hint. Define two partial evolution operators to represent the two kinds of evolution in Eqs. (1.14), one along a track and the other from track to track. Define a projection operator that sums the two densities and choose a suitable constant two-vector to facilitate computations. Put $\alpha = 0$ and derive the memory function without using projections, rather by repeated differentiation of the density. Now, with α non-zero, develop exact results involving projections in the manner of another, more complex, result introduced to you elsewhere in the book. Combining the two parts of this exercise, arrive at the exponential memory that results in the telegrapher's equation.

8.3   Chapter 8 in Summary

The technique of projection operators was explored further by showing a variety of examples of their use in non-memory contexts. These examples include tracing over quantum spaces to obtain correlation functions that appear naturally in the linear response theory of transport coefficients originated by Kubo, a use important to the theory of electrical resistivity. They include applications to the quantum control of dynamic localization relating to an understanding of internal structures in the evolution of a quantum dimer. They involve integration of classical variables to obtain the BBGKY (Born-Bogolubov-Green-Kirkwood-Yvon) hierarchy in classical statistical mechanics of gases, as well as use in the Torrey-Bloch equation in NMR microscopy, a subject of relevance to medical physics. They even apply to the simple railway-track problem of Chap. 1.

Spatial Memories and Granular Compaction

9

The science of granular materials is a subject that is at once profound, difficult, of extreme technological and human relevance, and fascinating from a scientific point of view. One realizes that sand is made of grains that are unquestionably a form of the solid state of matter. Pour it into a vessel and it takes the shape of the vessel, mocking you to regard it as a liquid. And twirl it at high speed in a rotating cylinder and you know it behaves, for most practical purposes, as a gas. If you press it in compacts, you know little of how stress distributes itself spatially in the granular material. If you make engine components from that compact without knowing well, and being able to manipulate to advantage, details of that stress distribution, you are liable to have the components crack after manufacture. Farmers need to know the statics of the grain they store in silos. The movement of pharmaceutical pills within factories where they are made begs to be understood properly. The movement of snow and earth on mountain tops leading to avalanches is a phenomenon of dire importance to humans. Every way you look at it, granular science is important, yet poorly understood. The interested scientist will find its problems and features described everywhere. See, for example, (Edwards and Oakeshott 1989; Jaeger et al. 1996b,a; Mehta and Barker 1994; Kadanoff 1999; de Gennes 1999; Duran 2012).

I have had occasion to participate in studies of patterns in granular flow,[1] segregation, and even aspects of manufacture in these investigations (Kenkre et al. 1996; Monetti et al. 2001; Vidales et al. 2001) but they lie outside the scope of this book. What does belong to the general theme here is our analysis of stress distribution, (Kenkre et al. 1998a; Scott et al. 1998; Kenkre 2001a,b), which is the only area I will touch upon here in this brief chapter. Some of the work reported was done in collaboration with Al Hurd, then of Sandia National Laboratory, and

---

[1] A part of Mark Endicott's dissertation and Anastasia Ieride's preliminary research dealt with this topic although their work with me is still in an unpublished state.

© The Author(s), under exclusive license to Springer Nature Switzerland AG 2021
V. M. (Nitant) Kenkre, *Memory Functions, Projection Operators, and the Defect Technique*, Lecture Notes in Physics 982,
https://doi.org/10.1007/978-3-030-68667-3_9

some in the context of the Ph.D. thesis of Joe Scott.[2] It turns out that almost nothing is known definitively about the so-called constitutive relations among the stresses in a granular compact. While it is relatively easy to measure stress at the surfaces of a compact, values of stress in the interior must often be deduced from indirect observations such as density distributions. Of physical significance are also some detailed remarks that are included in this chapter concerning the origin of spatial memory functions in granular materials.

9.1   Spatial Memories and Correlations in the Theory of Granular Materials

The subject of the following discussion is a method we developed to analyze stress distributions on the basis of spatially nonlocal equations for stress, thus involving "memory functions" that arise, not in time but in space. The method has two important ingredients. One is what is sometimes called in the literature the $t - z$ transformation: the singling out of one spatial direction in the granular material and treating it for the purpose of description as if it were time. Stress distribution in an n-dimensional system is then viewed as stress "propagation" in an $n - 1$-dimensional system. For instance, the study of the spatial variation of stress in a 3-dimensional die becomes equivalent to the study of the 'time-evolution' of disturbances in a 2-dimensional membrane. Geometrical changes in the shape along the chosen direction are reinterpreted as temporal changes in the extent of the region under consideration. While the $t - z$ transformation came up naturally in our analysis of stress distribution, it had also appeared in another, earlier, investigation (Bouchaud et al. 1995). Its advantage is that it simplifies the mathematical treatment considerably and provides physical intuition based on knowledge of initial value problems in other fields. Its disadvantage is that approximation procedures, that are normally used along with the transformation, place restrictive limitations on the applicability of the analysis.

The other ingredient of the analysis, particular to our approach, is a non-local formalism (Kenkre et al. 1998a; Scott et al. 1998; Kenkre 2001a) based on integro-differential equations of the Volterra type incorporating memory functions which characterize spatial correlations in the granular material. Its novel aspects useful for the problem of determining stress in granular compacts arise in the form of eigenvalue analysis as we will see below.

---

[2]Granular compacts was not the only common interest that Joe and I had. The story of Tarzan was another although we had no interest in Weissmuller or the movies. While Joe was an expert, the character in the novels was of great interest to the entire research group throughout my years as an advisor. Any of my ex-students will testify that we called our weekly research meetings *bundolos* after the word for 'kill' in the language of the apes as set forth by Edgar Rice Burroughs. The nomenclature originated from the fact no quarter was given or expected when we presented our research results to one another in those meetings.

Once one has selected the z-direction (to be thought of as time) as the direction of gravity and/or the applied stress, one proceeds to seek a closed evolution equation for the scalar field σ_{zz} which represents the zz-component of the stress tensor. The characteristic ingredient of our approach is the use of an evolution equation that is non-local in the depth z:

$$\frac{\partial \sigma_{zz}(x, y, z)}{\partial z} = D \int_0^z dz' \, \phi(z - z') \left[\frac{\partial^2 \sigma_{zz}(x, y, z')}{\partial x^2} + \frac{\partial^2 \sigma_{zz}(x, y, z')}{\partial y^2} \right]. \quad (9.1)$$

The function ϕ that connects the derivatives of the stress at various depths z is what we call the spatial memory function. Along with the multiplying factor D, it is indicative of the spatial correlations of the granular material which arise from the granularity (variations in shape and size of the grains) and other properties such as friction. Generally, the memory functions can depend on x and y as well, but we suppress that dependence here to focus on the essential features of our memory description. The idea of utilizing a memory function approach arose originally from the need to address curious features such as oscillations in the values of observed stress down the center line in compacts. See, as an example, (Aydin et al. 1994, 1996).

To prepare ourselves for the eigenvalue analysis directed at the oscillatory stress observations, let us first ask what Eq. (9.1) would predict for stress distribution in a situation without boundaries and for a memory function $\phi(z)$ that is a simple exponential $\alpha e^{-\alpha z}$ with $D = c^2/\alpha$. Equation (9.1) then yields the telegrapher's equation as we saw in Chap. 1. Indeed, if for the sake of simplicity we consider here only a two-dimensional system, we get from (9.1)

$$\frac{\partial^2 \sigma_{zz}(x, z)}{\partial z^2} + \alpha \frac{\partial \sigma_{zz}(x, z)}{\partial z} = c^2 \frac{\partial^2 \sigma_{zz}(x, z)}{\partial x^2}, \quad (9.2)$$

exactly as we obtained Eq. (1.8) from Eq. (1.7) in Chap. 1. The task of finding solutions proceeds in a manner identical to that in Chap. 1. Thus, if we apply at the "surface" $z = 0$ a delta-function stress, i.e., $\sigma_{zz}(x, 0) = \delta(x)$, the stress will "propagate" below the surface as described by

$$\sigma_{zz}(x, z) = e^{-\alpha z/2} \left[\frac{\delta(x + cz) + \delta(x - cz)}{2} + T(x, z) \right]. \quad (9.3)$$

In Eq. (9.3), $T(x, z)$ vanishes identically in the spatial region $cz \leq |x|$ and equals, otherwise,

$$T(x, z) = \left(\frac{\alpha}{4c} \right) \left[I_0 \left(\frac{\alpha}{2c} \sqrt{c^2 z^2 - x^2} \right) + \frac{cz}{\sqrt{c^2 z^2 - x^2}} I_1 \left(\frac{\alpha}{2c} \sqrt{c^2 z^2 - x^2} \right) \right], \quad (9.4)$$

the I's being modified Bessel functions of order 0 and 1.

This memory function analysis given by Kenkre et al. (1998a) provided a unification for two different and disparate treatments published in the literature: by Bouchaud et al. (1995) who predicted wave-like stress "propagation" forming the so-called light cones, and by Liu et al. (1995) who predicted random distribution of stress appropriate to diffusive propagation in our terms. The former is the $\alpha \to 0$ limit of our analysis whereas the latter comes about when $\alpha \to \infty$, $c \to \infty$, such that $c^2/\alpha = D$ is finite. The unification is clear from Fig. 9.1 (see Figure caption). The wave-like limit in (a) corresponds to Eq. (9.3) being reduced to

$$\sigma_{zz}(x, z) = (1/2)[\delta(x + cz) + \delta(x - cz)]$$

as $\alpha \to 0$. For non-vanishing but small α, our solution retains the light-cones feature but shows explicitly that there is non-vanishing stress distribution *within* the light cones. This stress is given by our term T shown in Eq. (9.4). In the limit that reduces our theory to the opposite extreme of Liu et al. (1995), the light cones spread out to coincide with the surface $z = 0$, and the entire region experiences stress:

$$\sigma_{zz}(x, z) = \frac{e^{-x^2/4Dz}}{\sqrt{4\pi Dz}}.$$

It goes without saying that the solution for any arbitrary prescribed distribution at the surface $z = 0$ is obtained from the Green function prescription

$$\sigma_{zz}(x, z) = \int dx' G(x - x', z)\sigma_{zz}(x', 0),$$

where $G(x, z)$ is the right hand side of Eq. (9.3).

9.2 Eigenvalue Analysis of Stress Distribution in Compacted Granular Material

Bouchaud et al. (1995) introduced the idea of light cones into the area of stress distribution and Liu et al. (1995) based their analysis on random walk considerations. However, neither set of investigators appear to have taken the problem seriously enough to carry out a boundary value treatment of the phenomenon as we will now proceed to perform. The former group focused attention on what may be termed "the ray optics limit" of the wave equation in their analysis of silo geometry while the latter concentrated their efforts on a mean field treatment of the stress distribution.

By contrast, our interest was (and is) in reproducing, if possible, actual reported observations of stress distributions in granular compacts exemplified by the work of (Aydin et al. 1994, 1996) in long pipes, at the very least, semiquantitatively. The convenient manner of doing this is to use (Kenkre et al. 1998a) the method of separation of variables by assuming a form for the stress,

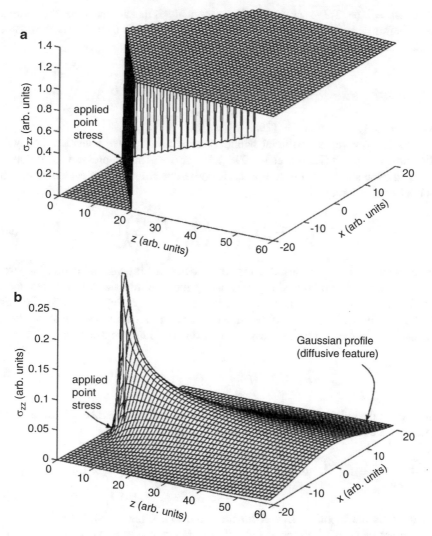

Fig. 9.1 Our unification of two disparate treatments of stress distribution in the literature brought about with the memory formalism. Plotted is the stress distribution in an unbounded medium for an applied δ-function stress. The wave limit of Bouchaud et al. (1995) showing the "light-cones" is clear in (**a**) where we take $c = 1$ and $\alpha = 0.005$ in arbitrary but consistent units. The diffusive limit of Liu et al. (1995) is similarly clear in the Gaussian profile in (**b**) where we have $D = 1$ and $\alpha \to \infty$ as well as $c \to \infty$. Adapted with permission from fig. 2 of ref. (Kenkre et al. 1998a); copyright (1998) by the American Physical Society

$$\sigma_{zz}(x, z) = \sigma_k (A_k \cos kx + B_k \sin kx) g_k(z), \tag{9.5a}$$

$$g_k(z) = e^{-\alpha z/2} \left[\cosh z\Omega_k + \frac{\alpha}{2\Omega_k} \sinh z\Omega_k \right], \tag{9.5b}$$

where $\Omega_k = \sqrt{(\alpha/2)^2 - (ck)^2}$. A preliminary analysis that assumes that the compact extends from $x = -L/2$ to $x = L/2$ and $\sigma_{zz}(\pm L/2, z) = 0$ so that only the cosines in Eq. (9.5) survive, leads to the evaluation of the A_k through

$$A_K = (2/L) \int_{-L/2}^{L/2} dx \, \cos kx \, \sigma_{zz}(x, 0),$$

with $m = 0, 1, 2, \ldots$ and $k = (2m + 1)(\pi/L)$.

This is, however, an artificial boundary condition that we considered only for illustrative purposes (Kenkre et al. 1998a). If a constant punch pressure p_0 is applied across the top surface of the compact, the centerline stress can be evaluated exactly in the Laplace domain as

$$\tilde{\sigma}_{zz} = \frac{p_0}{\epsilon} \left[1 - \text{sech} \left(\frac{L}{2c} \sqrt{\epsilon^2 + \epsilon\alpha} \right) \right]. \tag{9.6}$$

In the wave limit $\alpha = 0$, inversion is straightforward. It gives the centerline stress as a square wave, symbolized as $W(z)$, along the z-coordinate. It is constant at the applied value p_0 for $0 < z < L/2c$, flips to $-p_0$ for $L/2c < z < 3L/2c$, flips back to p_0 for $3L/2c < z < 5L/2c$, and continues alternating in this fashion. In the diffusive limit, the centerline stress distribution is evaluated exactly as

$$\sigma_{zz}(0, z) = 2p_0 \int_0^{1/2} dv \, \theta_1 \left(v \mid \frac{4Dz}{L^2} \right). \tag{9.7}$$

in terms of the elliptic theta function of the first kind. The general expression for the intermediate region, when one is not in either limit, is

$$\frac{\sigma_{zz}(0, z)}{p_0} = 1 + \int_0^z du \, e^{-(\alpha/2)u} \left[M(u) + \left(\frac{\alpha}{2} \right) \int_0^u ds \, I_1(s) M \left(\sqrt{u^2 - s^2} \right) \right]. \tag{9.8}$$

where I_1 is the modified Bessel function and $M(z)$, the derivative $dW(z)/dz$ of the square wave $W(z)$ described above, can be expressed as an infinite sum of δ functions centered at multiples of $L/2c$.

This illustrative analysis shows oscillations in the centerline stress in a clear and direct way. However, it contains unphysical elements which arise from the vanishing boundary conditions at the die walls because of the wave element in the evolution. To construct a realistic theory devoid of these unphysical elements, we took the stress at the wall surfaces of the compact, $\sigma_{zz}(\pm L/2, z)$, *not to vanish* but to be a given function $h(z)$ of the depth,

$$\sigma_{zz}(x, z)|_{x=\pm L/2} = p_0 h(z) \tag{9.9}$$

given by experiment. The solution of the telegrapher's equation with such initial *and* boundary conditions presents a quite unusual problem. It is analogous to propagation problems in which boundary conditions are themselves dependent on time (Farlow 1993).[3] To tackle it, we generalize Eqs. (9.5) and, for this context, take

$$\sigma_{zz}(x, z) = \sum_k A_k G_k(z) \cos kx$$

and solve for $G_k(z)$. The function $h(z)$ appearing in the solution is taken from experimental observations to be an exponentially decreasing function. Solutions for the stress distribution are found explicitly and shown to reproduce observed behavior in compaction experiments carried out in uranium dioxide. The observed stress oscillations are thus explained (Scott et al. 1998) on the basis of interference of reflected "stress waves".

On the basis of the telegrapher's equation, equivalently of an exponential *spatial memory*, our theory is thus able to describe the entire range of observed behavior from wavelike limit (with the light cones) to the opposite, i.e. diffusive extreme. To make the qualitative difference between these extreme limits clear, we present in Fig. 9.3 cases (a) and (c) (the leftmost and rightmost panels respectively) of Fig. 9.2 in the form of a 3-dimensional representation. The considerable difference in the qualitative variations of stress, labelled as wave-limit and diffusive-limit are clear in Fig. 9.3. The relevant experimental variation of stress in the form of a contour plot reported by Macleod and Marshall (1977) in uranium dioxide powders pressed at 160 MPa is shown in Fig. 9.4. The oscillations are clear in this contour plot of observations. Previous theories such as of Aydin et al. (1994) were unable to produce such oscillations (see the discussion in Scott et al. (1998)).

9.3 Physical Origin of the Memory Functions

There are several different ways one can understand the origin of memory function in the description of stress distribution. We present three:

9.3.1 Stochastic Origin and Connection to Depth-Dependent Correlations

Consider for simplicity a 2-dimensional granular compact (z along the vertical and x along the horizontal) consisting of weightless circular disks of a given radius arranged in perfect order. Let a vertical force be applied to one of the disks lying in the top layer of the compact. Newtonian laws of statics dictate that the consequent force distribution, equivalently stress distribution, is down two lines

---

[3]Note that, as a consequence of the $t - z$ transformation, z here is playing the role of time t.

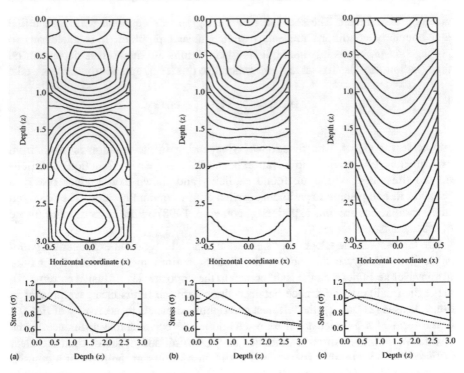

Fig. 9.2 Calculated stress distributions from the theory of Scott et al. (1998) showing the transition from fully wavelike stress propagation down the compact to diffusive propagation. Plots below the contour plots show the centerline stress (solid lines) relative to the exponential variation at the boundary (dotted line). Units are arbitrary. Adapted with permission from fig. 4 of ref. (Scott et al. 1998); copyright (1998) by the American Physical Society

in the compact, representative of the so-called "light cones". Viewed through the $t - z$ transformation, the representative point in the 1-dimensional space of x travels ballistically with constant speed which we will call c. Here, 'travel' obviously represents changes in the x-coordinate of the representative point with changes in its depth z.

If the array is now considered to be realistic, and therefore not perfectly periodic, we come upon irregularities stemming from changes in shape and size of the disks and/or the presence of friction. The speed c now changes from location to location. The path of the representative point is jagged: c is a *stochastic* variable. Restricting attention to its z-variation only, we can write an equation of the kind one encounters in the description of random motion,

$$\frac{dx}{dz} = c(z). \tag{9.10}$$

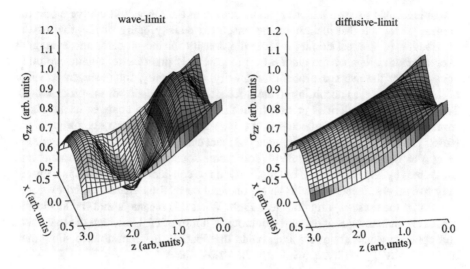

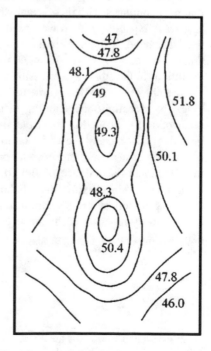

Fig. 9.3 Three-dimensional representations of the stress distribution calculated using our theory in the wave limit (left panel), and in the diffusive limit (right panel) to make clearer the oscillations predicted in the former and their absence in the latter. The two limits correspond precisely to the contour plots (**a**) and (**c**) in Fig. 9.2. Units are arbitrary. Stress "reflections" from the walls produce the oscillations that are clearly visible in the Figure. Adapted with permission from fig. 5 of ref. (Scott et al. 1998); copyright (1998) by the American Physical Society

Fig. 9.4 Contour plot of density distribution (in percent solid density) for cylindrical compacts of uranium dioxide powder pressed at 160 MPa taken from Macleod and Marshall (1977). The density distribution is taken to represent the stress distribution as commonly done in the literature. Adapted with permission from fig. 2 of ref. (Scott et al. 1998); Copyright (1998) by the American Physical Society

An ensemble of representative points started at the top would evolve along the various paths, the distribution of the ensemble density being descriptive of the distribution of stress. Defining a Liouville density for the process, and averaging over all realizations of the stochastic process, it is possible to obtain evolution equations for the average probability density, equivalently, for the average value of the stress $\sigma_{zz}(x, z)$ according to the stochastic characteristics of the process $c(z)$. The particular irregularities arising from the shape and size changes in the disks, from the roughness, are reflected in the $c(z)$ and thereby in the evolution of the stress. The irregularities in the granular compact change the direction and magnitude of $c(z)$ as one goes down the depth coordinate sometimes by small amounts and sometimes by large amounts. For simplicity, let us consider that $c(z)$ jumps between only two values, c and $-c$, with an exponential correlation characterized by depth l, and that the noise is a random telegraph. What this means is that the stochastic process is dichotomous and that the number of jumps of $c(z)$ between the two equal and opposite values follows a Poissonian distribution. It is possible to show then that the Liouville density, equivalently the stress, obeys

$$\frac{\partial \sigma_{zz}(x, y, z)}{\partial z} = c^2 \int_0^z dz' \, e^{-\alpha(z-z')} \left[\frac{\partial^2 \sigma_{zz}(x, y, z')}{\partial x^2} \right]. \tag{9.11}$$

The parameter $\alpha = 2/l$ describes the reciprocal of the depth scale over which the exponential correlation of the random process decays. If α vanishes, which means that $c(z)$ continues with its original value forever, the equation obeyed by $\sigma_{zz}(x, z)$ is a wave equation with wave speed c and corresponds to constant memory. On the other hand, if the correlation of c's decays infinitely rapidly, specifically such that $\alpha \to \infty$, $c \to \infty$, $c^2/\alpha = D$, the memory $\phi(z)$ equals a δ-function, and the evolution for the stress is a diffusion equation.

The above stochastic argument (Kenkre 2001b), while highly simplified, provides the essential understanding of the origin of memory functions and of their connection to correlation functions in the granular system. The roughness of the granular particles, and the variation in their shape and size, produce a decay in the correlation of $c(z)$, and this correlation directly gives rise to the memory equation, the functional dependence of the correlation function and the memory function being identical in the simplified dichotomous case. Generally, a stochastic process involving a sum of many random telegraphs may be approximately represented by a memory function in the stress equation that is the sum of a large number of exponentials.[4]

---

[4]I am indebted to Marek Kuś for the detailed demonstration. On the basis of his arguments, we have been able to present an analysis of a combination of random telegraphs for the quite different topic of charge mobility. See Kenkre et al. (1998b) where the technique is explained.

9.3.2 Origin of Spatial Memory Functions from Effective Medium Considerations

Even if you feel that the stress distribution problem is entirely diffusive in the sense that the depth correlation of the c's decays extremely rapidly (notice the clear demonstration that we have argued in (Scott et al. 1998) and in Sect. 9.3.1), memory functions can arise from granularity and disorder, even if the starting point is a diffusion equation, as we will now show.

Granularity of the material we consider demands that one replace x by a discrete index m. Randomness of shapes and sizes of the particles making up the granular material demands that the rates in the "evolution equation" be random functions. It makes sense, therefore, to start (Kenkre 2001a) with a diffusive but discrete description represented by

$$\frac{d\sigma_m(z)}{dz} = F_{m+1,m}[\sigma_{m+1}(z) - \sigma_m(z)] + F_{m,m-1}[\sigma_{m-1}(z) - \sigma_m(z)], \qquad (9.12)$$

Where σ_m is the value of σ_{zz} at site m, the latter being the discrete index representing the horizontal x (or y) coordinate. Generally m is a vector index, the evolution equation being appropriately modified. We continue to consider one horizontal dimension for the sake of simplicity.

In the latter part of this book, the reader will come across two chapters, (13 and 14), devoted entirely to the subject of effective medium theory (EMT). That EMT converts spatial disorder into a memory function to be used on an ordered lattice, specific recipes being available to calculate the memory function from disorder information such as the distribution of the rates F. In other words, Eq. (9.12) can be replaced by

$$\frac{d\sigma_m(z)}{dz} = \int_0^z dz' \, \mathcal{F}(z - z')[\sigma_{m+1}(z') + \sigma_{m-1}(z') - 2\sigma_m(z')]. \qquad (9.13)$$

The prescription to go from the random distribution function $\rho(f)$ that determines the rates F to the memory $\mathcal{F}(z)$ can be expressed in different ways (see Chap. 13). We give one of them here:

$$\int_0^\infty \frac{df \, \rho(f)}{f + \left[\frac{\tilde{\mathcal{F}}\zeta(\tilde{\mathcal{F}})}{1 - \tilde{\mathcal{F}}\zeta(\tilde{\mathcal{F}})}\right]} = \frac{1}{\tilde{\mathcal{F}}(\epsilon)} - \zeta(\tilde{\mathcal{F}}(\epsilon)). \qquad (9.14)$$

Here, $\tilde{\mathcal{F}}(z) = \int_0^\infty dz \, e^{-\epsilon z} \mathcal{F}(z)$ and ζ has a known dependence both on ϵ and on $\tilde{\mathcal{F}}$.

It is important to keep in mind that memory functions arising from such effective medium considerations are characterized by two "time" scales that is a fast z-decay and a slower z-decay and consequently lead to stress distributions that can be qualitatively different (Kenkre 2001a) from those predicted by the simple exponential memory explored in earlier sections.

9.3.3 Spatial Memories from Generalizing Constitutive Relations

A third elucidation of how spatial memories can come about in the description
of stress distributions is to be found in arguments presented in Bouchaud et al.
(1995) and Kenkre et al. (1998a) regarding generalizations of constitutive relations.
Typically, stress balance equations appearing from the so-called Cauchy relations
(Newtonian statics) are three in number but involve six independent quantities. They
need to be supplemented, therefore, by additional relations that go under the name
of constitutive or closure relations. Whether made explicit or not, they are *ad hoc*
in nature. The assumption made in the past was of proportionality between the
diagonal elements of the stress tensor (σ_{xx}, σ_{yy} and σ_{zz}) as well as the vanishing
of the components in the xy plane: $\sigma_{xy} = \sigma_{yx} = 0$. This last relation is not used
directly but only in the form of spatial derivatives (Bouchaud et al. 1995):

$$\frac{\partial \sigma_{xx}}{\partial x} = c^2 \frac{\partial \sigma_{zz}}{\partial x}, \tag{9.15a}$$

$$\frac{\partial \sigma_{yy}}{\partial y} = c^2 \frac{\partial \sigma_{zz}}{\partial y}. \tag{9.15b}$$

If we extend this existing constitutive relation by generalizing it to incorporate
the contributions of σ_{xz} and σ_{yz}, representing these contributions through the
addition of first-order terms in the sense of a Taylor's series expansion,

$$\frac{\partial \sigma_{xx}}{\partial x} = c^2 \frac{\partial \sigma_{zz}}{\partial x} + \alpha \sigma_{xz}, \tag{9.16a}$$

$$\frac{\partial \sigma_{yy}}{\partial y} = c^2 \frac{\partial \sigma_{zz}}{\partial y} + \alpha \sigma_{yz}, \tag{9.16b}$$

we are all set to obtain the exponential memory we have used in the discussion in
this chapter. All we need is to combine this extended form of the constitutive relation
with the standard Cauchy equations. An immediate consequence is

$$\sigma_{xz}(z) = -c^2 \int_0^z dz' e^{-\alpha(z-z')} \frac{\partial \sigma_{zz}}{\partial x}, \tag{9.17a}$$

$$\sigma_{yz}(z) = -c^2 \int_0^z dz' e^{-\alpha(z-z')} \frac{\partial \sigma_{zz}}{\partial y}. \tag{9.17b}$$

Kenkre et al. (1998a) have shown how Eqs. (9.17) lead without effort to the closed
equation (9.1) for σ_{zz} with exponential memory, $\phi(z) = \alpha \exp -\alpha z$. This is how we
arrive at the Eq. (9.2) that we have used for the analysis in this chapter.

It must not be overlooked that, although they might have some formalistic
(mathematical) appeal, this manner of obtaining the stress memory has less physics
in it than the other two ways we have commented on in this Section.

9.4 Chapter 9 in Summary

We ventured in this chapter to a particularly distant field in the application of memory functions. The topic treated was granular materials. The memories were not functions of time but of depth in a granular compact and played the role of time-dependent memories after a $t - z$ transformation. A telegrapher's equation analysis was constructed on the basis of a simple exponential approximation to the memory. An eigenvalue analysis was developed which allowed us to understand experimental observations on packed powders of uranium dioxide and similar systems which show oscillatory behavior in their density distributions. Those considerations were useful to the manufacture of technological items such as engine blocks. We commented on three different origins of the spatial memories. One was based on considerations of how stress is distributed as a result of variations of size and shape of the particles forming the material. Another was based on effective medium theory. A third appeared from simple but reasonable assumptions regarding constitutive relations.

Memories and Projections in Nonlinear Equations of Motion

10

The tools we have employed in conjunction with the development and applications of our formalism so far appear to be eminently those associated with *linear* equations, whether it is projection operators that we use or Fourier-Laplace transforms that we analyze with. One might be tempted to ask whether memory functions and projection techniques can live in harmony with nonlinearities. A surprising incursion of nonlinear entities in three separate situations forms the subject of the present chapter.

10.1 The Discrete Nonlinear Schrödinger Equation and the Related Density Matrix Equation

The discrete nonlinear Schrödinger equation is an object one encounters in polaron physics if the electron-phonon interaction is strong, in the study of optics as in waveguide phenomena, and in the form of the Gross-Pitaevsky equation in the analysis of Bose-Einstein condensates. It is characterized by a cubic nonlinearity. If we select the polaron context for purposes of discussion, the entity of interest is a quantum mechanical particle moving among sites m while interacting strongly with phonons or lattice vibrations such that its amplitude $c_m(t)$ to be at site m evolves through

$$i\frac{dc_m(t)}{dt} = V\left[c_{m+1}(t) + c_{m-1}(t)\right] - \chi|c_m(t)|^2 c_m(t). \tag{10.1}$$

Here V is the transfer interaction matrix element as in earlier situations we have encountered. However, χ introduces a novel feature. It arises from strong interactions with vibrations and is the coefficient of nonlinearity that represents the site energy lowering that the moving quasiparticle undergoes when it occupies any given site. Its source is a feedback process whose intensity can give rise to an abrupt transition in the motion of the particle. If the effect is small enough, the

© The Author(s), under exclusive license to Springer Nature Switzerland AG 2021
V. M. (Nitant) Kenkre, *Memory Functions, Projection Operators, and the Defect Technique*, Lecture Notes in Physics 982,
https://doi.org/10.1007/978-3-030-68667-3_10

energy lowering coming about from the occupation of a site causes the temporary destruction of resonance between site energies and merely reduces mobility. If the energy lowering crosses a certain threshold, the particle cannot bounce back to a resonant situation and finds itself self-trapped. It is thus a polaron. An entire field of research has built itself around this phenomenon since the days of the studies carried out by Holstein. Unfortunately, elucidation of those investigations lies outside the scope of this book. I will only peck at the physics of the system by discussing the GME and the memory function that arise in this system.

The amplitude equation (10.1) leads to the consequence that the density matrix $\rho_{mn}(t) = c_m^*(t)c_n(t)$ obeys (see, e.g., Kenkre and Campbell 1986)

$$i\frac{d\rho_{m,n}}{dt} = V\left(\rho_{m+1,n} + \rho_{m-1,n} - \rho_{m,n+1} - \rho_{m,n-1}\right) - \chi(\rho_{m,m} - \rho_{n,n})\rho_{m,n}.$$

(10.2)

What do we expect to happen if we apply the diagonalizing Zwanzig projections to this equation? We would think the familiar method of deriving the GME would not work because of the lack of linearity in the evolution.[1] Yet, let us reexpress the nonlinear von Neumann equation as

$$\frac{d\rho}{dt} = -iL_V\rho + iL_\chi\rho = -iL\rho$$

(10.3)

where the "Liouville" operator $L_V O = [V, O]$ imposes a commutation on any operator O as usual in the standard manner but

$$(L_\chi O)_{mn} = \chi(\rho_{mm} - \rho_{nn})O_{mn}.$$

(10.4)

If we now follow Zwanzig in applying the projection operator through $(\mathcal{P}O)_{mn} = O_{mm}\delta_{mn}$, we get the usual equations of motion for the diagonal part ρ' and the off-diagonal part ρ'', respectively,

$$\frac{d\rho'}{dt} = -i\mathcal{P}(L_V + L_\chi)\rho' - i\mathcal{P}(L_V + L_\chi)\rho'',$$

(10.5a)

$$\frac{d\rho''}{dt} = -i(1 - \mathcal{P})(L_V + L_\chi)\rho'' - i(1 - \mathcal{P})(L_V + L_\chi)\rho'.$$

(10.5b)

---

[1]This is what I loudly declared decades ago in a lecture I delivered to my class. Honglu Wu, who had started with me on his research was in the class. He approached me the next day with a GME he obtained via Zwanzig projections for this problem and I told him he had to have made an error. I soon saw he was right and I was wrong. On closer inspection, I found out what the subtlety was. The following discussion is, thus, the result of Wu's refusing to believe his teacher. It is a trait that is to be heartily encouraged in any student.

It is important that we have defined L_χ *not* through

$$(L_\chi O)_{mn} = \chi(O_{mm} - O_{nn})O_{mn}$$

for any O but through Eq. (10.4) which preserves the linearity of the projection operator. Our definition has the consequence that, while (10.5a) is a nonlinear equation for ρ', the diagonal part of the density matrix, (10.5b) is a *linear* equation for ρ'', the off-diagonal part. This is a consequence of the fact that Eq. (10.4) shows that the action of L_χ on the off-diagonal part ρ'' involves multiplication by elements of the *diagonal* part of the density matrix. We can thus solve Eq. (10.5b) and substitute the solution in Eq. (10.5a) through standard methods of linear differential equations precisely as taught to us by Zwanzig for his completely linear case. Assuming, as usually done, that an initial random phase condition applies, the result is, of course,

$$\frac{d\rho'}{dt} = -i\mathcal{P}L(t)\rho' - \mathcal{P}L(t)\int_0^t ds \left[e^{-i\int_s^t dz\,(1-\mathcal{P})L(z)} \right] (1-\mathcal{P})L(s)\rho'(s). \quad (10.6)$$

The GME can now be written down and has the usual form

$$\frac{dP_m(t)}{dt} = \int_0^t ds \sum_n [\mathcal{W}_{mn}(t,s)P_n(s) - \mathcal{W}_{nm}(t,s)P_m(s)]. \quad (10.7)$$

It differs from the usual GMEs that we have encountered in that the memories $\mathcal{W}(t,s)$ are *not* functions of the difference $t - s$ and, furthermore, that they depend on the probabilities.

10.1.1 Exact Evaluation for the Dimer and a Surprising Connection

For a two-state system, the disentangling of the projection operators in the GME expressions can be done exactly, i.e., through analytical means. Because the action of L_χ as defined in (10.4) involves the multiplication by the difference of the diagonal elements, L_χ produces a vanishing result when it acts on ρ'. This leads to

$$\langle 1|(1-\mathcal{P})L(s)\rho'(s)|2\rangle = -V[\rho_{11}(s) - \rho_{22}(s)]. \quad (10.8)$$

Arbitrary powers r of the operator $(1 - \mathcal{P})L(z)$ acting on $(1 - \mathcal{P})L(s)\rho'(s)$ yield the result

$$\langle 1| [(1-\mathcal{P})L(z)]^r (1-\mathcal{P})L(s)\rho'(s)|2\rangle = (-1)^{r+1}\chi^r V[\rho_{11}(s) - \rho_{22}(s)]$$

$$[\rho_{11}(z) - \rho_{22}(z)]^r, \quad (10.9)$$

which makes possible the exact evaluation of the series produced by the action of $\exp\left[-i \int_s^t dz\,(1-\mathcal{P})L(z)\right]$ on the operator $(1-\mathcal{P})L(s)\rho'(s)$. The result is a sum of two terms, one proportional to the sine and the other to the cosine of the expression $\int_s^t dz\,\chi\,[\rho_{11}(z) - \rho_{22}(z)]$. When $\mathcal{P}L(t)$ acts from the left, it makes the sine term vanish. The evaluated GME for the two-state system is, thus, exactly

$$\frac{d\rho_{11}(t)}{dt} = 2V^2 \int_0^t ds\,[\rho_{22}(s) - \rho_{11}(s)]\cos\left(\int_s^t dz\chi\,[\rho_{22}(z) - \rho_{11}(z)]\right),$$

$$(10.10\mathrm{a})$$

$$\frac{d\rho_{22}(t)}{dt} = 2V^2 \int_0^t ds\,[\rho_{11}(s) - \rho_{22}(s)]\cos\left(\int_s^t dz\chi\,[\rho_{11}(z) - \rho_{22}(z)]\right).$$

$$(10.10\mathrm{b})$$

Note that it is a simple dimer GME except for the fact that its memories are functions of the (time integrals of) probability differences, thus making the GME itself certainly nonlinear.

It is possible that the aspect of Eqs. (10.10) seems unfamiliar, or even strange, to you. But, as we showed in Wu and Kenkre (1989), where we also described this entire derivation, all you have to do is subtract the two equations from each other, write down the integro-differential equation obeyed by the probability difference $p = \rho_{11} - \rho_{22}$,

$$\frac{dp(t)}{dt} + 4V^2 \int_0^t ds\,p(s)\cos\left(\chi \int_s^t dz\,p(z)\right) = 0, \qquad (10.11)$$

⊛   and recognize the pendulum that is staring at you. If you still do not recognize the oscillating beast in the equation, I suggest you undertake it as an exercise and stop only when you have shown that the time dependence of the probability difference is given by an elliptic cosine. If you desire a hint, it is the transformation

$$\xi(t) = \int_0^t ds\,p(s) \qquad (10.12)$$

that will simplify the situation. It leads to the second order differential equation for the integral $\xi(t)$,

$$\frac{d^2\xi}{dt^2} + \left(\frac{4V^2}{\chi}\right)\sin(\chi\xi) = 0, \qquad (10.13)$$

which is identical to the evolution equation of the physical pendulum in which one has *not* made the small angle approximation that would simplify it to the simple harmonic motion of an oscillator.[2] Indeed, the explicit solution of (10.13) in terms of the inverse trigonometric and direct elliptic sine functions may already be known to you:

$$\xi(t) = (2/\chi)\sin^{-1}\left[\operatorname{sn}\left(\chi t/2|4V/\chi\right)\right].$$

Differentiation of this result gives

$$p(t) = \frac{d\xi}{dt} = \operatorname{cn}\left(2Vt|\chi/4V\right). \tag{10.14}$$

This result for the probability difference in a nonlinear dimer is part of the literature (see, e.g., Kenkre and Campbell 1986), is known to satisfy a simple cubic nonlinear differential equation with coefficients determined by initial conditions in the dimer (Kenkre and Tsironis 1987), and forms the basis of a simple description of polaron effects from nonlinearities arising from strong electron-phonon interactions.

10.1.2   Weak-Coupling GME for Motion on a Chain

Although the exact evaluation of the memory functions in the GME is possible only for the two-site system, an approximation scheme can be developed for the chain (Wu and Kenkre 1989). We will treat the nonlinearity exactly but will consider the intersite transfer relatively weak so that we can use a perturbation in its powers.

As we learned from Zwanzig for the linear GME, it is possible to replace here too, without approximation,

$$\mathcal{P}L(t)\int_0^t ds\, e^{-i\int_s^t dz\,(1-\mathcal{P})L(z)}L(s)\rho'(s) = \mathcal{P}L_V\int_0^t ds\, e^{-i\int_s^t dz\,(1-\mathcal{P})L(z)}L_V\rho'(s),$$

i.e., replace the first and last occurrence of the full L by L_V alone which, moreover, is independent of time. This follows from the nature of $\mathcal{P}$ and L_χ.

We are now ready to introduce the weak-coupling approximation. Making use of the fact that L_χ acting on an arbitrary operator produces an operator that is completely off-diagonal in the site-representation, we see that the lowest nonvanishing term in an expansion of the expression in the kernel above is

$$\mathcal{P}L_V\int_0^t ds\, e^{i\int_s^t dz\,L_\chi(z)}L_V\rho'(s).$$

We have ousted the projection operators in the exponent of the expression. The expansion of the exponential in powers of $i\int_s^t dz\,L_\chi(z)$ along with the use of the

---

[2]The equation has also appeared in the work of Cruzeiro-Hansson et al. (1988).

definition of L_χ allows us now to write, completely transparently,

$$\left[e^{i \int_s^t dz\, L_\chi(z)} O \right]_{mn} = e^{i \int_s^t dz\, \chi [\rho_{mm}(z) - \rho_{nn}(z)]} O_{mn}. \tag{10.15}$$

Using Eq. (10.15), we at once obtain the GME (10.7) for the chain, the memory functions being given, via the weak-coupling approximation, by

$$\mathcal{W}_{mn}(t, s) = 2V^2 (\delta_{m,n+1} + \delta_{m,n-1}) \cos\left(\chi \int_s^t dz\, [P_m(z) - P_n(z)] \right). \tag{10.16}$$

The procedure for the evaluation of the memory for this nonlinear problem is similar to the one carried out in Kenkre (1978a) and reported in Sect. 2.3.3 of the present book for the case of interaction with a reservoir described through a simple SLE of the Avakian type. The present procedure is more complicated because of the nonlinearity. This is also the reason why analytic evaluation is possible for the linear problem of Sect. 2.3.3 for the extended chain as well as for the dimer whereas it can be done only for the dimer in the present case. Notice that the starting evolution equation is *formally* the same, i.e., (10.3) for both problems. The definition of L_χ here is as given in Eq. (10.4) but in the problem of Sect. 2.3.3, it is simpler as given in Eq. (2.27). An application of the weak-coupling memory in Eq. (10.16) to study the phenomenon of self-trapping via the transfer rate is depicted in Fig. 10.1.[3]

10.1.3 Additional Results

Returning to the dimer, an exact evaluation can be given (Wu and Kenkre 1989) also for the initial condition term in the Zwanzig procedure for arbitrary placement. Defined as an initial driving term that appears in the GME arising from the condition that the density matrix is not initially diagonal in the 1,2 representation, it is well known that it is given by

$$\mathcal{I}(t) = -i \mathcal{P} L(t) e^{-i \int_0^t ds\, (1 - \mathcal{P}) L(s)} (1 - \mathcal{P}) \rho(0).$$

We gave an explicit evaluation in Wu and Kenkre (1989) in terms of the initial values of the density matrix element combinations

$$r_0 = [\rho_{12}(0) + \rho_{21}(0)], \quad q_0 = i[\rho_{12}(0) - \rho_{21}(0)].$$

The explicit expressions for $\mathcal{I}_1(t)$ and $\mathcal{I}_2(t)$ to be respectively added to the GME equations for $P_1(t)$ and $P_2(t)$ are

---

[3] The operators $(1 - \mathcal{P}) L_\chi$ and $(1 - \mathcal{P}) L_V$ commute in the reservoir problem of 2.3.3 but not here. The primary difference is that successive powers of $(1 - \mathcal{P}) L_\chi$ introduce simply multiplicative powers of the c-number α in 2.3.3 but products of different quantities in the nonlinear case here.

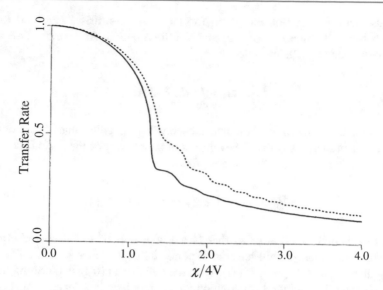

Fig. 10.1 Dependence of the transfer rate on the nonlinearity parameter $\chi/4V$ in the discrete nonlinear Schrödinger equation. The rate (normalized to 1) is defined as the reciprocal of the time taken for the mean square displacement to rise from 0 to a^2 where a is the lattice constant, from an initially localized occupation at the origin. The exact solution (solid line) is obtained by numerical means from the original density matrix equation (10.2) while the approximate solution (dotted line) is the result of the weak-coupling procedure, i.e., of the GME (10.16) (dashed line). The qualitative features of the former are reproduced pretty well by the latter. The difference is small except near the self-trapping transition where we also see the behavior change from "free" to "self-trapped" as the transfer rate changes from large to small values. Reprinted with permission from fig. 2 of Ref. Wu and Kenkre (1989); copyright (1989) by the American Physical Society

$$\mathcal{I}_1(t) = \left(-V\sqrt{1 - p_0^2}\right)\cos\left(\Delta_0 - \chi \int_0^t dz\, p(z)\right), \tag{10.17a}$$

$$\mathcal{I}_2(t) = \left(V\sqrt{1 - p_0^2}\right)\cos\left(\Delta_0 - \chi \int_0^t dz\, p(z)\right), \tag{10.17b}$$

satisfying, as they must, the sum rule $\mathcal{I}_1 + \mathcal{I}_2 = 0$ at all times t.

In writing Eqs. (10.17b), use has been made of the definition $\Delta_0 = \tan^{-1}(r_0/q_0)$ and of the identity $p_0^2 + q_0^2 + r_0^2 = 1$ where p_0 is the initial value p(0) of the difference of probabilities.[4] As might be expected, these expressions agree with the corresponding ones that had been derived earlier (Kenkre and Knox 1974a) for a dimer of non-degenerate sites with fixed energy difference.

---

[4]For those who might compare the expressions to the ones in the original publication, I have corrected a typo in Eqs. (A7), (A8) of Wu and Kenkre (1989) and rewritten the expressions here to make them more compact, adding, it is hoped, no further typographical errors.

It also turns out that an intriguing formal connection arises (Wu and Kenkre 1989) between the present problem and the Toda lattice. As an extension of the dimer definition (10.12), let us define

$$\xi_m(t) = \int_0^t dz\, P_m(z). \tag{10.18}$$

Let us consider an ensemble of members, each with a specific value of the nonlinear parameter χ in such a way that the evolution of the particle on the chain is described by the ensemble average of the memory functions,

$$\mathcal{W}^{eff}_{mn}(t, s) = \int d\chi\, \rho(\chi) \mathcal{W}_{mn}(t, s, \chi).$$

For a distribution function $\rho(\chi)$ that is a Lorentzian of width α, the effective memory function $\mathcal{W}^{eff}$ will have an exponential dependence on the modulus of the argument $\int_s^t dz\, [P_m(z) - P_n(z)]$. With the definition (10.18) describing the ξ's, we then obtain the evolution equation on the chain which is a generalization of the dimer equation (10.13):

$$\frac{d^2 \xi_m}{dt^2} = \left(\frac{\pi V^2}{\alpha}\right) \left(e^{-\alpha|\xi_{m+1}-\xi_m|} - e^{-\alpha|\xi_{m+1}-\xi_m|}\right). \tag{10.19}$$

Equation (10.19) is remarkably similar to the evolution equation for the Toda lattice which is (Toda 2012)

$$\frac{d^2 \xi_m}{dt^2} = (\text{constant}) \left(e^{-\alpha(\xi_{m+1}-\xi_m)} - e^{-\alpha(\xi_{m+1}-\xi_m)}\right). \tag{10.20}$$

The only difference between Eqs. (10.19) and (10.20) is that absolute values of the differences of the ξ's appear in the former while their actual values appear in the latter.[5]

Summarizing the application of diagonalizing projections to the discrete nonlinear Schrödinger equation (10.1) that we have described here, one can appreciate the surprise that it was possible to apply this patently linear technique despite the nonlinearity of the starting equation. One can state with confidence that we understand why the application works. One can see that the memory functions are certainly nonlinear in the probabilities in that they are functions of the (time integrals) of the memories. And one can use the results, exact for the dimer and approximate for the chain, for transport studies. In the exact application of the context of the two-site system, the passage from the probability difference p to its time integral ξ happens to be a well-known transformation that changes the cubic

---

[5]I must confess I have never figured out the significance of this unexpected connection. I hope that you, the reader, will assist me in understanding this strange state of affairs.

nonlinearity in an equation such as (10.1) to a sinusoidal nonlinearity characteristic of the equation for a physical pendulum, Eq. (10.13).

10.2 Nonlinear Waves in Reaction Diffusion Systems

Let us now replace the discrete nonlinear Schrödinger equation as the nonlinear evolution to study, by the so-called Fisher equation that has found increasing use in mathematical ecology.[6]

The Fisher equation typically describes non-interacting particles whose density is $u(x, t)$ executing a simple random walk (in 1-dimension for simplicity here), while simultaneously undergoing nonlinear processes described by an additive term, which supplements the diffusion equation that is representative of the random walk (Fisher 1937):

$$\frac{\partial u}{\partial t} = D\frac{\partial^2 u}{\partial x^2} + kf(u). \tag{10.21}$$

Our interest here is in exploring the consequences of replacing the simple random walk by motion of the particles which would involve a finite mean free path before collisions, i.e., of finite coherence in the motion described by an exponential memory $\phi(t) = \alpha \exp(-\alpha t)$ assumed in the transport part of the Fisher equation.

There are two distinct ways in which coherence can be introduced into the Fisher equation (10.21), one by injecting memory into the diffusion equation, converting it to the telegrapher's equation as we have done in Chap. 1, for instance in the railway-track problem, and then adding the nonlinear term subsequently:

$$\frac{\partial^2 u}{\partial t^2} + \alpha\frac{\partial u}{\partial t} = v^2\frac{\partial^2 u}{\partial x^2} + s^2 f(u). \tag{10.22}$$

The medium speed is taken to be v and the time between collisions $1/\alpha$ so $D = v^2/\alpha$. The particles are envisaged as moving coherently until they collide, all the time undergoing the nonlinear process.

The other way with the same physical idea would be to append a memory to the diffusion term of the Fisher equation (10.21) in the form in which the *entire* equation stands, and thus write

$$\frac{\partial u(x, t)}{\partial t} = D\int_0^t dt'\, \phi(t - t')\frac{\partial^2 u(x, t')}{\partial x^2} + kf(u), \tag{10.23}$$

---

[6]Numerous results have been worked out on the time evolution of populations of entities ranging from animals roaming on the terrain to bacteria growing in Petri dishes with the help of the Fisher equation. Indeed, a book has been published recently that is based, in considerable part, on consequences of that equation in the study of the spread of epidemics (Kenkre and Giuggioli 2020).

wherein the memory $\phi(t) = \alpha \exp(-\alpha t)$ just as in the other case. In the absence of the nonlinearity, both these ways produce the telegrapher's equation. However, with the nonlinearity already established, there is a difference. This second way results in

$$\frac{\partial^2 u}{\partial t^2} + \left[\alpha - kf'(u)\right]\frac{\partial u}{\partial t} = v^2\frac{\partial^2 u}{\partial x^2} + \alpha kf(u). \tag{10.24}$$

We have taken s to measure the strength of the nonlinearity, and to be related to k in the full random walk version (10.21) of our Fisher equation in such a way that, in the limit $\alpha \to \infty, s \to \infty$, we have $s^2/\alpha \to k$. This is precisely similar to the limit $\alpha \to \infty, v \to \infty$, such that $v^2/\alpha \to D$ that corresponds to the passage of the telegrapher's equation to the standard diffusion equation.

In a publication stimulated initially by some problems in granular science, Manne et al. (2000) indicated both these ways of introducing coherence into the Fisher equation but chose to explore only the first in detail. Subsequently, Abramson et al. (2001) took up the second way and showed similarities and differences that would come about in the predictions of the other coherent generalization of the Fisher equation.

Let us briefly inspect the consequences of each in turn.

10.2.1 Consequences of the Telegrapher's Equation with an Additive Nonlinear Term

The general problem is quite complicated because of the nonlinearity. Let us therefore restrict the analysis in two ways. First, let us choose to study only the phenomenon of traveling waves. What this means is that only solutions which are combinations of x and t in the form $z = x - ct$ are sought where c is the speed of the traveling wave. This will help us convert the partial differential equation problem to one involving only an ordinary differential equation in the variable z. Second, let us soften the difficulty presented by the actual functional form of the nonlinearity in the Fisher equation, the so-called 'logistic' nonlinearity, by replacing it by a piecewise linear approximation. This will remove the nonlinear nature of the problem altogether and allow us multiple linear problems where the price we have to pay is book-keeping the different solutions in the multiple regions.

Specifically, whereas the logistic nonlinearity $f(u)$ actually present in the Fisher equation is proportional to $u(b-u)$, we will approximate it as $f(u) = u/a$ for $u \le a$ and $f(u) = (b-u)/(b-a)$ for $u \ge a$. The speed of the traveling wave appearing in the solution $u(x, t) = U(x - ct) = U(z)$ is to be distinguished from v, the natural speed dictated by the medium at which *linear* waves travel in it. Equation (10.22) results in the ordinary differential equation in z-space:

$$m\frac{d^2U(z)}{dz^2} + 2\Gamma\frac{dU(z)}{dz} + k_1^2U(z) = 0, \qquad U \le a,$$

$$m\frac{d^2U(z)}{dz^2} + 2\Gamma\frac{dU(z)}{dz} + k_2^2\,(b - U(z)) = 0, \qquad U \geq a. \tag{10.25}$$

These can be regarded as describing the dynamics of a damped harmonic oscillator of fictitious mass m that arises from the difference in the wave-front speed and the linear medium speed. The frequency of the oscillator is a consequence of the strength and shape of the nonlinearity. Specifically in Eq. (10.25), $m = v^2 - c^2$, $\Gamma = c\alpha/2$, and k_1^2 and k_2^2 are the ratios of s^2 to a and $b - a$, respectively.

Manne et al. (2000) were able to extract a number of interesting conclusions from their analysis. One of them is related to a well-known property of the Fisher equation that traveling waves obeying it cannot have a speed below a minimum value which happens to be proportional to the square root of the diffusion constant D. The result obtained by Manne et al. (2000) is that the traveling waves must have a speed that exceeds or equals a minimum value given by

$$c_{min} = \frac{v}{\sqrt{1 + a(\alpha/2s)^2}}. \tag{10.26}$$

This result is a generalization of the Fisher result to finite correlation times of the memory function (equivalently to partially coherent transport within the Fisher equation). Indeed, the old expression can be actually recovered from Eq. (10.26) under the limit $\alpha \to \infty$, $s \to \infty$ such that $s^2/\alpha \to k$, and $v \to \infty$ such that $v^2/\alpha \to D$.

But the coherence feature we introduced into the Fisher equation also gives rise to a new result. We know that the speed limit c_{min} always exists no matter what the parameter values. But if, additionally $2s > \alpha\sqrt{b - a}$, then a new quantity makes it appearance. It is c_{osc} given by

$$c_{osc} = \frac{v}{\sqrt{1 - (b - a)(\alpha/2s)^2}}. \tag{10.27}$$

Following Manne et al. (2000), we call it the *oscillation* speed. Its significance is that if it is exceeded by the wavefront speed, the wave-front shape *exhibits spatial oscillations*.

Such oscillations cannot appear in the original Fisher equation. This is so because, in the diffusive limit which the Fisher equation represents, $\alpha \to \infty$, $s \to \infty$, $s^2/\alpha \to k$, $v^2/\alpha \to D$, the quantity c_{osc} becomes infinitely large. What happens in the opposite transport limit? As $\alpha \to 0$, all three, c_{osc}, c_{min}, and v, respectively, the oscillatory speed, the minimum speed and even the medium speed, become identical to one another, and the shape always exhibits oscillations. These shape oscillations are a signature of the coherence in the Fisher equation introduced by the non-vanishing mean free path, equivalently our exponential memory (Fig. 10.2).

Many more results of an interesting nature await the interested reader in the publication of Manne et al. (2000) whose original motivation for the research stemmed surprisingly from the possibility that the combined effect of (a spatial

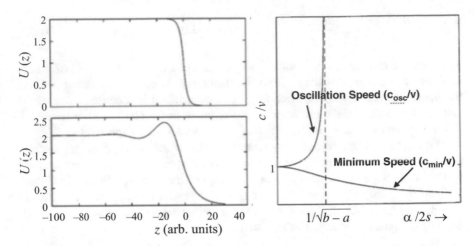

Fig. 10.2 Finite coherence consequences on transport, specifically on the shape and speed of wave-fronts, in the modified Fisher equation. Left panel shows wavefronts with different front speeds: $c = 9$ (top) and $c = 18$ (bottom) but identical (arbitrary) parameter values $v = 10$, $\alpha = 5$, $s = 3$, $a = 1$, $b = 2$. Coherence is strong in the bottom case and leads to visible oscillations in the front shape not present in the top case. Right panel shows the dependence on the damping-nonlinearity parameter $\alpha/2s$ of the characteristic limiting values c_{min}, and c_{osc} of the speed of the nonlinear waves, expressed in units of the medium speed v. While c_{min} decreases monotonically with increasing $\alpha/2s$, c_{osc} diverges as $\alpha/2s$ approaches a characteristic value beyond which the speeds is infinite. Adapted with permission from figs. 2 and 4 of Ref. Manne et al. (2000); copyright (2000) by the American Physical Society

rather than temporary) memory and nonlinearity in stress distributions might be related to arches in granular compacts (Kenkre et al. 1996, 1998a; Scott et al. 1998).

Because our interest in this chapter is, however, only in exploring what effects memory functions have when blended with nonlinear equations, let us now turn to the alternative procedure to incorporate coherence into the Fisher equation via memories that we mentioned at the beginning of this section.

10.2.2 Alternative Way of Introducing Partial Coherence into the Fisher Equation

The passage from the diffusive Fisher equation (10.21) to the partially coherent counterpart (10.22) occurred by introducing memory first and thereby introducing wave-like behavior of motion first and only then adding the nonlinearity. What if we added memory directly into the Fisher equation? This would convert Eq. (10.21) into Eq. (10.23) which leads to the second-order differential equation (10.24) which is more complicated than Eq. (10.22). The added complication comes from the fact that the damping coefficient (that multiplies the first order derivative in (10.24) is

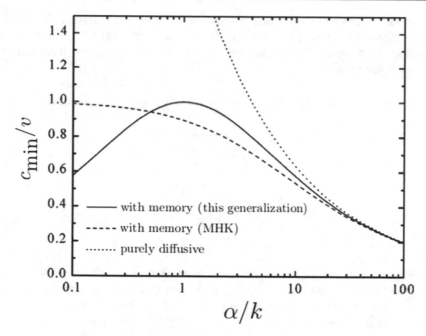

Fig. 10.3 Dependence of the minimum speed of nonlinear waves expressed as a ratio to the medium speed on the parameter α/k which is the ratio of the scattering rate to a nonlinearity parameter. Lower rates of α/k represent more coherent motion. See text. Adapted with permission from fig. 1 of Abramson et al. (2001); copyright (2001) by the American Physical Society

not constant as in (10.22) but nonlinear in u. Furthermore it can cross through zero and become negative causing *anti-damping*!

Abramson et al. (2001) carried out a careful analysis of the consequences of this second manner of introducing coherence into the Fisher equation. Relative to the results of the first way of introducing coherence, they found the loss of one feature and the gain of another. They showed that the interesting oscillating shapes in wavefronts predicted in Manne et al. (2000) disappear if Eq. (10.24) is the governing equation. Figure 10.3 makes clear the other result. As in Manne et al. (2000), it is found that traveling waves must have speeds exceeding a minimum value c_{min}. However, its dependence on the incoherence α in the transport (expressed in units of the nonlinearity k) is quite different from that predicted in Manne et al. (2000). In the latter study the dependence is monotonic, always decreasing as the motion becomes more incoherent. In Abramson et al. (2001) the dependence is non-monotonic, rising to a peak before decreasing as α/k increases. We see the contrasting behaviors in Fig. 10.3 where the dependence of the minimum speed in the pure diffusive Fisher case (no coherence) is also shown for comparison. The Fisher equation without any coherence, called "purely diffusive", is the dotted line. The dashed line labelled "MHK" refers to Eq. (10.22) after the names of the authors of Manne et al. (2000) and corresponds to that way of introducing coherence into the Fisher equation.The solid line (referred to as "this generalization" in the plot represents the present equation (10.23). The three respective expressions for v/c_{min}, the ratio of the medium speed to the minimum speed, clearly increase in complexity and are given by

$$\frac{y}{2},$$

$$\sqrt{1 + \left(\frac{y}{2}\right)^2},$$

$$\sqrt{1 + \left(\frac{y - \frac{1}{y}}{2}\right)^2},$$

where $y = \sqrt{\alpha/k}$.

What we can conclude from this brief inspection of the effects of blending a nonlinear equation and memory (finite coherence), is that nonlinear waves with lower speed than those present in the pure diffusive Fisher equation are allowed and, in addition, there is no divergence for fully coherent behavior ($\alpha/k \rightarrow 0$). The generalization in Eq. (10.24) also predicts sharply different behavior for $\alpha < k$. Indeed, profound consequences in the nature of traveling waves emerge because of the *antidamping* possibility that the damping coefficient

$$\left[\alpha - kf'(u)\right]$$

of Eq. (10.24) can go negative. We refer the reader to Abramson et al. (2001) for details.

10.3   Influence Functions in the Fisher Equation and Pattern Formation

The Fisher equation (10.21) written out explicitly by substituting the quadratic logistic nonlinearity in $f(u)$ is

$$\frac{\partial u(x,t)}{\partial t} = D\frac{\partial^2 u(x,t)}{\partial x^2} + A\,u(x,t) - B\,u^2(x,t). \tag{10.28}$$

The quadratic term represents competition as members of the population, whose density u signifies, compete with one another for resources. If, instead of considering competing members to lie at the same location, we take them to be separated by some distance as they compete, one being at x and the other at y (we are continuing to consider a 1-dimensional system), we might want to write down for their evolution,

$$\frac{\partial u(x,t)}{\partial t} = D\frac{\partial^2 u(x,t)}{\partial x^2} + A\,u(x,t) - B\,u(x,t)\int_\Omega f_\sigma(x,y)u(y,t)dy.$$

(10.29)

Perhaps surprisingly for the uninitiated investigator, the interplay of nonlinearity with spatial nonlocality represented by Eq. (10.29) results in the formation of patterns. Let us briefly touch upon this subject as a peripheral member of the topics of this book.

The generalization of the Fisher equation we have carried out here is through an *influence function* characterized by a range σ and normalized in the domain Ω under study, that describes an interaction between members at two points x and y. The physical origin of such non-local aspects in the competition interaction is perhaps easy to justify. For instance, in the case of bacteria, the diffusion of nutrients and/or the release of toxic substances can cause non-locality in the interaction. While non-local competition has been mentioned earlier (Lee et al. 2001; Mogilner and Edelstein-Keshet 1999), the formation of patterns caused by influence functions as we reported in Fuentes et al. (2003) seemed not to have been known earlier. We will briefly touch upon it below.[7] In this section we are thus interested, not in memory functions in time but in nonlocality in space, specifically the blend of such nonlocality (of influence functions) with nonlinearity (of the Fisher equation). This is, therefore, similar to our adventures in Chap. 9 where (although we did not study patterns) our interest lay in the effects of spatial nonlocality.

The Fisher equation has been studied so thoroughly that every one was/is familiar with the fact that for the extreme when the influence function is completely local, $f_\sigma(x,y) = \delta(x-y)$, nothing in the way of patterns happens. It was also easy to prove (Kenkre 2003), that no patterns form for the opposite extreme of a constant $f_\sigma(x,y)$ throughout the domain. For instance, integrating Eq. (10.29) over all space leads to the explicit time-dependent solution of the integral of the density over all space, and substituting it back into the equation one obtains, with $\bar{u}_0$ denoting the initial value of that integral, the complete solution for the space and time dependent density as

$$u(x,t) = \left[e^{-At} + \frac{B\,\bar{u}_0}{A}\left(1 - e^{-At}\right)\right]^{-1}\int \Psi(x-y,t)u_0(y)dy.$$

(10.30)

Here $\Psi(z,t)$ is the standard Gaussian propagator $(4\pi Dt)^{-1/2}\exp(-z^2/4Dt)$ of the diffusion equation. There are clearly no patterns in this Gaussian form. The reader should use the derivation of Eq. (10.30) from Eq. (10.29) as an exercise.

---

[7] Suspicious of the possibility that pattern formation might be in the air, but approaching the subject with little familiarity of the subject of pattern formation, I worked with two collaborators I assumed more learned in the area (Miguel Fuentes and Marcelo Kuperman of Bariloche) and we decided to carry out numerical explorations first and only then attempt to gain some understanding.

10.3.1 Patterns Arising from Long-Range Competition Interactions: Theory

Although our own work (Fuentes et al. 2003) was first experimental (in a numerical sense), and only later involved analytical study, let us here reverse the order and learn a bit first about what one can say about patterns in this system (Fuentes et al. 2004). A lot has been theorized generally in the field of patterns. The reader is referred to famous reviews such as Cross and Hohenberg (1993); Newell (1997) and books such as Nicolis (1995) and Rabinovich et al. (2000).

The technique to approach the formation of patterns is standard. One looks for steady state homogeneous solutions of the system under study. Those solutions represent an absence of patterns. One considers a departure from such solutions and analyzes whether such departures have an innate tendency to disappear with the passage of time. If they do tend to go away, we have ascertained the stability of the uninteresting homogeneous solutions: there are no patterns. If they show the potential to grow, there *may* be patterns that form. The idea is, thus, to investigate a necessary, not sufficient, condition for pattern formation.

To find the steady-state homogeneous solutions u_0, we put to zero the temporal and spatial derivatives in the governing equation. Thus, from Eq. (10.29), we get $u_0 = A/B$ as the non-trivial answer. If we consider a small deviation from the homogeneous steady-state solution,

$$u(x, t) = u_0 + \epsilon \cos(kx) \exp(\varphi t), \tag{10.31}$$

where the amplitude ϵ is small enough that its powers (higher than the first-order) can be neglected,[8] substitution in Eq. (10.29) yields the dispersion relation

$$\varphi = -Dk^2 - A\mathcal{F}_\sigma. \tag{10.32}$$

You should note that the difficulties of nonlinearity are removed by considering small enough departures so the neglect of powers is justified. Here, the function $\mathcal{F}_\sigma = \int_\Omega \cos(kz) f_\sigma(z)$ is the cosine Fourier transform of the influence function.

Since positive φ would make the deviations grow while negative φ would tend to make them disappear, steady-state patterns require that $\lambda = 2\pi/k$, the wavelength of the kth mode of the Fourier expansion of the pattern, satisfies

$$\lambda > 2\pi \sqrt{\frac{D}{-A\mathcal{F}_\sigma}}. \tag{10.33}$$

The inequality (10.33) means that a necessary condition for steady-state patterns to appear is that the Fourier transform of the influence function at the wavelength under consideration should be *negative* and that its magnitude should be large enough. This condition may be reworded as requiring that 2π times the "effective diffusion length" should be smaller than the wavelength for the pattern to occur. By

---

[8]We use this symbol to maintain consistency with the literature, braving any possible confusion with our usage for the Laplace variable.

the effective diffusion constant we mean D divided by $-\mathcal{F}_\sigma(\lambda)$, which is a factor decided by the influence function, and by the diffusion length we mean the distance traversed diffusively in a time interval $1/A$ on the order of the inverse of the growth rate. The reader is reminded here that A is the growth rate in the nonlinear equation.

This suggests, and makes clear why, we should expect no patterns for a smooth influence function such as a Gaussian in an infinite domain: the Fourier transform would remain positive throughout and (10.33) could not be possibly satisfied. On the other hand, we know that a cutoff in the influence function produces oscillations in the Fourier transform of the Gaussian which can make the transform go negative for certain wavelengths. This would make it possible to satisfy (10.33).

Figure 10.4, which depicts the plot of the dispersion relation for a simple example, should make this clear. The example considered is a square influence function made from step functions,

$$f_\sigma(z) = (1/2w)\Theta(w - z)\Theta(w + z). \tag{10.34}$$

Patterns appear for two of the three cases shown for certain values of wavenumber. The cutoff range is measured by w from its center. The dispersion relation in terms of the dimensionless growth rate φ' and the dimensionless wavenumber $k' = k\sqrt{D/A}$,

$$\varphi' = -\frac{\sin(k'\eta)}{k'\eta} - k', \tag{10.35}$$

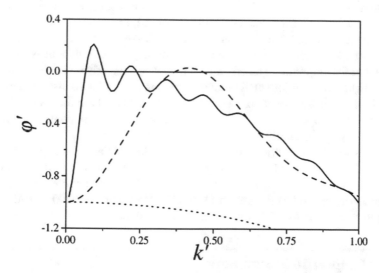

Fig. 10.4 Dispersion relation between the dimensionless growth exponent φ' and wavenumber k' for different values of $\eta = w/\sqrt{D/A}$ (see text): 50 (solid line), 10 (dashed line) and 2 (dotted line). Patterns appear in two of the three cases for those values of k' for which φ' is positive. Reprinted with permission from fig. 1 of Fuentes et al. (2004); Copyright (2004) American Chemical Society

is plotted for three values of the ratio η of the cutoff width to the diffusion length. This ratio,

$$\eta = \frac{w}{\sqrt{D/A}}, \tag{10.36}$$

is the crucial quantity that controls the formation of the pattern.

10.3.2 Confirming the Theory: Results from Numerical Experiments

Armed with the essential knowledge gathered above about what to look for, we could now look for patterns as was done through numerical experiments carried out by Fuentes et al. (2003) who studied many different influence functions $f_\sigma(x, y)$ dependent only on the spatial difference $z = x - y$.

They utilized a precise definition of the width, which they termed Σ, which could be used for any influence function:

$$\frac{1}{\Sigma} = \sqrt{-\left[\frac{d^2 \ln(f_\sigma(z))}{dz^2}\right]_{z=0}}.$$

Thus, Σ equals what is called the Gaussian width for a Gaussian influence function but is infinite for a square function. What is found (see Fig. 10.5) is in accord with the theoretical analysis above. If we use a pure square influence function (infinite Σ), patterns of nontrivial amplitude appear for all values of the cut-off interval of the square. For a Gaussian influence function (finite Σ), the patterns exhibit a curious feature in the periodic boundary condition case. Two critical values of the width Σ are seen separating trivial patterns with vanishing amplitude from nontrivial patterns with substantial amplitude. The critical width depends linearly on the domain size L (with a slight deviation for small domains) making it possible to display the results conveniently with Σ/L on the x-axis. The results displayed are for a system size L of 100 sites with $D = 1 \times 10^{-3}$, $A = 1$, $B = 1$, in arbitrary but consistent units for a system obeying Eq. (10.29).

We thus see that rich effects emerge from the interaction of spatial nonlocality of influence functions and nonlinearity. The reader interested in pursuing the area might consult some of our work in Clerc et al. (2005, 2010) that covers other types of nonlinearity as well as Fuentes et al. (2003), Fuentes et al. (2004), and Kenkre (2003) in which only the Fisher nonlinearity is considered.

10.4   Chapter 10 in Summary

Projections are typically applied to linear equations of motion such as the von Neumann or the Liouville equation for the evolution of the quantum density matrix. This chapter showed how their use can be extended, surprisingly, to systems

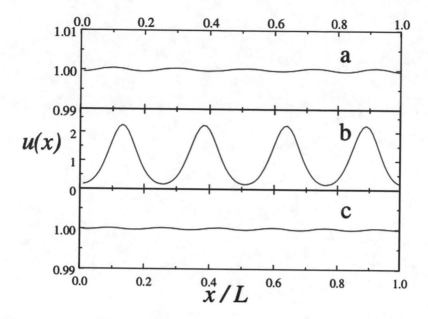

Fig. 10.5 Steady state patterns formed as a result of a blend of nonlinearity and spatial nonlocality of the influence function for a Gaussian influence function and periodic boundary conditions, and the dependence of the patterns on the cutoff-width relative to the domain size Σ/L. This ratio equals 0.205 in (**a**), 0.245 in (**b**), and 12 in (**c**). And $u(x)$ is plotted in units of a/b, and x in units of L. Reprinted with permission from fig. 1 of Ref. Fuentes et al. (2003); copyright (2003) by the American Physical Society

which obey *nonlinear* equations of motion, although it would appear such was not envisaged in the original construction by Zwanzig. The first example treated was the discrete nonlinear Schrödinger equation which has a cubic nonlinearity in the quantum mechanical amplitude, or equivalently, its consequence, the nonlinear von Neumann equation for the density matrix. A connection seemed to arise naturally but unexpectedly in this analysis to the evolution of the physical pendulum. The second example was to nonlinear waves that appear in reaction diffusion equations of a type exemplified by the Fisher equation. The third example had to do with non-local connections not in time but in space, which are better termed "influence functions" rather than memory functions. They facilitate the exploration of interesting phenomena in theoretical ecology.

The Montroll Defect Technique and Its Application to Molecular Crystals

11

11.1 Introduction: Need to Develop the Defect Technique

It appears to be a reasonable argument that, if we wish to learn about the nature and magnitude of the motion of excitons, we should place a few of them at some known location, and then pick them up at some other location whose distance from the first is known, and measure the time elapsed. This is precisely the idea used for decades in the field of sensitized luminescence. Dopant molecules such as tetracene are introduced into a host crystal such as anthracene. Typically, the two molecules absorb (and emit) in different regions of the spectra. Shining light of an appropriate frequency to excite anthracene electronically produces these excitations, or Frenkel excitons, that move about through resonance in the host crystal of anthracene, are captured by the guest molecules of tetracene, and may eventually radiate, i.e., give off light in quite a different part of the spectrum, given that the electronic structure of tetracene differs from that of anthracene.

Even though the host system (without dopant molecules) may possess translational invariance (as a crystal of anthracene does), that symmetry would be lost by the introduction of the dopant molecules which are energetically different from the molecules forming the host. This makes it necessary to develop new mathematical tools for the analysis of the transport of excitons or the transfer of energy. This is the reason we undertake our study of the Defect Technique. It was originated by Montroll and Potts (1955). It was popularized in recent times by various articles by Montroll and West (1979) and others (Montroll 1964, 1969; Rubin and Weiss 1982; Hughes 1995; Redner 2001). We explain the technique in the next section along with extensions and developments we found compelled to add to it. The emphasis in the present chapter is on describing successes and failures we encountered in its application to molecular crystals and their associated phenomena. Applications in other systems are the content of Chap. 12.

© The Author(s), under exclusive license to Springer Nature Switzerland AG 2021
V. M. (Nitant) Kenkre, *Memory Functions, Projection Operators, and the Defect Technique*, Lecture Notes in Physics 982,
https://doi.org/10.1007/978-3-030-68667-3_11

The experimental method is natural to the study of sensitized luminescence of aromatic hydrocarbon crystals and is also of importance in investigations of photosynthesis in which there is transfer of excitation from antenna molecules to the reaction center.

11.2 Overview of the Defect Technique and General Results

In a translationally invariant host, such as a crystal, given the initial probability distribution among locations n, viz. $P_n(0)$, of a particle such as a Frenkel exciton, the occupation probability at a later time t can be typically expressed as

$$P_m(t) = \sum_n \Pi_{m-n}(t) P_n(0) \tag{11.1}$$

in which $\Pi_{m-n}(t)$ is the probability propagator which expresses the probability that the particle is at m at time t, given that it is at n at time 0. It is a function of the single quantity $m - n$ rather than of m, and n separately because of the translational invariance, which also makes it relatively easy to determine the actual propagator from the structure of the governing equation.

A simple but realistic example of such motion is in a 3-dimensional cubic lattice whose sites are denoted by integers m_x, m_y, and m_z in the three Cartesian directions taking all values, negative, positive and zero, and where the motion is via nearest neighbor interactions represented by hopping rates F_x, F_y, and F_z, respectively. The propagators mentioned in Eq. (11.1) above are, in this case, products of modified Bessel functions and decaying exponentials as can be easily shown via discrete Fourier transforms:

$$\Pi_{\mathbf{m-n}}(t) = I_{m_x-n_x}(2F_x t) I_{m_y-n_y}(2F_y t) I_{m_z-n_z}(2F_z t) e^{-2(F_x+F_y+F_z)t}. \tag{11.2}$$

We might want to write a GME or a Master equation as the governing entity depending on the degree of coherence in the transport. Let us assume the motion is incoherent for now. Let us, therefore, assume a Master equation. We will mention how to treat coherence consequences of memories in a GME as we proceed. Generally, all these considerations apply for any dimensionality, m, n being vectors in the appropriate number of dimensions but, for simplicity, let us also take the system to be 1-dimensional and the transfer rates F to be nearest neighbor in character, except where noted otherwise.

11.2.1 Trapping at a Single Site

If we consider that the moving particle can disappear out of the host at a rate c if it lands on a site r, the governing equation is

$$\frac{dP_m}{dt} = F\left(P_{m+1} + P_{m-1} - 2P_m\right) - \delta_{m,r}cP_r. \tag{11.3}$$

Whereas in the absence of the trap at location r, the probability at all times would be given by Eq. (11.1), we have to determine now the effect of it trickling out at r as described by Eq. (11.3).

Before we explain the defect technique which provides the methodology for the solution of the problem presented by Eq. (11.3), let us describe what calculable quantities we are after. All information would be present in the probabilities $P_m(t)$. But there is a simpler quantity which is their sum over all the sites. This can be said to be the survival probability $Q(t) = \sum_m P_m(t)$, equivalently the (normalized) number of excitons n_H still in the host crystal.[1] We will use the latter notation, which is found in the sensitized luminescence literature in the present chapter and use the former in the next. Obviously, n_H would be 1 in the absence of trapping but less than 1 in its presence and dependent on the motion rate F, the capture rate c, the location of the trap r and the initial distribution. It is obviously given by

$$n_H(t) = 1 - c\int_0^t dt' P_r(t'). \tag{11.4}$$

Linearity of the system allows us to treat the capture term as a driving term and write, in the Laplace domain where tildes denote Laplace transforms and ϵ is the Laplace variable,

$$\widetilde{P}_m(\epsilon) = \sum_n \widetilde{\Pi}_{m-n}(\epsilon)\left[P_n(0) - \delta_{n,r}c\widetilde{P}_r(\epsilon)\right] = \widetilde{\eta}_m(\epsilon) - c\widetilde{\Pi}_{m-r}(\epsilon)\widetilde{P}_r(\epsilon), \tag{11.5}$$

where η_m is the homogeneous solution in the absence of the trap for the given initial condition, viz. the right hand side of Eq. (11.1). This is the standard Green function approach based on the knowledge of the propagators Π, but (11.5) is hardly a solution of the problem since it gives P_m in terms of the yet unknown P_r on the right hand side. To turn it into a true solution, the defect technique puts $m = r$ in Eq. (11.5), solves for P_r,

$$\widetilde{P}_r(\epsilon) = \frac{\widetilde{\eta}_r}{1 + c\widetilde{\Pi}_0}, \tag{11.6}$$

and then puts the solution back into Eq. (11.5) to give the probability at *any* site as

$$\widetilde{P}_m = \widetilde{\eta}_m - \frac{\widetilde{\Pi}_{m-r}\widetilde{\eta}_r}{(1/c) + \widetilde{\Pi}_0}. \tag{11.7}$$

---

[1]The symbols n and H refer to 'number' and 'host' (crystal), respectively.

Note the even simpler expression that (11.7) reduces to, if the trapping is *perfect*, which means if c is infinite, since then $(1/c)$ vanishes; the trap probability P_r in (11.6) in such a case is itself identically zero.[2]

What do we get if we sum Eq. (11.7) over all sites and thereby evaluate n_H in the presence of the trap? The propagators Π_{m-r} as well the homogeneous solution η_m for any initial distribution refer to the defect-less system and must therefore add up to 1 in the time domain when summed over all sites m. The sum of their Laplace transforms is, thus, $1/\epsilon$. Consequently, n_H is given in the Laplace domain as

$$\widetilde{n_H}(\epsilon) = \frac{1}{\epsilon}\left(1 - \frac{\widetilde{\eta}_r}{(1/c) + \widetilde{\Pi}_0}\right). \tag{11.8}$$

In the time domain, inspection reveals that Eq. (11.8) is equivalent to

$$\frac{dn_H(t)}{dt} = -\int_0^t ds\, \mathcal{M}(t - s)\eta_r(s), \tag{11.9}$$

showing that the exciton number in the host is decaying at a "rate" described by the convolution of a memory $\mathcal{M}(t)$ with what the probability would be at the trap site for the given initial condition of the problem but in the absence of the trap. The Laplace transform of the memory $\mathcal{M}(t)$,

$$\widetilde{\mathcal{M}}(\epsilon) = \frac{1}{(1/c) + \widetilde{\Pi}_0}, \tag{11.10}$$

has a very interesting structure which seems to suggest that one should think of the processes of motion and capture as occurring sequentially, at least for the purposes of this expression, with the overall trapping time being the sum of the times taken by each separate process. The motion time appears to be represented by the time given by the Laplace transform of the self-propagator (roughly a measure of the time a particle spends at the site of its initial occupation in the absence of the trap),

$$\int_0^\infty dt\, e^{-\epsilon t} \Pi_0(t),$$

and the capture process naturally takes the time given simply by the reciprocal of the capture rate, $1/c$. We will find this interpretation, formal as it is, to be of great significance in our subsequent analysis in this chapter and the next: you should already understand the tremendous errors that one could make while interpreting experiments thinking that the capture time is tiny in systems where it is the motion time that is actually negligible. More of that to come in the later part of the chapter.

---

[2]It is this extreme limit that is accessed by the so-called First Passage Time procedures.

This, in essence, is the defect technique in its simplest version (illustrated for one capturing site). The solution is explicit in the Laplace domain and is given in terms of known host propagators Π_m including the self-propagator Π_0 which describes the probability of excitation of an initially excited site in the trap-less host, the initial condition and consequently the homogeneous solution η, and the capture rate c. An inversion of the transform is necessary to get the results back into the time domain. The straightforward method for inversion is the evaluation of an integral on the Bromwich contour as is known from Laplace transform theory. However, practically, such inversion is typically done in one of three ways: simply looking up tables of transforms and identifying expressions, performing an asymptotic analysis, or doing a numerical inversion which is quite reliable for functions that do not oscillate widely.

11.2.2 How Laplace Inversion may be Avoided in Some Situations

However, there happens to be a situation in this area of investigation when the Laplace inversion need not even be attempted. A Frenkel exciton typically decays, spewing out a photon in the process and the experimental procedure often includes the collection of these emitted photons. Because the photons emerge at different frequencies according to whether they come from the host or the trap molecules, detecting them with frequency resolution can give information as to whether the excitons did, or did not, get transferred from the host to the guest molecules. A simple way to represent this radiative decay of the excitons is to append a term $P_m(t)/\tau$ to the left hand side of an equation such as Eq. (11.3), with τ as the lifetime of the exciton against radiative decay.[3] Let us put a prime on the P's to represent the probabilities in this situation:[4]

$$\frac{dP'_m}{dt} + \frac{P'_m(t)}{\tau} = F\left(P'_{m+1} + P'_{m-1} - 2P'_m\right) - \delta_{m,r}cP'_r. \tag{11.11}$$

We notice at once that the $P'(t)$'s are exactly related to the solutions $P(t)$ of Eq. (11.3) via

$$P'_m(t) = P_m(t)e^{-t/\tau}$$

which means that, in the Laplace domain, the connection is

---

[3]The actual situation may require that we distinguish between a host lifetime τ_H and a guest lifetime τ_G. For the moment, let us pretend that the latter is infinite and denote the first without label.

[4]All the indices m here, including r, refer to sites in the host in this simple model, the actual guest molecules being considered as being external to the system of all (say N) host sites.

$$\widetilde{P'_m}(\epsilon) = \widetilde{P_m}\left(\epsilon + \frac{1}{\tau}\right).$$

It should certainly be clear that the ratio of the total number of photons that is collected in a detector throughout all time, to the number of excitons put in into the host initially (for instance through the process of illumination) is a physical quantity that is relevant to how much of trapping at defect sites r actually occurred. This quantity is called the quantum yield (Wolf 1968b; Wolf and Port 1976) and has been the target of much experimentation.[5] Also clear should be that this quantity is the integral over all time of the term $P'_m(t)/\tau$ in Eq. (11.11) summed over all sites. It is given by

$$\sum_m \int_0^\infty dt\, \frac{P'_m(t)}{\tau} = \sum_m \int_0^\infty dt\, \frac{P_m(t)e^{-t/\tau}}{\tau} = \frac{1}{\tau}\widetilde{n}_H\left(\frac{1}{\tau}\right).$$

On substituting the expression found in Eq. (11.8), one sees that this is

$$\frac{1}{\tau}\widetilde{n}_H\left(\frac{1}{\tau}\right) = 1 - \frac{\widetilde{\eta}_r\left(\frac{1}{\tau}\right)}{(1/c) + \widetilde{\Pi}_0\left(\frac{1}{\tau}\right)}. \tag{11.12}$$

In order to find the quantum yield from Eq. (11.11) that contains the radiative decay term, it is thus sufficient to consider Eq. (11.3) that does *not* contain the decay term, calculate from it the survival probability $n_H(t)$, and simply divide its Laplace transform evaluated at $\epsilon = 1/\tau$ by τ to get the quantum yield. No inversion of the Laplace transform is necessary! We will have occasion to make extensive use of this simplification in later parts of this chapter.

11.2.3 Trapping at More than 1 Site: Exercise for the Reader

If capture can occur at two sites, at locations r and s, the starting equation is

$$\frac{dP_m}{dt} = F\left(P_{m+1} + P_{m-1} - 2P_m\right) - c\left(\delta_{m,r}P_r + \delta_{m,s}P_s\right). \tag{11.13}$$

Following the procedure sketched for a single site, we arrive at the need for the solution of two simultaneous equations, equivalently the matrix equation

---

[5]Fifty years ago some of us felt a bit annoyed that the word "quantum" had been unnecessarily appended to "yield". In today's world, however, when everything is pushed as being "quantum", from "information" to "leap", the usage fits right in.

$$\begin{pmatrix} 1 + c\widetilde{\Pi}_0 & c\widetilde{\Pi}_{r-s} \\ c\widetilde{\Pi}_{s-r} & 1 + c\widetilde{\Pi}_0 \end{pmatrix} \begin{pmatrix} \widetilde{P}_r \\ \widetilde{P}_s \end{pmatrix} = \begin{pmatrix} \widetilde{\eta}_r \\ \widetilde{\eta}_s \end{pmatrix}. \tag{11.14}$$

Evaluation of the 2×2 determinant and obtaining from it all the $\widetilde{P}_m(\epsilon)$ as for a single trap are straightforward if slightly tedious. The solution can be then used for subsequent inspection of matters such as the effect of the distance between the two trap sites. They are recommended for the reader who wishes to gain some experience with such manipulations.[6] However, the primary lesson to be learnt from this example is about the practical usefulness of the defect technique. It is always possible to use the technique in principle but the *size* of the defect, i.e., the number of sites involved which break the translational invariance of the original (defect-less) system, can limit the usefulness of the method, if applied blindly.

To appreciate the snag we have arrived at, consider an aromatic hydrocarbon crystal such as anthracene into which we have doped a small concentration of tetracene guest molecules: a standard experimental situation in sensitized luminescence (Wolf 1968b; Wolf and Port 1976). Let us assume that we are considering a chain and there is a guest molecule (trap) for every 1000 host molecules. If the host sample has 10^{23} sites arranged periodically, the determinant we would have to evaluate in the counterpart of Eq. (11.14) would be 10^{20} elements on each side! Not even the best computers available could begin to dream of tackling this problem.

How should we proceed? In the dilute system presented above, in which one encounters a guest molecule for every thousand host ones, we might consider a host crystal of 1000 sites with a single trap, put periodic boundary conditions for simplicity, and evaluate the 1-trap problem. This strategy was used by Kenkre and Wong (1981) to analyze the effect of exciton coherence on quantum yield. While quite adequate for low concentrations of dopants, the procedure can land us in trouble when pressed for higher concentrations. Better methods can be developed for this scenario as was done by Kenkre and Parris (1983).

11.2.4 Coherence Effects on Sensitized Luminescence

In order to understand the effect of coherence on the yield, one starts with a equation such as

$$\frac{d\rho_{mn}(t)}{dt} = -i \sum_s (V_{ms}\rho_{sn} - V_{ns}\rho_{sm}) - (1 - \delta_{m,n})\alpha\rho_{mn},$$

where the V's are obtained from the bandwidth of the Frenkel exciton, and α is the decoherence rate as discussed in Chap. 4, and calculates the Laplace transform of the self-propagator $\widetilde{\Pi}_0(\epsilon)$ evaluated at $\epsilon = 1/\tau$. For a tightbinding 1-dimensional

---

[6]An additional bonus this exercise will confer upon the compliant reader is preparation for another (even more important) exercise assigned in Chap. 13 where effective medium theory is discussed.

model of Frenkel excitons with nearest neighbor intersite matrix elements V, the above equation becomes the simple SLE discussed in Chap. 4 (associated sometimes with the name of Avakian). From it, Kenkre and Wong (1981) calculated for the present purpose the Laplace transform of its self-propagator and found it to be a sum of three terms. The first is

$$\frac{\alpha}{\sqrt{(\epsilon^2 + 2\epsilon\alpha)(\epsilon^2 + 2\epsilon\alpha + 16V^2)}},$$

the second is

$$\frac{(2/\pi)\mathbf{K}(k)}{\sqrt{((\epsilon + \alpha)^2 + 16V^2)}}$$

where $k = 4V\left[(\epsilon + \alpha)^2 + 16V^2\right]^{-1}$, and the third is

$$\frac{(2/\pi)\alpha^2 \mathbf{\Pi}(a^2, k)}{\sqrt{(\epsilon + \alpha)^2 + 16V^2}\sqrt{\epsilon^2 + 2\epsilon\alpha + 16V^2}},$$

where a^2 has nothing to do with any lattice constant but equals $16V^2(\epsilon^2 + 2\epsilon\alpha + 16V^2)^{-1}$. Here, $\mathbf{K}$ and $\mathbf{\Pi}$ are complete elliptic integrals of the first and third kinds, respectively.[7]

By substituting the expression for the Laplace-transform of the selfpropagator evaluated at $\epsilon = 1/\tau$ in Eq. (11.12), Wong was able to display visually the effect of coherence on the quantum yield in sensitized luminescence both for the simple model evident in the above discussion and more complicated versions such as the "substitutional" model. In the latter, the exciton does not leave the system but merely enters the guest molecule at r, trapping and detrapping rates being considered explicitly. The reader is referred to the original publication (Kenkre and Wong 1981) for all details.

---

[7]I must confess I have come across complete elliptic integrals of the first and second kind elsewhere in my researches but never one of the third kind, this being the only place, courtesy of Wong; the derivation here is his work. We had lengthy discussions during his dissertation work whether expressing integrals that have to be otherwise evaluated numerically in terms of known special functions was progress or only lateral activity. Our tentative conclusion was that it was progress if the special functions were highly studied and familiar ones such as sines and cosines. For instance, it is significant that, for the trigonometric functions mentioned, we already know without doing numerical work that they are bounded by ± 1, and oscillate equally on both sides of 0. The reader should enquire of herself whether it is important/useful or not to identify a complicated integral as a complete elliptic integral of the third kind.

11.3   High Defect Concentration: The ν-Function Approach

Solubility restrictions make high trap concentrations impossible to achieve in many aromatic hydrocarbon crystals (e.g., tetracene-doped anthracene in which the relative trap concentration cannot exceed 10^{-4}). The subterfuge used by Kenkre and Wong (1981) works well to calculate sensitized luminescence observables in such systems. However, systems also exist in which such restrictions are not present and relative trap concentrations can be made to approach unity. It is then necessary to develop serious modifications in the theory.

In the presence of many traps, the last term in Eq. (11.3) must be replaced by $-\sum_{r}' \delta_{m,r} c P_m$ where the primed summation is over all the sites from which the excitations leak out into the trap molecules. Evaluation of any quantity requires now the evaluation of a determinant of enormous size decided by the number of trap-influenced host sites as we have seen in Sect. 11.2.3. Equation (11.5) is now replaced by

$$\widetilde{P}_m(\epsilon) = \widetilde{\eta}_m(\epsilon) - c \sum_{s}' \widetilde{\Pi}_{m-s}(\epsilon) \widetilde{P}_s(\epsilon) \qquad (11.15)$$

where the label s denotes a trap-influenced site. It is not possible to proceed from here in the same manner as that used for one or a few traps, unless one undertakes the impossible task of evaluating the determinant of a size that could be gigantic.

However, one may use a non-standard approximation as follows (Kenkre 1982). Define a new function

$$\nu_s(t) = \sum_{r}' \Pi_{r-s}(t) \qquad (11.16)$$

so that putting $m = r$ in Eq. (11.15), where r is a trap-influenced site, and summing over r, yield the exact result

$$\sum_{r}' \widetilde{P}_r(\epsilon) = \sum_{r}' \widetilde{\eta}_r(\epsilon) - c \sum_{s,r}' \widetilde{\nu}_s(\epsilon) \widetilde{P}_s(\epsilon).$$

Now make the approximation that, in that result, ν_s *is considered in an average or ensemble-averaged sense* so that its dependence on s is dropped. If this can be done, we at once have

$$\sum_{r}' \widetilde{P}_r(\epsilon) = \frac{\sum_{r}' \widetilde{\eta}_r}{1 + c\widetilde{\nu}(\epsilon)} \qquad (11.17)$$

as the respective many-trap generalization of Eq. (11.6) and

$$\widetilde{n_H}(\epsilon) = \frac{1}{\epsilon}\left(1 - \frac{\sum'_r \widetilde{\eta}_r}{(1/c) + \widetilde{v}(\epsilon)}\right) \tag{11.18}$$

as the many-tap generalization of Eq. (11.8). Equation (11.17) describes the probability of the exciton in the entire trap-influenced region of the host, and Eq. (11.18) the probability still left in the host as earlier, both in the Laplace domain.

Whether the introduction of the v-function helps at all is obviously dependent on whether calculating it "in an average sense" is possible in concrete situations. We now show that it is, in a number of relevant cases.

11.3.1 Calculation of the v-Function in Specific Cases

In an infinite lattice, which we will take 1-dimensional for simplicity, and with nearest neighbor incoherent rates F as usual, the propagators are $\Pi_m(t) = e^{-2Ft} I_m(2Ft)$. From their Laplace-transforms, we can state that, if the traps are themselves placed periodically, v_s automatically sheds its s-dependence and becomes

$$\widetilde{v}(\epsilon) = \sum_r' \widetilde{\Pi}_{r-s}(\epsilon) = \frac{1}{\sqrt{\epsilon^2 + 4\epsilon F}} \sum_m \left[\frac{\epsilon + 2F - \sqrt{\epsilon^2 + 4\epsilon F}}{2F}\right]^{|m|/\rho}. \tag{11.19}$$

Indeed, as the m-summation is over all integers, positive and negative, and zero as well, one has the closed-form result (Kenkre 1982)

$$\widetilde{v}(\epsilon) = \frac{1}{\epsilon}\frac{\tanh(\xi/2)}{\tanh(\xi/2\rho)}, \tag{11.20}$$

where ρ is the ratio of the number of trap-influenced sites to all the sites, and ξ is defined through[8]

$$\cosh\xi = 1 + \frac{\epsilon}{2F}. \tag{11.21}$$

It is easy to see from Abelian theorems (taking the limits of ϵ as it tends to zero and infinity respectively) that $v(t)$ starts initially at 1 and tends to the ratio ρ. This is obvious from the meaning of the function and displayed clearly in the plot of Fig. 11.1 which also shows how the function $v(t)$ differs from the self-propagator $\Pi_0(t)$ used in its place earlier. It was natural for me to suggest from this exact limiting result a simplified approximate expression (Kenkre 1982) in the general

---

[8]I learned the idea of defining this inverse hyperbolic function of a combination of the Laplace variable and the transfer rate from the work of Katja Lindenberg and her collaborators in their analysis of motion on finite linear segments. See Lakatos-Lindenberg et al. (1972).

Fig. 11.1 Time-dependence of the $\nu(t)$-function compared with that of the self-propagator $\Pi_0(t)$ showing that the former tends to the concentration ρ (here taken arbitrarily to be 0.2) but the latter vanishes at long times. Adapted with permission from fig. 1 of reference Kenkre and Parris (1983); copyright (1983) by the American Physical Society

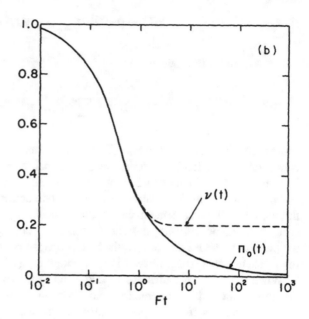

case when the traps are placed randomly rather than periodically: $\nu(t) \approx \rho + (1 - \rho)e^{-\Gamma t}$ where Γ describes in some phenomenological manner the motion of the exciton. The "correct" expression, however, is $\nu(t) = \rho + (1 - \rho)\Pi_0(t)$ which is exact when the trap placement is random.[9]

Parris also showed Kenkre and Parris (1983) by effectively coiling the infinite chain as many times as required that, if the trap-influenced host sites form a superlattice (being placed every $1/\rho$ sites in the 1-dimensional case), the ν-function in the infinite chain exactly equals the self-propagator Π_0 in a smaller ring of $1/\rho$ sites obeying periodic boundary conditions.

Several of these results can be easily derived as corollaries of a general result as follows. Let us express $\nu_r(t) = \sum'_s \Pi_{s-r}(t)$ as

$$\nu_r(t) = \sum_s \sum_m \Pi_{m-r}(t)\delta_{m,s} = \sum_s \sum_m \Pi_m(t)\delta_{m,s-r},$$

noting that the m-summation is unrestricted but the primed s-summation is over trap-influenced sites only. Then, the r-independent ν that is obtained from ν_r through an ensemble average, may be immediately written as

---

[9]This result was due to Paul Parris who proved it to me along with several other results when I showed him the ν-function idea. I remember Paul as one of the sharpest persons I have ever met who had a remarkable background in art before he began working as my Ph. D. student. I thought of him as a *Mentat*, a reference that will be understandable to those of us who are fans of Isaac Asimov's Foundation series.

$$v(t) = \langle v_r(t) \rangle = \sum_m \Pi_m(t) p_m. \tag{11.22}$$

The new quantity introduced here,

$$p_m = \left\langle \sum_s {}' \delta_{m,s-r} \right\rangle, \tag{11.23}$$

is the probability that the mth host site is trap-influenced, given that the 0th host site is trap-influenced. In other words, it is the trap pair correlation function!

Equations (11.22) and (11.23), obtained in Kenkre and Parris (1983), are useful and powerful results. The nature of trap-trap interactions, if present, might decide the nature of their placement within the host during the doping process, i.e., to specific forms of their pair correlation function. The nature of exciton motion in the pure host decides the form of the host propagators. The two together determine the v-function in a straightforward way through Eq. (11.22).

When the trap molecules are sprinkled into the host crystal, they would sit at random positions if no interactions exist among those traps, and form a periodic lattice themselves if the opposite extreme applies, i.e., if very strong repulsive attractions exist among them. Let us treat these in turn.[10]

No Interactions Among Traps, Random Placement

In the absence of trap-trap interactions, the trap-influenced host sites are located completely randomly. The probability of finding a trap-influenced host site at m given that there is one at site 0, which we have denoted by p_m, is clearly a constant for $m \neq 0$. The constant equals $(N' - 1)/(N - 1)$ and cannot be distinguished from the concentration $\rho \equiv N'/N$ in the limit that N and N' are both large which we will assume here. Then the trap correlation function is given by

$$p_m = \delta_{m,0} + \rho(1 - \delta_{m,0}). \tag{11.24}$$

When substituted in the extreme right hand side of Eq. (11.22), it yields the v-function for this random placement:

$$v(t) = \Pi_0(t) + [1 - \Pi_0(t)]\rho = \rho + (1 - \rho)\Pi_0(t). \tag{11.25}$$

The behavior of $v(t)$ at the initial time, as well as at infinite time, is clearly as seen in Fig. 11.1.

---

[10]It must not be forgotten here that, in the simple model under discussion, we do not consider additional trap molecules sitting (perhaps substitutionally) in the host crystal. When we refer to the location of a trap molecule, it actually means that we refer to the location of a trap-influenced host molecule.

Strong Repulsive Interactions Among Traps, Periodic Placement
The opposite limit of extreme interactions (of a repulsive nature) would cause the trap-influenced host sites to be as far as possible from one another and form a lattice by themselves. In a 1-dimensional system such as the one we are considering here (for simplicity), we would then have a superlattice of the trap-influenced host sites, with p_m equaling 1 whenever m equals nN/N', equivalently n/ρ for large N and N':

$$p_m = \sum_n \delta_{m,\,(n/\rho)}. \tag{11.26}$$

Substitution in Eq. (11.22) as before, yields now the ν-function for this periodic placement as

$$\nu(t) = \sum_n \Pi_{(n/\rho)}. \tag{11.27}$$

In the summations in Eqs. (11.26) and (11.27), n takes values 0, 1, 2, ..., $N' - 1$.

The Parris result mentioned earlier about the relationship of ν and the selfpropagator for a smaller (periodic) lattice of $1/\rho$ sites, which one could symbolically express as

$$\nu = \Pi_0^{(1/\rho)},$$

the superscript (obviously, it is not a power) referring to the size of the smaller lattice, can be recovered by the application of discrete Fourier transforms. Because ⊛ it would help one to understand how to treat lattices within lattices, it is suggested as an exercise for the reader.[11]

A large variety of additional interesting issues ranging from time dependence of sensitized luminescence to the exploration of the Kenkre-Wong coherence considerations in the high concentration regime were addressed comprehensively in the article (Kenkre and Parris 1983); I especially recommend it to the reader.[12]

11.4   Ising Model Analysis of Arbitrary Magnitude and Sign of Trap-Trap Interactions

We have seen the consequences of the respective cases of random and periodic trap placement arising from extremes of interaction among them. In two publications

---

[11]The detailed solution is available in Section IV of (Kenkre and Parris 1983).

[12]I recall a kind colleague, Gert Zumofen, approaching me during a conference after that paper was published and saying to me in mock frustration, "There is nothing you and your student have now left for the rest of us to do." While that statement bespeaks of Gert's generosity more than anything else, there is surely a lot that Paul Parris and I succeeded in packing into that publication.

(Kenkre et al. 1984; Parris et al. 1984), we addressed the consequences of arbitrary magnitude and sign of the interaction among trap-influenced sites through an Ising model (equivalently, lattice gas) treatment. The analysis is in two parts, one in which it can be shown that the pair correlation function between trap-influenced sites is given by

$$p_m = \delta_{m,0} + (1 - \delta_{m,0})[\rho + (1 - \rho)x^{|m|}], \qquad (11.28)$$

and the other in which the consequences of substituting this expression in Eq. (11.22) are explored.

The quantity x in Eq. (11.28) is defined through

$$x = \frac{\sqrt{1 - 4\rho(1 - \rho)(1 - E)} - 1}{\sqrt{1 - 4\rho(1 - \rho)(1 - E)} + 1}.$$

The quantity E equals $e^{\Delta/k_B T}$ as we will see below, where T, the temperature during the process of doping, is certainly always positive but Δ can be of either sign. Consequently, E is always positive but can be 1, smaller than 1, or greater than 1. The three cases correspond, respectively, to vanishing, repulsive, and attractive interactions. As Fig. 11.2 shows, the pair correlation function has distinct behaviors in the three cases, the repulsive case being the most interesting as it leads to oscillations.

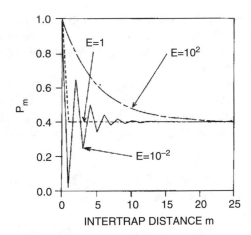

Fig. 11.2 Pair correlation function of the trap-influenced host sites from Eq. (11.28) plotted as a function of distance in number of sites for three cases of interaction represented by the dimensionless parameter E taking the values 10^2, 10^{-2} and 1 as shown. They represent, respectively, attractive, repulsive, and vanishing interaction. The last of these is denoted by a dashed line. The value of the concentration ρ has been set arbitrarily to 0.4. Reprinted with permission from fig. 1 of reference Kenkre et al. (1984); copyright (1984) by Elsevier Publishing

11.4.1 Calculation of p_m from Ising Model/Lattice Gas Considerations

The *modus operandi* of calculating the pair correlation function p_m consists of representing the molecular crystal as a lattice gas, and exploiting the well-known analogy between the latter and the Ising model. Let us refer to the host sites in the crystal as "occupied" (in the language of the lattice gas) if they are trap-influenced and "unoccupied" if they are not. Let us characterize the interaction between any two trap-influenced host sites by an energy that equals infinity if the two coincide, $-\Delta$ if they are nearest neighbors, and 0 in all other cases. The corresponding Ising model has the Hamiltonian (see any textbook on statistical mechanics, e.g., Pathria (1972)),

$$-J \sum_{i,j}' \sigma_i \sigma_j - B \sum_i \sigma_i$$

in standard notation. The spins σ can take on values ± 1. Through the analogy we are using, the respective correspondence is to the host site being or not being trap-influenced.

If in the Ising model, $M = < \sigma_m >$ denotes the magnetization per spin, we can set up a bridge from the spin system to the doped crystal as shown in Table 11.1. As shown in detail in Kenkre et al. (1984), one can deduce from the table that p_m is given by

$$p_m = 1 - \frac{1- < \sigma_0 \sigma_m >}{4\rho}. \tag{11.29}$$

On the other hand, it is known from Ising model analysis that

$$< \sigma_0 \sigma_m > = \frac{\sinh^2(B/k_B T) + e^{-4J/k_B T} x^{|m|}}{\sinh^2(B/k_B T) + e^{-4J/k_B T}},$$

where $\beta = 1/k_B T$ and x is given as

$$x = \frac{M - \tanh(B/k_B T)}{M + \tanh(B/k_B T)}.$$

Table 11.1 Correspondence between the Ising model and the doped crystal

Ising model	Doped crystal
Spin up	Host site trap-influenced
Spin down	Host site not trap-influenced
4J	Δ
Magnetization M	2ρ-1

The elimination of the magnetization from these expressions is done with the help of the additional Ising model result

$$M = \frac{\sinh(B/k_BT)}{\sqrt{\sinh^2(B/k_BT) + e^{-4J/k_BT}}}.$$

This elimination finally yields

$$p_m = 1 - (1 - \rho)(1 - x^{|m|})$$

which we rewrite in the form of Eq. (11.28).

A plot of p_m versus the distance between a given site and one m sites distant from it is given in Fig. 11.2. Interesting oscillations arise from repulsive interactions. For vanishing interactions, one recovers the earlier result in (11.24).

11.4.2 Exploration of $\nu(t)$ and Sensitized Luminescence Predictions

There are a large number of interesting consequences that emerge from the analysis on the form of the $\nu(t)$ function, including exact analytical expressions that have transparent consequences on luminescence predictions. In the interests of space limitation in the book, I display here only one exact expression for ν and one plot of relevance to sensitized luminescence, and encourage the reader to consult both the original publication (Kenkre et al. 1984) and an exact independent calculation (Parris et al. 1984) that succeeded it.

The exact expression is for ν in the Laplace domain. It is

$$\epsilon\tilde{\nu}(\epsilon) = \rho + (1 - \rho)\left[\epsilon\tilde{\Pi}_0(\epsilon)\right]\left[\frac{1 + x\tilde{h}(\epsilon)}{1 - x\tilde{h}(\epsilon)}\right]. \tag{11.30}$$

where

$$\tilde{h}(\epsilon) = (1/2F)\left[\epsilon + 2F - \sqrt{\epsilon(\epsilon + 4F)}\right] = e^{-\xi}, \tag{11.31}$$

and one may remind oneself that $\tilde{\Pi}_0(\epsilon)$ exactly equals $\tanh(\xi/2)$. The meaning of ξ is as (11.21) in the Lakatos-Lindenberg propagator expression. It equals $\cosh^{-1}[1 + (\epsilon/2F)]$.

The plot that might whet the reader's appetite is a combination of two graphs from Kenkre et al. (1984), see Fig. 11.3. Both graphs show a dependence on the relative trap concentration $\rho = N'/N$. The left panel shows the key quantity $\epsilon\tilde{\nu}(\epsilon)$ evaluated at $\epsilon = 1/\tau$. The right panel is the quantum yield of the guest ϕ_G. The three lines in each panel correspond to three respective values of the cooperative interaction between traps that lead to their placement within the host. The middle line in each panel represents no interaction ($E = 1$). The top line in the left panel and the bottom line in the right panel correspond to attractive interactions ($E = 100$) while the bottom line in the left panel and the top line in the right panel correspond to repulsive interactions ($E = 1/100$). Attractive interactions lead

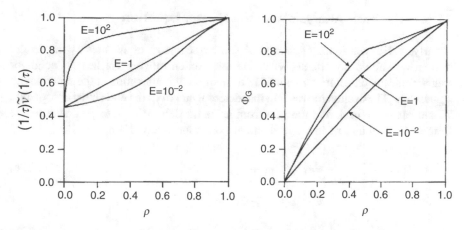

Fig. 11.3 Effect of cooperative trap interactions shown through the dependence of the key quantity $(1/\tau)\widetilde{v}(1/\tau)$ in the left panel and the guest quantum yield ϕ_G in the right panel on the concentration ρ for three cases of the interaction ($E = 10^2$, 1 and 10^{-2} as shown explicitly. The middle curve in each case represents absence of interactions and the other two curves are flipped in each panel relative to the other. Adapted with permission from figs. 4 and 5 of reference Kenkre et al. (1984); copyright (1984) by Elsevier Publishing

to cluster formations of trap-influenced host sites and to less efficient transfer of excitation to the guest in the assumed instance of uniform initial illumination.

11.5 End-Detectors in Simpson Geometry

During an assignment he had to complete, Seshadri[13] discovered there was a discrepancy of several orders of magnitude in the reported values of the singlet diffusion constant of Frenkel excitons in anthracene at room temperature, all from experiments performed in what was called Simpson geometry. I also saw the discrepancy commented on in an excellent review article by Powell and Soos (1975) that was being widely read at that time. Obviously, this was a serious enough issue that one had to understand and resolve and so I undertook an analysis of the experiments involved. The observations had been reported by Simpson (1956), Gallus and Wolf (1966) and Takahashi and Tomura (1971). The quantum yield (or exciton flux) was measured by creating excitons at one end of the host crystal through illumination and detected at the other end where they were captured by a detector layer of another material. The distance traversed by the excitons in these experiments is, thus, a given and measurable quantity rather than random as in the bulk sensitized luminescence experiments we have been analyzing so far.

---

[13]He was my first ever Ph.D. student and he went on to do some fine work on the Montroll-Shuler equation for vibrational relaxation, a subject quite removed from the one under discussion here.

11.5.1 The Basic Analysis with the Master Equation

Clearly, a 1-dimensional *finite* crystal (linear segment) of N sites is the model system we should investigate. Visualize the exciton moving via nearest neighbor transfer rates F on a line segment of N host sites communicating at the far end with a guest site. The exciton moves into that detector site from the end (Nth) site via rate $\mathcal{F}$ and can come back into the host from it via the detrapping rate f. The radiative lifetimes are τ_H and τ_G in the host and guest , respectively. Then,

$$\frac{dP_m}{dt} + \frac{P_m}{\tau_H} = F(P_{m+1} + P_{m-1} - 2P_m), \tag{11.32a}$$

$$N - 1 \geq m \geq 2,$$

$$\frac{dP_1}{dt} + \frac{P_1}{\tau_H} = F(P_2 - P_1), \tag{11.32b}$$

$$\frac{dP_N}{dt} + \frac{P_N}{\tau_H} = F(P_{N-1} - P_N) + f P_G - \mathcal{F} P_N, \tag{11.32c}$$

$$\frac{dP_G}{dt} + \frac{P_G}{\tau_G} = \mathcal{F} P_N - f P_G. \tag{11.32d}$$

The first equation above describes what is going on in the bulk of the host, the second and third equations what happens at the left and right ends respectively, and the fourth is a statement of what happens in the detector. The host system is not translationally invariant but the propagators , that now have two labels instead of a single difference one, are known from the work in Lakatos-Lindenberg et al. (1972). While the details of the procedure followed to solve the equations are best learned from the original publication (Kenkre and Wong 1980), it should be sufficient to state that the primary quantity to calculate is the guest quantum yield given by

$$\phi_G = \int_0^\infty dt \, \frac{P_G(t)}{\tau_G} e^{-t/\tau_H},$$

and the results are

$$\phi_G = \phi_G^s \left[\frac{(\mathcal{F}/F)}{(\mathcal{F}/F) + c} \right], \tag{11.33a}$$

$$\phi_G^s = \mathrm{sech}\left[\xi' \left(N - \frac{1}{2} \right) \right] \mathcal{I}(\kappa), \tag{11.33b}$$

$$c = \left[\frac{1 + f\tau_G}{\sqrt{F\tau_H}} \right] \sinh(N\xi') \mathrm{sech}\left[\xi' \left(N - \frac{1}{2} \right) \right], \tag{11.33c}$$

$$
\mathcal{I}(\kappa) = \frac{1}{2}\left[\frac{1 - e^{-\kappa}}{1 - e^{-\kappa N}}\right]\left[\frac{e^{\xi'/2}(1 - e^{-N(\kappa - \xi')})}{1 - e^{-(\kappa - \xi')}} + \frac{e^{-\xi'/2}(1 - e^{-N(\kappa + \xi')})}{1 - e^{-(\kappa + \xi')}}\right].
$$

$$(11.33d)$$

The quantity ξ' is defined as ξ, given in Eq. (11.21), but evaluated at $\epsilon = 1/\tau_H$.

These results provide a possible explanation of the large discrepancies in the reported values of the exciton diffusion by correcting several features of the conclusions drawn previously in the literature. The first equation here, (11.33a), clearly shows that the crucial quantity measured, the guest yield ϕ_G, is not simply ϕ_G^s as was thought and used in former interpretations of the observations but depends crucially on the value of the rate $\mathcal{F}$ relative to the host transfer rate F. Different detectors were used in different measurements. Furthermore, the sensitivity of this dependence is decided by c. Allowing for the detrapping rate f as we did (unlike all those who analyzed the observations earlier) shows that the c can be large and thus the actual yield ϕ_G can be quite different from (and smaller than) the saturation value ϕ_G^s assumed by all interpretations earlier. The latter is only the limit of the former for infinite $\mathcal{F}/F$ and is given in the second equation, (11.33b). The explicit expression for the sensitivity factor c is in the third equation, (11.33c). The initial condition is decided by the absorption coefficient κ through the function $\mathcal{I}(\kappa)$ given in (11.33d) and appears multiplicatively in the second equation, i.e., (11.33b).

To make the argument visually clear, let us inspect Fig. 11.4 as provided in Kenkre and Wong (1980), in which the quantum yield is plotted as a function of the ratio $\mathcal{F}/F$ for three values of the diffusion constant (equivalently of F given that one can call Fa^2 the diffusion constant D.) The three values are in the ratio 1:5:10 as we go upwards in the figure. The saturation value of the yield, ϕ_G^s, which has been taken as the yield itself in previous interpretations, is shown for each of the three cases. Because $\phi_G^s > \phi_G$, there are errors in previous interpretations.

One can use Fig. 11.4 to determine F and from it the diffusion constant in the host, if the host-to-detector rate $\mathcal{F}$ is known. In addition to obtaining the latter independently from the Förster spectral prescription, we suggested two methods of extracting F without knowing $\mathcal{F}$. One was to evaluate F from the dependence of the initial distribution of the excitons on the penetration length of the excitons (Kenkre 1981c). The other was to make time-dependent measurements retaining the Simpson geometry (Parris and Kenkre 1982). We comment on these in turn.

11.5.2 Suggested Experiment to Determine D from Variation of the Penetration Length

A careful inspection of Eq. (11.33) shows that the effect on ϕ_G of the initial variation of the exciton density or probability, which can be brought about by varying κ,

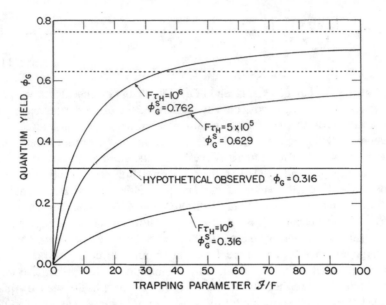

Fig. 11.4 Quantum yield ϕ_G plotted as a function of the ratio of the host-to-detector rate $\mathcal{F}$ to the host transfer rate F for three values of $F\tau_H$ as shown. See text. Each curve corresponds, thus, to a value of the diffusion constant. Dotted lines represent the limiting value of the yield for perfect absorbers. The solid straight line shows a hypothetical observed yield of 0.316. The value of $f\tau_G$ is taken to be 10^4. Adapted with permission from fig. 1 of reference Kenkre and Wong (1980); copyright (1980) by the American Physical Society

$$P_m(0) = e^{-\kappa(m-1)} \left[\sum_{m=1}^{N} e^{-\kappa(m-1)} \right]^{-1},$$

appears *solely* in $\mathcal{I}(\kappa)$ as given Eq. (11.33d). This term, while dependent on F, is independent of detector quantities $\mathcal{F}$ and f. This prompted me to propose, in Kenkre (1981c), observation of the variation of the quantum yield with changes in the wavelength of excitation expressed as changes in the penetration length $l_p = a/\kappa$. The latter is the ratio of the lattice constant and the absorption coefficient. The proposal can be understood from Fig. 11.5. The ratio of the yield to its value for uniform illumination in which the entire crystal is excited homogeneously (by letting $\kappa \to 0$) is plotted as a function of the penetration length l_p for reasonable values relevant to the experiment. The three curves shown correspond to three assumed values of the diffusion constant of the exciton in the host crystal, equivalently three values of F. The values vary over two orders of magnitude. The penetration length dependence is weak (strong) if the diffusion constant is large (small) which is clear from the expressions but also from straightforward considerations as well: a fast exciton bent on reaching the detector end cares little whether it is initially close to the detector or far!

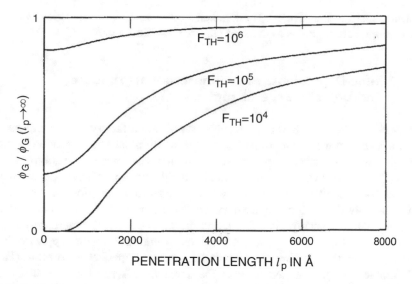

Fig. 11.5 The detector quantum yield is plotted as a function of the penetration length for three values of the exciton diffusion constant as shown. The yield is normalized to its value for uniform illumination. The sample length is $5000A$ and the lattice distance a is $5A$. Reprinted with permission from fig. 2 of reference Kenkre (1981c); copyright (1981) by Elsevier Publishing

Not shown in Fig. 11.5 is a graphical demonstration of the sensitivity of the yield to variation of l_P for slowly diffusing excitons and lack of sensitivity for rapidly diffusing ones. Such a graphical display was provided in the original publication by plotting the ratio of the yields for very large and vanishing κ respectively. This ratio is easily seen to be given by

$$\lim_{\kappa \to \infty} \phi_G / \lim_{\kappa \to 0} \phi_G = \frac{N}{\sinh(N\xi')} \frac{\cosh(\xi'/2)}{\sqrt{F\tau_H}}.$$

With a sample length taken to be $0.5\,\mu$m and the lattice constant a to be $5A$, for a diffusion constant D of $2.5\,\text{cm}^2$/s which here would correspond to $F\tau_H = 10^7$, the values of the yield for extreme l_p values bear a ratio of 0.98. However, for $D = 2.5 \times 10^{-2}\,\text{cm}^2$/s, which corresponds to $F\tau_H = 10^5$, the ratio is 0.27. And for $D = 2.5 \times 10^{-3}\,\text{cm}^2$/s, equivalently $F\tau_H = 10^4$, the ratio is less than 10^{-3}. To put it differently, observation of the yield as a function of the penetration length can show a variation over a factor of 1000 for $D \approx 2 \times 10^{-3}\,\text{cm}^2$/s, but only a factor of 1 for $D \approx 2\,\text{cm}^2$/s. Little observed variation is thus indicative of a high value of D and a great deal of variation represents slow exciton transport.[14]

---

[14]We are, needless to say, making these estimates in the approximate range of reported values of D and other parameters to emphasize experimental relevance.

That such experiments can be undertaken from a practical point of view was clear from reported observations as those of Haarer and Castro (1976).[15]

11.5.3 Time-Dependence of Luminescence in Proposed End-Detector Experiments

Detectors that we encounter in molecular crystals are a far cry from being *perfect* absorbers and yet the assumption of perfect absorption has been erroneously made through decades of interpretation of sensitized luminescence observations. This is a message that began to form in my mind through years of analyzing experiments and one that I am attempting to convey to the reader in the previous sections while simultaneously describing proposed experiments that could beat the situation in principle. Another such proposal is based on the hope that the multiple items of information that *time dependence* of the detector emission might supply could be used in lieu of the unknown information regarding imperfect absorption (Parris and Kenkre 1982). The geometry is the same as described in the discussion above and the equations of motion are also the same, i.e., Eq. (11.32) set out in Kenkre and Wong (1981). What we now calculate is the guest intensity $n_G(t)$ by obtaining its Laplace transform through the previous analysis and then inverting it numerically via the Stehfest algorithm procedure (Stehfest 1970). Let us make two simplifications. Let us take $\tau_H = \tau_G = \tau$, and assume that the initial illumination is homogeneous; in other words, the entire host crystal, from site 1 to site N, is initially excited.

The detector intensity is given in the Laplace domain by

$$\widetilde{n_G}(\epsilon) = \frac{(\mathcal{F}/N\epsilon')}{\epsilon'' + f + \epsilon''\mathcal{F}\widetilde{\Pi}_{NN}(\epsilon')} \tag{11.34}$$

where $\epsilon' = \epsilon + (1/\tau_H)$ and $\epsilon'' = \epsilon + (1/\tau_G)$ and we have used the Lakatos-Lindenberg propagator expression

$$\widetilde{\Pi}_{NN}(\epsilon) = \left(\frac{\cosh(\xi/2)}{F}\right)\left(\frac{\cosh[\xi(N - (1/2))]}{\sinh(\xi)\sinh(N\xi)}\right).$$

As earlier, the definition of ξ is $\cosh\xi = 1 + \epsilon/2F$. By virtue of our simplification assumption, we take $\tau_H = \tau_G$ and therefore $\epsilon' = \epsilon''$ below.

The result of the Laplace inversion is in the two accompanying plots. In Fig. 11.6, curve c corresponds to perfect absorption and is measurably different from the other two curves a and b which represent finite values of $\mathcal{F}/F$, 5 and 12 respectively. The

---

[15]Unfortunately, I have not heard of experiments undertaken on the basis of the idea proposed. A theoretical extension (Gülen 1988) appears, however, to have been stimulated by the proposal.

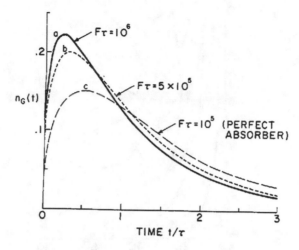

Fig. 11.6 Time dependence of the detector luminescence intensity plotted in units of the radiative lifetime τ. The three curves are for three respective values of the diffusive constant $D = Fa^2$, (1, 0.5, and 0.1 cm^2/s) equivalently of three values of the intersite host transfer rate F which in units of $1/\tau$ are as shown. The lattice constant is $5A$. Values of the trapping rate into the detector are different for the three curves but adjusted so that the quantum yield for each of the three is 0.316. Reprinted with permission from fig.1 of reference Parris and Kenkre (1982); copyright (1982) by Elsevier Publishing

selected value 0.316 of the yield kept constant in order to construct the curves is one borrowed from one of the reported experiments and is otherwise arbitrary.

Our recipe for extracting the value of the diffusion constant without an a priori knowledge of $\mathcal{F}$ is explained fully in the caption of Fig. 11.7. As an example of its implementation, note that, if the observed yield were (arbitrarily) 0.5, and the observed peak time in units of τ were 0.4, one would focus on the point of intersection of the two relevant curves. That point is marked in Fig. 11.7 by an arrow. The abscissa of that point would give us the additional information we need: $\mathcal{F}/F$ equals about 33. That value would then be used in an expression such as in Eq. (11.33a) shown earlier to obtain the host diffusion constant.

11.6 A Surprise from a Combined Study of Observations of Capture and Annihilation

Doping a host with guest molecules and measuring the energy transfer rate k from the host to the guest featuring in an assumed equation such as

$$\frac{dn_H(t)}{dt} = -\frac{n_H(t)}{\tau_H} - kn_H(t) \tag{11.35}$$

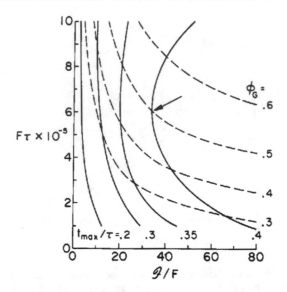

Fig. 11.7 How the host transfer rate F can be extracted without knowledge of the trapping rate into the detector. Each of the four solid curves corresponds to a value of t_{max}/τ, the peak time in units of the radiative lifetime as indicated, and each of the four dotted lines to a value of the quantum yield ϕ_G. From an experimental curve such as in Fig. 11.6, one notes the values of these two quantities. On a plot such as the present one, one identifies the corresponding solid line and dashed line and notes the point of intersection. The abscissa at that intersection yields the ratio $\mathcal{F}/F$. One then uses this in the earlier analysis to get F. Reprinted with permission from fig. 2 of reference Parris and Kenkre (1982); copyright (1982) by Elsevier Publishing

for the host exciton density $n_H(t)$, is not the only experimental probe used.[16] The mutual annihilation of excitons in pure (undoped) crystals of the host is also used for the purpose. The underlying equation is assumed to be

$$\frac{dn_H(t)}{dt} = -\frac{n_H(t)}{\tau_H} - \gamma n_H^2(t),\tag{11.36}$$

and k and γ extracted routinely from the observations of $n_H(t)$, either from its time dependence or from steady state quantities such as the quantum yield.

---

[16]Taking advantage of a program that encouraged international collaborations between experimentalists and theorists, I persuaded Dankward Schmid to join me in undertaking a sustained investigation into sensitized luminescence in molecular crystals. He was at the University of Stuttgart when we started our collaboration but moved to take up a Chair at the University of Düsseldorf where he was during most of our collaboration. I learned from him the nuances of the assumptions made in the interpretation of experiments in the field of sensitized luminescence. He was an expert in the subject and came from the impeccable Stuttgart tradition in this area, having learned the subject from Hans C. Wolf, known for decades as an authority in the field.

What is remarkable is that both k and γ are believed to be proportional to the exciton diffusion constant D via a prescription vaguely thought to come from the well-known and often-quoted Smoluchowski or Chandrasekhar recipe for coagulation (Chandrasekhar 1943):

$$\gamma = v\gamma' = 8\pi R_d D, \tag{11.37a}$$

$$k = \rho v k' = 4\pi R_c D. \tag{11.37b}$$

Here v is the volume per molecule of the host crystal, ρ is the relative guest concentration, and R_d and R_c the radii of influence over which the elementary processes of mutual destruction and capture occur, respectively. The quantities $1/\gamma'$ and $1/k'$ are measured in s.

Equation (11.37) are used extensively (Powell and Soos 1975) to extract diffusion constants for excitons in molecular crystals from experiment. Assumptions considered reasonable are made on the radii of influence R_d and R_c. This procedure of interpreting data has led in some cases to alarming values of D for reasonable values of the radii of influence, and vice versa. It has led some to abandon the expressions in favor of time-dependent (Powell and Soos 1975) energy transfer rates $k(t)$. Yet, observations in aromatic hydrocarbon crystals seldom show such time dependence.

There is no escape from a paradox that faces us when we use prescriptions such as in (11.37). Perusal of experiments as in Benderskii et al. (1978), Broude et al. (1978), Auweter et al. (1979), Braun et al. (1982) leads to the unequivocal conclusion that in naphthalene and anthracene, the annihilation constant γ is essentially T-independent in the range $4\,\mathrm{K} < T < 300\,\mathrm{K}$ while the energy transfer rate k has a rather *pronounced* dependence on T. See Fig. 11.8.

Unless one follows the unsatisfactory procedure of attaching to R_d and R_c dependences on T invented artificially to produce the observed result, there seems to be no way of getting out of this conundrum. The observed value of γ remains 4×10^{-11} cm^3/s in naphthalene and 1×10^{-8} cm^3/s in anthracene throughout the temperature range indicated. The observed value of k in the same range undergoes a large change. Thus, k', in the same T-range, decreases from 3×10^{12} s^{-1} at 4 K to 2×10^{11} s^{-1} at 300 K in naphthalene and from 1.7×10^{13} s^{-1} at 4 K to 3×10^{12} s^{-1} at 300 K in anthracene. These are changes of an order of magnitude.

How do we resolve this paradox? Easily, it turns out, provided we pay heed to expressions such as Eq. (11.10) that have naturally emerged from our defect technique analysis of sensitized luminescence (Kenkre and Wong 1980, 1981; Kenkre and Parris 1983) and of annihilation as well (Kenkre 1980; Kenkre and Reineker 1982). Simply put, what our derivations have resulted in, is

$$\gamma' = \left(\frac{1}{d} + \frac{1}{M} \right)^{-1}, \tag{11.38a}$$

$$k' = \left(\frac{1}{c} + \frac{1}{M} \right)^{-1}. \tag{11.38b}$$

Fig. 11.8 Annihilation and energy transfer times, respectively $(\gamma')^{-1}$ and $(k')^{-1}$, expressed in seconds, plotted logarithmically against the temperature for naphthalene and anthracene. Clearly, it is impossible that each is proportional to the diffusion constant D via T-independent constants as is normally assumed from the Chandrasekhar recipe. Reprinted with permission from fig.1 of reference Kenkre and Schmid (1983); copyright (1983) by Elsevier Publishing

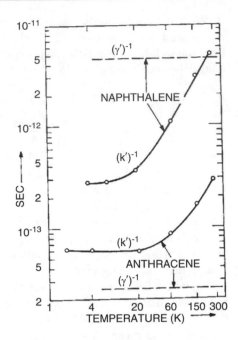

If one assumes perfect absorption and perfect mutual destruction in the two kinds of experiment, c and d are infinite, and it is impossible for k' to be strongly temperature-dependent in the same range for the same crystal while γ' remains constant. Yet, consider a hypothetical situation in which d is small enough relative to M so that $\gamma' \approx d$ but c large enough relative to M so that $k' \approx M$. In such an extreme case γ' and k' need have little to do with each other given that the first is determined totally by the process of mutual destruction while the second is controlled completely by exciton motion through the host! In this hypothetical case, M could be strongly *T-dependent* and d quite T-independent.[17]

There is only a paradox here if one makes the unjustified assumption of perfect rates for the reaction part of the reaction-diffusion process. By "reaction" event is meant the event that is not motion (absorption in the case of capture, mutual destruction in the case of annihilation).

The detailed analysis along the above lines was provided in a paper that we wrote (Kenkre and Schmid 1983) along with a procedure to extract bounds, not actual values, for the exciton diffusion constants. We hope the reader will examine that analysis and the additional information discussed in that publication. Its primary message should be inescapable in addition to being quite clear. In general, we cannot make the perfect reaction rate assumption. The existence of the additional detail of the process of reaction accounts for the large discrepancy in reported diffusion

---

[17]The choice of the subscripts d and c in this and earlier discussions refers to 'destruction' and 'capture', respectively.

constants in the literature. It also accounts for the present fiasco, transparent in Fig. 11.8, of the T-dependence of the energy transfer rate while at the same time having to contend with the T-independence of the annihilation constant.

11.7 Is Sensitized Luminescence an Appropriate Probe of Exciton Motion in Molecular Crystals?

After spending about a decade examining existing experiments and their interpretations, as well as attempting to construct, when possible, proposed observations on the basis of theoretical calculations, we expressed, in two publications, our conclusions about the question posed in the title of this section. One was a preliminary paper (Kenkre and Parris 1984) and another a comprehensive one (Kenkre et al. 1985a).[18] Let us inspect the conclusions in summary form.

As a procedure to measure the amount and nature of the motion of Frenkel excitons, sensitized luminescence as well as its associated observation of mutual annihilation, are based on a reaction-diffusion phenomenon. There are two parts to such a phenomenon, one in which the excitons move from one location to another in the host, and one in which the reaction occurs. The second part is the transfer to the detector in the case of sensitized luminescence. In the annihilation scenario, because the excitons themselves serve as detectors of other excitons, it is the final process of mutual destruction. Given the usual (almost universal) assumption that the reaction process is short-range in space,[19] the two processes are visualized as happening sequentially. If the reaction is long-range, modifications in our visualization and a revamping of most theoretical interpretations will be necessary.

With this conception of the procedure in mind, it should be straightforward to understand that the measurement process will pick out the slower of the two sequential processes. If excitons move very fast, they will arrive at the detector locations immediately after the experiment is begun; what the measurement will reflect in that case is the slow process: capture or destruction. If excitons move very slowly with respect to the detection process, it will be the motion characteristics that will determine the overall features of the experiment. Quantitatively, the situation is described by equations such as (11.38). The totally unjustified assumption, that the reaction process is always the faster one of the two, indeed *infinitely fast*, has been used essentially in all interpretations of experiment in this field.

It is interesting to ruminate upon the source of this habitual assumption and the reason for what appears to be a huge inertia to change the habit. I believe there are two sources. For most, it seems to have come from the great success of the

---

[18]The work was a collaboration that was initially begun with Yiu-man Wong, and then continued with Paul Parris and Dankward Schmid.

[19]The possibility certainly exists that in some systems long-range reaction processes (e.g., for annihilation) may be at work. This was brought to our attention by H. C. Wolf. In such a case the arguments presented need to be modified.

coagulation theory of Smoluchowski as represented, for example in Chandrasekhar (1943). The perfect absorber assumption might well work for that process but there is no reason it should work in every phenomenon imaginable. As we have seen, it certainly does not, in pure molecular crystals.[20]

I happen to believe there is another source for the practice. Assume that a theorist is studying motion effects in an arbitrarily general context, not directed at a specific experimental system. It would be natural for her to minimize the consequences of an extraneous process such as reaction so the focus can be placed on the motion itself. The perfect absorber assumption achieves that focus because the faster (reaction) process drops out of consideration. This could well be the reason for the concentration of investigators on the so-called "first passage time" procedures. As soon as the moving particle reaches a location, it is taken to be removed from consideration. That allows the investigator to analyze the motion process without distraction.

A theorist who is interested in the analytic study of the process of motion by itself, has this luxury. One whose interest lies in determining from nature the movement characteristic of a given particle, does not. The latter kind of investigator cannot take an a priori stand irrespective of the actual state of affairs.

As a remedy for the malady present in the field of molecular crystals, we suggested various new experimental schemes as explained in this chapter. They included dependence on the penetration length of excitons into the host crystal, watching for the time dependence of the detector emission, even independent calculations of the reaction (e.g., capture) parameters so that observations could be interpreted without making the perfect absorption ansatz. The improper practice has led to an orders-of-magnitude disparity in quantities such as the exciton diffusion constant in the same material at the same temperature on the one hand, and the absurd situation related to Fig. 11.8 on the other.

The detailed exposition in Kenkre et al. (1985a) has a rather complete discussion of the various studies we attempted over the years to address the problem of extraction of motion parameters of the excitons. While it is best to refer to that publication for those details, it might be useful to show here from that work two items. The first is the result of an attempt by Parris and Kenkre (1984) to address measurements (Dlott et al. 1977) of the time-dependence of the X-trap phosphorescence in 1,2,4,5-tetrachlorobenzene (TCB), and the second is a table of information that can be legitimately gathered from sensitized luminescence observations in pure molecular crystals.

The plot in Fig. 11.9 shows the first. Results of a best-fitting procedure based on two *widely differing* theories are seen to be equally compatible with the data (on triplet excitons at the very low temperature 1.25 K) as stated in the caption. The dashed line is based on completely incoherent exciton motion with infinite capture rate. The solid line is based on largely coherent motion with a small enough capture

---

[20]In mixed crystals such as those studied by Kopelman and his collaborators and reported in Agranovich and Hochstrasser (1983), exciton motion is slowed down by irregularities in the underlying lattice. It is quite possible that in those systems the measurements are motion-controlled and that our concerns do not apply in those systems.

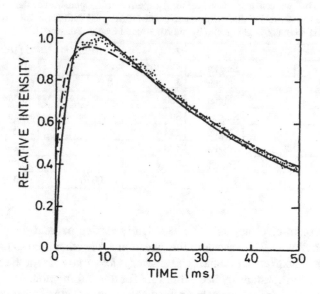

Fig. 11.9 Time dependence of the normalized X-trap phosphorescence in undoped TCB at 1.25 K along with lines fitting the data equally well on the basis of two quite different theories. The solid curve is based on a capture-limited theory, the capture rate deduced being $c = 6 \times 10^3$ s^{-1}. The dashed curve is based on a motion-limited theory, which means that c is infinite, the transfer rate deduced being $F = 3.3 \times 10^{10}$ s^{-1}. This latter curve is based on the idea that exciton motion is incoherent. By contrast, the former curve is based on the motion being quite coherent. What is additionally striking is that it presupposes that exciton motion characteristics drop out of the experiment because the situation is capture-controlled. Reprinted with permission from fig.1 of reference Parris and Kenkre (1984); copyright (1984) by Elsevier Publishing

rate so that the experiment seems capture-limited! Surely, if you are a proponent of incoherent motion, as Dlott et al. (1977) were in the publication in which they reported their data, you can deduce the triplet exciton diffusion constant from the observations. You merely multiply the deduced F (see caption of Fig. 11.9) by the square of the lattice constant to obtain D. If you are a believer that the triplet motion in this experiment was coherent, as Van Strien et al. (1982) would be on the basis of their spin echo experiments, you can claim compatibility with a value of the scattering rate from the experiment. But the experiment cannot serve to distinguish between these totally different kinds of motion because information necessary to establish such distinction is absent. Whereas in Chap. 6 we saw various theories that claimed to explain observations drop away one by one leading to the conclusion that none was applicable, here we find that both theories work and therefore, without further input, we can accept neither.

If the sensitized luminescence experiments in pure molecular crystals have such problems in determining definitive values for exciton diffusion constant and/or scattering rates, is there anything at all that we can deduce from them? The

Table 11.2 What we can learn from existing experiments in pure molecular crystals: lower bounds for the transfer rate F and diffusion constant D. Adapted with permission from reference Kenkre et al. (1985a); copyright (1985) by the American Physical Society

Host	Guest	Type	T (K)	F (10^{12} s^{-1})	D (10^{-2} cm^2/s)
Anthracene[a]	T	Singlet	5 − 300	>56	>15
Naphthalene[b]	A	Singlet	4	>3.9	>1
Naphthalene[b]	A	Singlet	300	>0.21	>0.054
p-Terphenyl[c]	T	Singlet	80	>1.5	>0.36
p-Terphenyl[c]	T	Singlet	250	>9	>2.2
Tetracene[d]	P	Singlet	170	>33	>8.2
Naphthalene[e]	β	Triplet	6	>0.016	>0.004
Naphthalene[e]	β	Triplet	16	>0.0076	>0.002

experiments themselves appear to be excellently carried out and there is a wealth of data in the reports. The answer to this question is the other item to take from the discussion in Kenkre et al. (1985a). The answer is *lower bounds* for the transfer rate F or the diffusion constant D. See Table 11.2 for this information.

The superscripts in the host column show the sources (published experiments) as follows: (a) is Braun et al. (1982), (b) is Auweter et al. (1979), (c) is Jones et al. (1981), (d) is Campillo et al. (1977), and (e) is Gentry and Kopelman (1983). In the guest column, T stands for tetracene, A for anthracene, P for pentacene, and β for β-methyl-naphthalene.

11.8 Chapter 11 in Summary

Experiments to probe the nature of a system may be either of the gentle kind examined in Chap. 5 or such that they modify the given system substantially. An example of the latter situation is when guest molecules such as those of tetracene are placed within an aromatic hydrocarbon crystal like anthracene to ascertain the nature of the motion of excitations in the latter. Such doping can alter the original system profoundly and require new methods of solution of the equations governing motion. The defect technique was constructed for such purposes in this chapter along lines first set out by Montroll and his collaborators. It was explained in the context of a crystal. An extension to study effects of the coherence of the motion when the detectors are placed at random positions within the crystal was explained next. It was then generalized to address high concentrations of defects via the so-called ν-function formalism. A theory was constructed to interpret measurements of the exciton diffusion constant in observations with detectors placed at the end of a crystal in what is sometimes referred to as Simpson geometry. It offered an explanation of the large discrepancy in reported values of the diffusion constant. Several experimental schemes were developed theoretically to be used along with proposed experiment. One such involved varying the penetration length of the excitons in the sample and another dealt with the time dependence of detector emission. The temperature dependence and magnitude of the diffusion constant, when extracted from annihilation experiments on the one hand and sensitized luminescence on the other, were found to be in stark conflict when interpreted in the usual way. The paradox was resolved with the theory developed in this chapter. The question of whether it is the motion or the reaction process that limits the reaction-diffusion phenomenon in a given experiment was investigated carefully and conclusions were presented.

The Defect Technique in the Continuum

12

It would be wrong to misinterpret the discussion in the last chapter to mean that the defect technique is generally unsuccessful as a tool in investigations. It is powerful and should be mastered and utilized. The discussion at the end of Chap. 11 simply showed how in pure molecular crystals (which happens to be one of the systems of interest to us), when used without the additional information (such as the rates of capture or similar reaction processes), it can lead to inaccurate interpretation. Even there, one can retrieve information concerning bounds of motion parameters rather than the parameters themselves.

Additional analysis based on the defect technique is given in the present chapter, with focus on equations of motion set out in the continuum rather than on discrete lattices. The discussion is in four parts. First, we will obtain the continuum counterparts of the Simpson geometry results obtained in the last chapter to show how such a procedure works. Next, we will see how to formulate the defect formalism in continuous space from the beginning, bypassing the discrete lattice description, and will present a catalog[1] of some useful results (Spendier and Kenkre 2013). Then we will discuss trapping in the presence of motion directed towards an attractive center, by using the Smoluchowski equation rather than the diffusion equation (Spendier et al. 2013). This theory will serve us in generalized form in Chap. 15 in the context of the transmission of infection in epidemics (Kenkre and Sugaya 2014; Sugaya and Kenkre 2018). Finally, I will mention, briefly, an application of the defect lattice in reverse in that it is aimed at understanding aggregation rather than depletion of a source of particles.

---

[1] Several of these results were derived by Kathrin Spendier in collaboration with me as part of her Ph.D. dissertation work (Spendier 2012). Others were compiled by her from sources in the literature and cited appropriately as noted. She was an exceptional student equally at ease with theoretical research and experimental work.

© The Author(s), under exclusive license to Springer Nature Switzerland AG 2021
V. M. (Nitant) Kenkre, *Memory Functions, Projection Operators, and the Defect Technique*, Lecture Notes in Physics 982,
https://doi.org/10.1007/978-3-030-68667-3_12

12.1   Continuum Limit of the Simpson Geometry Analysis

Some investigators make defect technique calculations on discrete lattices whenever possible, and then take the vanishing limit of the lattice constant a to pass to the continuum. Let us inspect a representative example to understand how it is done by recovering from our discrete lattice analysis the continuum result that used to be utilized earlier in interpretations of experiments such as in Simpson (1956).

In Chap. 11, we obtained Eqs. (11.33) for the detector quantum yield ϕ_G, working on a discrete lattice representing the host crystal. Let us begin by noting that, from the meaning of ξ' given immediately below Eqs. (11.33), the multiplying factor $1/\sqrt{F\tau_H}$ that appears in Eq. (11.33c) for c, exactly equals $2\sinh(\xi'/2)$, and that, in the limit of vanishing ξ', it would be identical to ξ' itself.[2]

The continuum limit consists of letting the lattice constant a vanish. However, we also let F and $\mathcal{F}$, as well as the number of sites N, tend at the same time to infinity, such that

$$\lim_{a\to 0} Fa^2 = D, \tag{12.1a}$$

$$\lim_{a\to 0} \mathcal{F}a = \mathcal{D}, \tag{12.1b}$$

$$\lim_{a\to 0} Na = L, \tag{12.1c}$$

where D is the diffusion constant and L the length of the sample. We also have $k = \lim_{a\to 0} \kappa/a$, and introduce the exciton probability density via $P(x,t) = \lim_{a\to 0} P_m(t)/a$, so that we can write, for its initial value,

$$P(x,0) = \frac{ke^{-kx}}{1 - e^{-kL}}.$$

Now, straightforward manipulations reduce (11.33), in the continuum limit, to

$$\phi_G = \frac{\lim_{\mathcal{D}\to\infty} \phi_G}{1 + \left(\frac{1+f\tau_G}{\mathcal{D}\tau_H}\right) l_D \tanh\left(\frac{L}{l_D}\right)}. \tag{12.2}$$

---

[2]For the rare reader who decides to compare expressions to those in the original publication, here is a note so confusion can be avoided. In this book, ξ' refers to what one gets by evaluating ξ at $\epsilon = 1/\tau_H$. This is opposite to the usage in the original publication and has been done here for convenience.

The ratio of the sample length L to the diffusion length $l_D = \sqrt{D\tau_H}$ makes its natural appearance here.[3] The numerator in the expression for the yield in Eq. (12.2) is the limit of the yield as $\mathcal{D} \to \infty$:

$$\lim_{\mathcal{D}\to\infty} \phi_G = \frac{k^2 l_D^2}{k^2 l_D^2 - 1} \left[\operatorname{sech}\left(\frac{L}{l_D}\right) - e^{-kL}\left(1 + \frac{\tanh(L/l_D)}{kl_D}\right) \right] \frac{1}{1 - e^{-kL}}.$$

(12.3)

Equation (12.3) in essence is the expression used in early analyses of the Simpson geometry experiments (Simpson 1956; Gallus and Wolf 1966; Takahashi and Tomura 1971) obtained on the basis of the diffusion equation (continuum approximation) and the implicit assumption that $\mathcal{D} \to \infty$.

12.2 Formalism in the Continuum and Some Useful Results

It is also straightforward to develop the defect technique directly from the diffusion equation if we happen to be in the continuum limit from the very beginning. If there is a single defect located at the point x_r, the starting equation is

$$\frac{\partial P(x, t)}{\partial t} = D\frac{\partial^2 P(x, t)}{\partial x^2} - C_1\delta(x - x_r)P(x, t).$$

(12.4)

Let us use a capital C here to remind ourselves that we are in the continuum and the subscript 1 to emphasize the number of dimensions. Generally, we will have C_d in d dimensions and replace the second spatial derivative in the right hand side by the Laplacian $\nabla^2 P$ in the appropriate number of dimensions. Note that C_1 has the dimensions of a velocity, C_2 those of a diffusion constant, and generally C_d those of a hyperarea (an obvious generalization of an area from the usual 2 to d dimensions) divided by time.

A moment's reflection shows that all of the results obtained for discrete lattices apply in the continuum with only the judicious replacement of some sums by integrals where necessary. The 1-dimensional propagator is now

$$\Pi\left(x, x', t\right) = \frac{1}{\sqrt{4\pi Dt}} \exp\left[-\frac{\left(x - x'\right)^2}{4Dt}\right]$$

(12.5)

which leads to the self-propagator

$$\Pi\left(0, 0, t\right) = \frac{1}{\sqrt{4\pi Dt}}.$$

(12.6)

---

[3]The diffusion length derives its name from the fact that, ignoring factors of order of unity, it is the distance the exciton would cover in the host in its radiative lifetime, while moving with the diffusion constant D.

The survival probability $Q(t) = \int_{-\infty}^{+\infty} dx \, P(x, t)$ of a particle placed initially at a point a distance x_0 away from a perfect absorber at the origin is known widely as being given by an error function, having been calculated by many in various contexts (Jaeger and Carslaw 1959; Spouge 1988; Redner and Ben-Avraham 1990; Redner 2001; Blythe and Bray 2003). It is given by

$$Q(t) = \operatorname{erf}\left(\frac{1}{2}\sqrt{\frac{\tau_1}{t}}\right),$$ (12.7)

where $\tau_1 = x_0^2/D$ is the diffusion time, i.e., the time taken by the diffusing particle in the trapless system to arrive from its initial location to the trap.

The counterpart of this result for arbitrary capture (imperfect absorption) is not known as well. Indeed, misleading statements that such results do not exist have been sometimes made in the literature. Therefore, when Kathrin and I derived this result in our own work, we had intended to publish it. However, we found that a number of investigators equal to those who had given the perfect absorber result had also derived and published it (Jaeger and Carslaw 1959; Rodriguez et al. 1993; Ben-Naim et al. 1993; Redner 2001). But certainly the general result deserves to be known, at least so it can dispel some incorrect reports, and is displayed here:

$$Q(t) = \operatorname{erf}\left(\frac{1}{2}\sqrt{\frac{\tau_1}{t}}\right) + e^{\frac{1}{\xi_1} + \frac{1}{\xi_1^2}\left(\frac{t}{\tau_1}\right)} \operatorname{erfc}\left(\frac{1}{2}\sqrt{\frac{\tau_1}{t}} + \frac{1}{\xi_1}\sqrt{\frac{t}{\tau_1}}\right).$$ (12.8)

The right hand side of Eq. (12.8) is perhaps surprising in that it appears as a *sum* of the perfect absorber result and a term that is proportional to a complementary error function. The argument of the latter is a sum of two terms, one of which is proportional to the reciprocal of the square root of t and the other proportional to the square root of t itself. The imperfectness of the absorption is reflected in the appearance of a non-zero $\xi_1 = 2D/(C_1 x_0)$. The reduction to the perfect absorber result is immediate on taking the limit $C_1 \to \infty$.

(✱)   It is not difficult to derive Eq. (12.8). The reader is invited to do the exercise by starting with (or deriving) two results from Laplace transform theory: a general one regarding scaling that, for any constant B, the transform of $f(Bt)$ is obtained by replacing ϵ by ϵ/B in the transform of $f(t)$ and then dividing the result by B; and a specific result, that, given a constant A, the transform of $(1/\sqrt{t}) \exp\left(-A^2/4t\right)$ equals $(\sqrt{\pi/\epsilon}) \exp\left(-A\sqrt{\epsilon}\right)$. Applying these to Eqs. (12.5) and (12.6) leads to

$$\widetilde{Q}(\epsilon) = \frac{1}{\epsilon}\left[1 - \frac{\exp\left(-\sqrt{\epsilon \tau_1}\right)}{1 + \xi_1 \sqrt{\epsilon \tau_1}}\right].$$ (12.9)

Laplace inversion of this expression is facilitated by the use of the scaling property of transforms again and leads to the general result (12.8). It is suggested that the reader should include within the exercise showing graphically that the expression does have the correct limiting behaviors at $t = 0$ and as $t \to \infty$, and that the additional term in the imperfect absorber expression simply increases to a maximum and drops to zero as t increases without limit.

12.3 Additional Results in 1-Dimensional Systems

Since the problem is linear, the superposition principle can be applied to obtain a solution for any distribution the particle may initially occupy.

12.3.1 Arbitrary Initial Conditions and Perfect Absorption

In the special case of perfect absorption, the principle of superposition states that

$$Q(t) = \int_0^\infty \rho(x_0) \, \text{erf}\left(\frac{x_0}{\sqrt{4Dt}}\right) dx_0, \tag{12.10}$$

where $\rho(x_0)$ is the initial distribution of point particles. We are considering here only the half-space from 0 to ∞.

Among results published in the literature for specific distributions (and perfect absorption) are the case of an initial random distribution for which $\rho(x_0)$ is of Poisson form $\rho(x_0) = c \exp(-cx_0)$, where c is an arbitrary constant,

$$Q(t) = \exp(c^2 Dt)\text{erfc}\left(c\sqrt{Dt}\right), \tag{12.11}$$

calculated by Torney and McConnell (1983) as well as by Sancho et al. (2007), and the case of the Rayleigh distribution, also known as a biased Gaussian and expressed as $\rho(x_0) = x_0 \exp\left(-x_0^2/(2\sigma^2)\right)/\sigma^2$, where σ describes the width of the distribution,

$$Q(t) = \frac{\sigma}{\sqrt{2Dt + \sigma^2}}. \tag{12.12}$$

The latter result has been reported by Doering and Ben-Avraham (1988). The selection of these particular cases comes from the fact that the first distribution that we have shown as an example drops away from its maximum at the origin while the second rises from a vanishing value at the origin to a peak and then falls away at large distances.

How about arbitrary capture rates (imperfect absorption) and arbitrary initial distributions? Below are some useful manipulations it might be interesting to share. Although the principle of superposition is obviously not restricted to perfect absorption, multiplying the general expression given in Eq. (12.8) by an initial distribution and integrating the product over x_0 lead to further manipulations that are algebraically tedious. To simplify, one can apply the superposition principle in the Laplace domain and attempt to derive a prescription that does not require the computation of Laplace inversions.

12.3.2 Arbitrary Initial Conditions and Imperfect Absorption

We start with $\widetilde{Q}(\epsilon)$ given in Eq. (12.9) showing the explicit dependence on the initial location x_0:

$$\widetilde{Q}(\epsilon) = \frac{1}{\epsilon} \left[1 - \frac{e^{-\sqrt{\epsilon/D}x_0}}{\frac{\sqrt{4\epsilon D}}{C_1} + 1} \right].$$

On averaging this result over the initial distribution $\rho(x_0)$, i.e., by performing an integral such as in (12.10), we see that the Laplace transform of dQ/dt equals

$$\left(\frac{p}{p + \sqrt{\epsilon/D}} \right) \int_0^\infty \rho(x_0) e^{-x_0\sqrt{\epsilon/D}} dx_0$$

where $p = C_1/2D$. This expression can be interpreted as the product of two Laplace transforms each with $\sqrt{\epsilon/D}$ as the Laplace variable. The first is the transform of an exponential, the second of the distribution ρ. If you now recall that, if $f(t)$ is the Laplace inverse of $\widetilde{f}(\epsilon)$, the inverse of $\widetilde{f}(\sqrt{\epsilon})$ is

$$\frac{1}{2\sqrt{\pi}} t^{-3/2} \int_0^\infty u e^{-\frac{u^2}{4t}} f(u) du,$$

and combine that result with the scaling rule of Laplace transforms, you arrive at our final result (Spendier and Kenkre 2013) for the rate of disappearance of the survival probability:

$$\frac{dQ(t)}{dt} = -\frac{C_1}{4\sqrt{\pi}} (Dt)^{-3/2} \int_0^\infty x e^{-\frac{x^2}{4Dt}} H(x) dx, \qquad (12.13)$$

where

$$H(x) = \int_0^x \rho(x_0) e^{-p(x-x_0)} dx_0 \qquad (12.14)$$

and $p = C_1/(2D)$.

Equation (12.13) is a prescription that allows us to calculate the particle survival probability for any initial distribution of non-interacting particles which diffuse in the presence of a single stationary trap. The two examples given at the beginning of this subsection can be recovered as particular cases.

12.4   Results for Symmetrical Systems in Higher Dimensions

Numerous real-world applications in systems whose dimensions are higher than 1, involve situations characterized by high symmetry wherein only the radial coordinate enters into consideration. For such, analytic results can be developed starting with the general result for the Laplace transform of the survival probability

$$\tilde{Q}(\epsilon) = \frac{1}{\epsilon}\left[1 - \frac{\sum_r' \tilde{\eta}(x_r, \epsilon)}{(1/C_d) + \sum_r' \tilde{\Pi}(x_r, x_r, \epsilon)}\right], \tag{12.15}$$

that follows from a generalization of Eq. (12.4) to include multiple traps and arbitrary dimensions. Here, surely, the summations may be replaced appropriately by integrations whenever necessary. Let us examine a few such results in the following for 2-dimensional and 3-dimensional systems of high symmetry in the continua.

12.4.1 Results for 2-Dimensional Systems

Let us consider a radially symmetric situation in which the trapping region is the circumference of a circle (disk) of radius R and the origin as its center. In cartesian coordinates, the 2-dimensional propagators are simply products of Gaussian 1-dimensional propagators (for isotropic diffusion as we will assume here) and, therefore, independent of the angular polar coordinate θ. In calculating sums or integrals of propagators in an expression such as (12.15), we can take one of the angles to be 0. The cartesian coordinates of the two points are $R\cos\theta$, $R\sin\theta$ and R, 0 respectively. We are, therefore, led to the evaluation of integrals such as

$$\int_0^{2\pi} \frac{e^{-\frac{R^2[(1-\cos\theta)^2+\sin^2\theta]}{4Dt}}}{4\pi Dt}\, d\theta,$$

which can be performed easily in terms of the I_0 Bessel function of argument $R^2/2Dt$. In 1-dimension, there would be no integral and the denominator would have a square root of t.

Inspection of (12.15) shows that we must calculate two quantities both of which are obtainable by evaluating

$$\int_0^{2\pi}\int_0^{\infty} G\left(\vec{r}, \vec{r}', t\right) P\left(\vec{r}', 0\right) r'\, dr'\, d\theta',$$

given the Green function

$$G\left(\vec{r},\vec{r}',t\right) = \frac{1}{4\pi Dt}e^{-\frac{|\vec{r}-\vec{r}'|^2}{4Dt}},\tag{12.16}$$

where $|\vec{r}-\vec{r}'|^2 = (x-x')^2 + (y-y')^2$, and the rotationally symmetric initial condition $P(\vec{r},0) = \delta(r-R_0)/(2\pi r)$. Carrying out the integration over r', and realizing that the left integral is the I_0 Bessel function, we get the important result

$$P\left(R,r,t\right) = \frac{1}{4\pi Dt}e^{-\frac{r^2+R^2}{4Dt}}I_0\left(\frac{rR}{2Dt}\right).\tag{12.17}$$

This is the probability density for the particle to occupy the circumference of radius R given that it initially occupied that of radius r. The result may also be seen in Jaeger and Carslaw (1959).

Circular Trap of Finite Radius

We now analyze an initial circular symmetric distribution of diffusing non-interacting point particles initially at R_0. The circular trap is centered at the origin and has a radius R, where $R < R_0$ (Fig. 12.1).

In our result derived above, we put $r = R$ to get

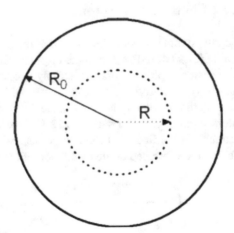

Fig. 12.1 Representation of a high-symmetry situation both in 2-d and 3-d. The inner region (dotted) is a circle in the 2-d case and a sphere in the 3-d case, has radius R, and is a trapping region. The outer region (solid) is also a circle in the 2-d case and a sphere in the 3-d case, has radius R_0, and is the region where the moving particles are initially placed uniformly. These regions are circumferences of circles in the 2-d case and surfaces of spheres in the 3-d case. There is no angular dependence in the problems considered so there is perfect radial symmetry. Reprinted with permission from fig. 1 of Spendier and Kenkre (2013); Copyright (2013) by the American Chemical Society

$$P(R, R, t) = \frac{1}{4\pi Dt} e^{-\frac{R^2}{2Dt}} I_0\left(\frac{R^2}{2Dt}\right),$$

and $r = R_0$ to get

$$P(R, R_0, t) = \frac{1}{4\pi Dt} e^{-\frac{R^2+R_0^2}{4Dt}} I_0\left(\frac{RR_0}{2Dt}\right).$$

In order to get the survival probability, we now need only Laplace-transform both expressions and substitute in Eq. (12.15). This yields

$$\tilde{Q}(\epsilon) = \frac{1}{\epsilon}\left[1 - \frac{\frac{1}{2\pi D}K_0\left(\sqrt{\frac{\epsilon}{\gamma_0}}\right) I_0\left(\sqrt{\frac{\epsilon}{\epsilon_0}}\right)}{\frac{1}{C_2} + \frac{1}{2\pi D}K_0\left(\sqrt{\frac{\epsilon}{\epsilon_0}}\right) I_0\left(\sqrt{\frac{\epsilon}{\epsilon_0}}\right)}\right], \tag{12.18}$$

where $\epsilon_0 = D/R^2$ and $\gamma_0 = D/R_0^2$. For perfect absorption, $C_2 \to \infty$ in Eq. (12.18). This gives

$$\tilde{Q}(\epsilon) = \frac{1}{\epsilon}\left[1 - \frac{K_0\left(\sqrt{\frac{\epsilon}{\gamma_0}}\right)}{K_0\left(\sqrt{\frac{\epsilon}{\epsilon_0}}\right)}\right]. \tag{12.19}$$

Here, I_0 and K_0 are the zero-order modified Bessel functions of the first and second kind, respectively. Our arbitrary capture rate (imperfect absorption) result is new whereas the perfect absorption result is known (Jaeger and Carslaw 1959) and can be obtained as a limit of ours in straightforward fashion. Both require numerical inversion to return to the time domain.

Asymptotic analysis can be carried out (Spendier and Kenkre 2013) on some of these expressions for long and short times. For short times, approximating the zero-order modified Bessel functions of the first and second kind in the Laplace domain as $K_0(z) \approx \sqrt{\pi/(2z)}\exp(-z)$ and $I_0(z) \approx \sqrt{1/(2\pi z)}\exp(z)$, one writes, for the finite reaction case,

$$\tilde{Q}(\epsilon) = \frac{1}{\epsilon}\left[1 - \sqrt{\frac{R}{R_0}}\frac{e^{-\sqrt{\epsilon\tau_2}}}{\xi_2\sqrt{\epsilon\tau_2} + 1}\right]. \tag{12.20}$$

Here, $\tau_2 = (R_0 - R)^2/D$ and $\xi_2 = 4\pi DR/[C_2(R_0 - R)]$ are 2-d quantities associated with motion and capture respectively. Equation (12.20) can be inverted exactly giving

$$Q(t) = 1 - \sqrt{\frac{R}{R_0}} \left[\text{erfc} \left(\frac{1}{2} \sqrt{\frac{\tau_2}{t}} \right) - e^{\frac{1}{\xi_2} + \frac{1}{\xi_2^2} \frac{t}{\tau_2}} \text{erfc} \left(\frac{1}{2} \sqrt{\frac{\tau_2}{t}} + \frac{1}{\xi_2} \sqrt{\frac{t}{\tau_2}} \right) \right].$$

(12.21)

This result appears to be new but is similar to Eq. (12.8) which is the *1-dimensional result* for a point trap and initial delta-function condition. It is straightforward to recover from our analysis other known results such as a short-time approximation for perfect absorption (Taitelbaum 1991),

$$Q(t) = 1 - \sqrt{\frac{R}{R_0}} \text{erfc} \left(\frac{1}{2} \sqrt{\frac{\tau_2}{t}} \right).$$

(12.22)

To obtain expressions for the long-time limit, one can follow Ritchie and Sakakura (1956) or Taitelbaum (1991) to obtain

$$Q(t) = \frac{2 \left[\ln \left(\frac{R_0}{R} \right) + \frac{2\pi D}{C_2} \right]}{\ln \left(\frac{4Dt}{R^2} \right) + \frac{4\pi D}{C_2} - 2E_\gamma}$$

(12.23)

for imperfect absorption. If absorption is perfect, $C_2 \to \infty$, and one recovers the result also mentioned in Ritchie and Sakakura (1956), with $E_\gamma = 0.57722...$ as the Euler's constant,

$$Q(t) = \frac{2 \ln \left(\frac{R_0}{R} \right)}{\ln \left(\frac{4Dt}{R^2} \right) - 2E_\gamma}.$$

(12.24)

Infinite Line Trap

Consider an infinite line of traps along the y-axis from $-\infty$ to ∞ through $x = 0$ and initial point particles placed on an infinite line from $-\infty < y < \infty$ through $x = x_0$. This problem can be solved by using the expression for a trapping ring of radius R and an initial radially symmetric distribution of point particles given in Eq. (12.18) by replacing R_0 with at $R_0 = R + x_0$

$$\tilde{Q}(\epsilon) = \frac{1}{\epsilon} \left[1 - \frac{\frac{1}{2\pi D} K_0 \left([R + x_0] \sqrt{\frac{\epsilon}{D}} \right) I_0 \left(R \sqrt{\frac{\epsilon}{D}} \right)}{\frac{1}{C_2} + \frac{1}{2\pi D} K_0 \left(R \sqrt{\frac{\epsilon}{D}} \right) I_0 \left(R \sqrt{\frac{\epsilon}{D}} \right)} \right].$$

(12.25)

As $R \to \infty$, the arguments of both $K_0(z)$ and $I_0(z)$ tend to infinity

$$I_0 \left(R \sqrt{\frac{\epsilon}{D}} \right) K_0 \left(R \sqrt{\frac{\epsilon}{D}} \right) \simeq \frac{\sqrt{D}}{2R \sqrt{\epsilon}},$$

$$I_0\left(R\sqrt{\frac{\epsilon}{D}}\right)K_0\left((R+x_0)\sqrt{\frac{\epsilon}{D}}\right) \simeq \frac{\sqrt{D}e^{-x_0\sqrt{\frac{\epsilon}{D}}}}{2R\left(1+\frac{x_0}{R}\right)\sqrt{\epsilon}}.$$

Substituting these expressions into Eq. (12.25) with $C_2 = \pi R C_1$, and taking the limit as $R \to \infty$, we obtain

$$Q(t) = e^{\frac{1}{\xi_1}+\frac{1}{\xi_1^2}\left(\frac{t}{\tau_1}\right)}\text{erfc}\left(\frac{1}{2}\sqrt{\frac{\tau_1}{t}}+\frac{1}{\xi_1}\sqrt{\frac{t}{\tau_1}}\right)+\text{erf}\left(\frac{1}{2}\sqrt{\frac{\tau_1}{t}}\right). \tag{12.26}$$

As one might expect from the physics of the limiting process, this result is identical to the expression obtained for a single trap at the origin and an initial particle placed at $x = x_0$, Eq. (12.8). Ben-Naim et al. (1993) have previously pointed out that the infinite 1-d trapping system with an imperfect trap is equivalent to a semiinfinite 1-d diffusion system. This equivalence has been used by Park et al. (2002) to explain results in a photobleaching experiment resulting from an infinite line trap.

Finite Line Trap: Open Trapping Surface
Each of the trapping surfaces considered so far has been a closed surface. What if we analyze an open surface, for instance a finite trap segment extending from $-l \le y \le l$ and passing through $x = 0$? The self-propagator is computed from $e^{-(y-y_0)^2/(4Dt)}/(4\pi Dt)$ by integrating y and y_0 from $-l$ to l and dividing by $2l$ for appropriate normalization:

$$P(0,0,t) = \frac{1}{4\pi Dt}\int_{-l}^{l}\int_{-l}^{l}e^{-\frac{(y-y')^2}{4Dt}}dy'dy$$

$$= \frac{1}{\sqrt{4\pi Dt}}\text{erf}\left(\frac{l}{\sqrt{Dt}}\right)-\frac{1}{\pi l}e^{-\frac{l^2}{2Dt}}\sinh\left(\frac{l^2}{2Dt}\right).$$

Under the limit $l \to \infty$, we obtain the same self-propagator as for an infinite line of traps. The first term in the above expression cannot be transformed exactly. However, for the second term, the exact Laplace transform can be found in Roberts and Kaufman (1966). If the particles are initially placed on a line from $-l$ to l through $x = x_0$ to the right of the trapping line, we can compute the homogeneous solution at the trap site:

$$P(0,x_0,t) = \frac{e^{-\frac{x_0^2}{4Dt}}}{8\pi l Dt}\int_{-l}^{l}\int_{-l}^{l}e^{-\frac{(y-y')^2}{4Dt}}dydy'$$

$$= \frac{1}{\sqrt{4\pi Dt}}e^{-\frac{x_0^2}{4Dt}}\text{erf}\left(\frac{l}{\sqrt{Dt}}\right)-\frac{1}{\pi l}e^{-\frac{2l^2+x_0^2}{4Dt}}\sinh\left(\frac{l^2}{2Dt}\right). \tag{12.27}$$

Spendier succeeded in arriving at the result for the survival probability for this problem as (with $\chi = 4l^2/D$),

$$\tilde{Q}(\epsilon) = \frac{1}{\epsilon}\left[1 - \frac{\int_0^\infty e^{-\frac{\tau_1}{4t}-\epsilon t}\operatorname{erf}\left(\sqrt{\frac{\chi}{4t}}\right)\frac{dt}{\sqrt{t}} - \frac{4}{\sqrt{\pi\chi\epsilon}}\left[\frac{\sqrt{\tau_1}}{2}K_1\left(\sqrt{\tau_1\epsilon}\right) - \frac{\sqrt{\tau_1+\chi}}{2}K_1\left(\sqrt{(\tau_1+\chi)\epsilon}\right)\right]}{\sqrt{\pi\tau_1}\xi_1 + \int_0^\infty e^{-\epsilon t}\operatorname{erf}\left(\frac{1}{2}\sqrt{\frac{\chi}{t}}\right)\frac{dt}{\sqrt{t}} - \frac{2}{\epsilon\sqrt{\pi\chi}} + \frac{2}{\sqrt{\epsilon\pi}}K_1\left(\sqrt{\epsilon\chi}\right)}\right].$$

(12.28)

The K_1's are modified Bessel functions of the second kind and of order 1. We believe Eq. (12.28) is not a result one finds in the literature, precisely because open trapping surfaces make the situation more complex due to the missing symmetry in the problem. Important to remember is that, as long as we obtain an expression in Laplace domain, it is possible to invert the solution numerically.

12.4.2 Results in 3-d Systems

The Green function in three dimensions is a product of three 1-d Gaussian propagators and written in spherical polar coordinates as

$$G\left(\vec{r},\vec{r}',t\right) = \frac{1}{(4\pi Dt)^{3/2}}e^{-\frac{|\vec{r}-\vec{r}'|^2}{4Dt}} \tag{12.29}$$

where $|\vec{r} - \vec{r}'|^2 = (x - x_0)^2 + (y - y_0)^2 + (z - z_0)^2$. For a spherically symmetric initial condition, $P(\vec{r},0) = \delta(r - R_0)/(4\pi r^2)$, the two quantities of interest to us are obtained by evaluating

$$\int_0^\pi \int_0^{2\pi} \int_0^\infty G\left(\vec{r},\vec{r}',t\right)P\left(\vec{r},0\right)r'^2\sin\theta dr'd\theta d\phi.$$

Their consequence is

$$P(R,r;t) = \frac{1}{8\pi Rr\sqrt{\pi Dt}}\left(e^{-\frac{(r-R)^2}{4Dt}} - e^{-\frac{(r+R)^2}{4Dt}}\right), \tag{12.30}$$

which is in agreement with earlier results as given in Jaeger and Carslaw (1959).

Spherical Trap of Finite Extent
Because of its important biological applications in the process of passive diffusion, let us consider the trapping problem of a spherical trapping shell and an initial spherical distribution of point particles that may represent a situation in which small molecules diffuse across a cell membrane. See Fig. 12.1. For example, in the process of photosynthesis, oxygen molecules may be absorbed by oxygen-evolving complexes embedded in the thylakoid membrane (Herrmann 1999) while undergoing passive diffusion through the membrane. In such a system, one might

be interested in the total amount of unbound oxygen, which is a measure of energy production in this process.

To solve this problem, we follow our previous methodology. The probability densities on the spherical surfaces of radius R and R_0 are calculated as

$$P(R, R, t) = \frac{1}{8\pi R^2 \sqrt{\pi Dt}} \left(1 - e^{-\frac{R^2}{Dt}}\right),$$

$$P(R, R_0, t) = \frac{1}{8\pi R R_0 \sqrt{\pi Dt}} \left(e^{-\frac{(R_0-R)^2}{4Dt}} - e^{-\frac{(R_0+R)^2}{4Dt}}\right).$$

Both expressions can be Laplace-transformed exactly (Roberts and Kaufman 1966) and the particle survival probability in Laplace domain for finite reaction is

$$\tilde{Q}(\epsilon) = \frac{1}{\epsilon} \left[1 - \frac{\frac{1}{4\pi R R_0 \sqrt{D\epsilon}} e^{-\sqrt{\frac{\epsilon}{\gamma_0}}} \sinh\sqrt{\frac{\epsilon}{\epsilon_0}}}{\frac{1}{C_3} + \frac{1}{4\pi R^2 \sqrt{D\epsilon}} e^{-\sqrt{\frac{\epsilon}{\epsilon_0}}} \sinh\sqrt{\frac{\epsilon}{\epsilon_0}}}\right], \tag{12.31}$$

where $\epsilon_0 = D/R^2$ and $\gamma_0 = D/R_0^2$. This is another of the useful expressions that Spendier derived and used for numerical inversions but was unable to invert analytically.

The instantaneous reaction limit, for $C_3 \to \infty$, is

$$\tilde{Q}(\epsilon) = \frac{1}{\epsilon}\left(1 - \frac{R}{R_0} e^{-\sqrt{\epsilon}\left(\sqrt{\frac{1}{\gamma_0}} - \sqrt{\frac{1}{\epsilon_0}}\right)}\right). \tag{12.32}$$

It can be inverted exactly (Roberts and Kaufman 1966), giving in the time domain,

$$Q(t) = 1 - \frac{R}{R_0} \text{erfc}\left(\frac{1}{2}\sqrt{\frac{\tau_3}{t}}\right), \tag{12.33}$$

where $\tau_3 = \tau_2 = (R_0 - R)^2/D$ in agreement with Jaeger and Carslaw (1959) and Rice (1985).

A straightforward expansion of Eq. (12.31) about $\epsilon = 0$, followed by Laplace inversion after retaining a few terms (Doetsch 1971), gives

$$Q(t) = 1 - \frac{RC_3}{R_0(4\pi DR + C_3)} + \frac{1}{\sqrt{t}}\left(\frac{RC_3}{\sqrt{D}(4\pi DR + C_3)} - \frac{C_3^2 R}{\sqrt{D}(4\pi D + C_3)^2}\right). \tag{12.34}$$

As $t \to \infty$, this result reduces to

$$\lim_{t \to \infty} Q(t) = 1 - \frac{RC_3}{R_0(4\pi DR + C_3)}, \tag{12.35}$$

which further yields, in the perfect absorber limit,

$$\lim_{t \to \infty} Q(t) = 1 - \frac{R}{R_0}. \tag{12.36}$$

Both these results are also to be found in Rice (1985). What this means is that, for finite and instantaneous reaction in the 3-dimensional system, the survival probability will never reach zero. This result is expected in light of the famous feature that random walkers escape in 3-d: the probability of a diffusing particle reaching any specific point (including the starting point) as time approaches infinity is less than 1 in 3-dimensions.

Infinite Sheet of Traps

Similarly, one can also obtain an expression for an infinite sheet in 3-dimensions. We again start with the solution for a spherical trap given in (12.31) and replace $R_0 = R + x_0$ to obtain

$$\tilde{Q}(\epsilon) = \frac{1}{\epsilon} \left[1 - \frac{\frac{1}{4\pi R^2(1+\frac{x_0}{R})\sqrt{D\epsilon}} e^{-x_0\sqrt{\frac{\epsilon}{D}}}}{\frac{1}{2\pi R^2 C_1} + \frac{1}{4\pi R^2 \sqrt{D\epsilon}}} \right] \tag{12.37}$$

with $2\pi R^2 C_1 = C_3$. In the limit as $R \to \infty$, it is easy to show that (12.37) becomes the 1-dimensional result given in Eq. (12.8) after Laplace inversion.

Trapping Ring in 3-d

Consider, as our final example, a trapping ring of radius R centered at the origin in the x, y plane ($z = 0$ or $\phi = \pi/2$) and an initial point particle at $(0, 0, z)$ above the ring on the z-axis. Tedious but straightforward calculations allow us to obtain a new expression for the survival probability in the Laplace domain,

$$\tilde{Q}(\epsilon) = \frac{1}{\epsilon} - \frac{e^{-\sqrt{(R^2+z^2)\epsilon/D}}}{\epsilon 4\pi D\sqrt{R^2 + z^2}}$$

$$\times \left\{ \frac{1}{C_3} + \frac{1}{(4\pi D)^{3/2}} \left[\frac{4R\epsilon}{\sqrt{\pi D}} \,_2F_3\,(A) - \frac{\sqrt{2D}}{R} \left[2E_\gamma + \ln\left(\frac{16R^2}{D}\right) \right] \right. \right.$$

$$\left. \left. -2\sqrt{\pi\epsilon} \,_1F_2\,(B) \right] \right\}^{-1}, \tag{12.38}$$

where $_pF_q$ is the generalized hypergeometric function defined in Abramowitz and Stegun (1965) with $A = 1, 1; 1, \frac{3}{2}, \frac{3}{2}, 2; \frac{R^2\epsilon}{D}$ and $B = \frac{1}{2}; 1, \frac{3}{2}; \frac{R^2\epsilon}{D}$. This expression cannot be inverted directly and must be evaluated through numerical procedures.

There are many other useful discussions in Spendier and Kenkre (2013) that the reader might find interesting as well as instructive, particularly regarding the fascinating topic of the relationship of continuum boundary condition studies to discrete sink term analysis.

12.5 Defect Technique with the Smoluchowski Equation

Typically, reaction-diffusion studies assume that, in the absence of the reaction phenomenon, the motion is translationally invariant. Our interest in the present section is to extend these studies by going beyond that assumption and analyzing systems in which a particle diffuses in a harmonic potential and undergoes capture at a given rate when it arrives at a fixed trap. The diffusion means that the particle performs a random walk. The attraction to the potential center means that a tendency to be tethered to a fixed point is present in addition. The location of the trap is generally arbitrary. e.g., not necessarily at the attractive center; this feature introduces richness in the consequences of the reaction-diffusion phenomenon and uncovers novel results.

Systems characterized by the above features occur often in nature. Molecular forces confine moving entities in various physical and chemical situations in sensitized luminescence and photosynthesis. Funneling phenomena attract excitations in photosynthetic antennae and molecular crystals and aggregates. Particle diffusion in a harmonic field occurs also in biophysical studies on DNA stretching with optical tweezers (Wang et al. 1997; Lindner et al. 2013). Yet another relevant area is electrostatic steering in enzyme ligand binding (Wade et al. 1998; Livesay et al. 2003). To bind at an enzymes active site, a ligand must diffuse or be transported to the enzyme surface, and, if the binding site is buried, the ligand must diffuse through the protein to reach it. Enhancement of this diffusion can be achieved by attractive electrostatic interactions between the substrate and the protein binding site. On a more macroscopic scale, animals feel a driving force pointing towards their *nest* (Ebeling et al. 1999) with the consequent emergence of home ranges (Kenkre and Giuggioli 2020; Giuggioli et al. 2006a). Transmission of infection in terrains where infected animals (such as rodents) interact with susceptible ones under the action of space confinement provides a related and more complex area of study.

A striking consequence of our analysis is that, depending on the relative position of the attractive center and the trap, a particle placed initially at some location may be affected *either* favorably *or* unfavorably during different stages of its motion as far as the efficiency of the trapping phenomenon is concerned. In a translationally invariant system, all that is relevant is the initial distance between the particle and reactive site, the manner of motion that occurs in between, and the rate at which the reaction occurs. The problem is rendered considerably richer in the presence of an attractive potential as we shall see below.

Our discussion below follows an analysis (Spendier et al. 2013) performed relatively recently in collaboration primarily with Kathrin Spendier.[4] Our study of the literature has uncovered no major previous advances in reaction-diffusion theory in the presence of a potential, prior to our work. Of the two relevant articles we have found, Drazer et al. (2000) and Bagchi et al. (1983), the former has no position-dependence in the capture which is represented via a constant term, the emphasis being on anomalous diffusion, and in the latter, only perfect absorption is treated and that only for a centrally located trap, which, as we shall show below, results in a relatively featureless case.

As in the rest of the chapter, we are generally interested in the particle survival probability $Q(t) = \int_{-\infty}^{+\infty} dx\, P(x, t)$ but our point of departure is the evolution equation

$$\frac{\partial P(x, t)}{\partial t} = \frac{\partial}{\partial x}\left(\gamma x P(x, t) + D\frac{\partial P(x, t)}{\partial x}\right) - C_1 \delta(x - x_r)P(x, t). \qquad (12.39)$$

While D is the familiar diffusion constant, the new parameter here is the rate at which the particle tends to return to the harmonic potential center (taken to be the origin without loss of generality). That rate is γ.

12.5.1 Propagator and Survival Probability

The propagator of this equation, i.e., the solution for $P(x, 0) = \delta(x - x_0)$, to be denoted by the symbol $\Pi(x, x_0, t)$, can be obtained by Fourier transforming the equation and solving the resulting first order partial differential equation by the method of characteristics. See, e.g., Risken (1984) and Reichl (2009):

$$\Pi(x, x_0, t) = \frac{e^{-\frac{(x - x_0 e^{-\gamma t})^2}{4D\mathcal{T}(t)}}}{\sqrt{4\pi D\mathcal{T}(t)}}, \qquad (12.40)$$

where $\mathcal{T}$ is a function of time given by

$$\mathcal{T}(t) = \frac{1 - e^{-2\gamma t}}{2\gamma}. \qquad (12.41)$$

The solution shows transparently that, wherever it is initially placed, the particle tends to move to the origin at rate γ but, as a result of the diffusion that it also

---

[4]This work was continued in a generalization to ecologically motivated problems (Kenkre and Sugaya 2014; Sugaya and Kenkre 2018) in collaboration with Satomi Sugaya. We will have occasion to return to that work, which was about transmission of infection in epidemics, in Chap. 15.

undergoes, ends up, in the steady state, occupying a Gaussian of width proportional to the square root of the ratio of the diffusion constant to γ.

The use of the standard defect technique yields the probability density in the Laplace domain in the usual manner. Because we will concern ourselves here with a single trap located at x_r and initial placement at x_0, the counterpart of the survival probability equation (12.15) we write here is

$$\widetilde{Q}(\epsilon) = \frac{1}{\epsilon}\left[1 - \left(\frac{\widetilde{\Pi}(x_r, x_0)}{(1/C_1) + \widetilde{\Pi}(x_r, x_r)}\right)\right], \tag{12.42}$$

where (and henceforth) we drop the specification of ϵ explicitly in the arguments in the right hand side. The starting point for the calculations we present below is the conjunction of Eq. (12.42) with Eq. (12.40).

12.5.2 Centrally Placed Trap: Analytic Solution

The complexity of the Smoluchowski propagator Eq. (12.40) makes it difficult or impossible to obtain analytic expressions in most cases. If, however, the trap is located at the attractive center of the potential, $x_r = 0$, progress can be made because the Laplace transforms of the propagators appearing in $\widetilde{Q}(\epsilon)$ can be computed explicitly in terms of the Whittaker function. The survival probability for this case is, in the Laplace domain,

$$\widetilde{Q}(\epsilon) = \frac{1}{\epsilon}\left[1 - \left(\frac{\widetilde{\Pi}(0, x_0)}{(1/C_1) + \widetilde{\Pi}(0, 0)}\right)\right]. \tag{12.43}$$

Putting $x_r = 0$ in the general expression for the Laplace transform of the Smoluchowski propagator Eq. (12.40),

$$\int_0^\infty dt \left[\frac{e^{-\frac{(x_r - x_0 e^{-\gamma t})^2}{4D\mathcal{T}(t)}}}{\sqrt{4\pi D\mathcal{T}(t)}}\right] e^{-\epsilon t},$$

one finds it impossible to give a convenient general inversion formula. However, from a table of Laplace transforms (Roberts and Kaufman 1966) or otherwise, one obtains particular results such as that

$$\widetilde{\Pi}(0, x_0) = \frac{(\gamma\tau_1)^{-1/4}}{2\sqrt{\pi}\,\sigma\gamma}e^{(\gamma\tau_1/2)}\Gamma\left(\frac{\epsilon}{2\gamma}\right)W_{\frac{1}{4}-\frac{\epsilon}{2\gamma},\frac{1}{4}}(\gamma\tau_1). \tag{12.44}$$

Here, $\tau_1 = x_0^2/2D$ is the time the particle would take to move via pure diffusion from its initial location to the attractive center. The dimensionless quantity $\gamma\tau_1$ is therefore the ratio of that diffusion time to $1/\gamma$, the time characteristic of motion resulting purely from the pull of the potential. On defining $\sigma = \sqrt{2D/\gamma}$ which is the width of the equilibrium distribution of the trap-less Smoluchowski equation, we see that $\gamma \tau_1$ can be given another physical interpretation: it is identical to the square of the initial location of the particle to the Smoluchowski equilibrium width:

$$\gamma \tau_1 = \left(\frac{x_0}{\sigma}\right)^2. \tag{12.45}$$

The W in Eq. (12.44) is the Whittaker W-function defined in Abramowitz and Stegun (1965) as

$$W_{\kappa,\mu}(z) = e^{-\frac{z}{2}} z^{\frac{1}{2}+\mu} U\left(\frac{1}{2} + \mu - \kappa, 1 + 2\mu, z\right), \; |argz| < \pi$$

in terms of the confluent hypergeometric function

$$U(a, b, c) = \frac{1}{\Gamma(a)} \int_0^\infty e^{-ct} t^{a-1} (1+t)^{b-a-1} \, dt.$$

The transform of the other propagator in Eq. (12.43) is even easier to calculate. One puts $x_0 = 0$ in the above expression, or directly computes the integral

$$\frac{1}{\sigma \sqrt{\pi}} \int_0^\infty dt \, \frac{e^{-\epsilon t}}{\sqrt{1 - e^{-2\gamma t}}},$$

to get

$$\tilde{\Pi}(0, 0) = \frac{1}{\epsilon \sigma} \frac{\Gamma\left(\frac{\epsilon}{2\gamma} + 1\right)}{\Gamma\left(\frac{\epsilon}{2\gamma} + \frac{1}{2}\right)} = \frac{1}{2\gamma \sigma \sqrt{\pi}} B\left(\frac{\epsilon}{2\gamma}, \frac{1}{2}\right), \tag{12.46}$$

where $\Gamma(n)$ is the Gamma function and $B(z, w) = \Gamma(z)\Gamma(w)/\Gamma(z+w)$ is the Beta function.

Substitution of Eqs. (12.44) and (12.46) into the prescription given in Eq. (12.43) provides an exact expression for the total survival probability in the Laplace domain:

$$\tilde{Q}(\epsilon) = \frac{1}{\epsilon} \left[1 - \frac{(\gamma \tau_1)^{-1/4} e^{(\gamma \tau_1/2)} \Gamma\left(\frac{\epsilon}{2\gamma}\right) W_{\frac{1}{4} - \frac{\epsilon}{2\gamma}, \frac{1}{4}}(\gamma \tau_1)}{\xi + B\left(\frac{\epsilon}{2\gamma}, \frac{1}{2}\right)} \right]. \tag{12.47}$$

The denominator of the second term in the square brackets is a sum of a dimensionless motion quantity that appears in the form of the Beta function, and a dimensionless capture parameter (not to be confused with the ξ defined in the beginning of this chapter and most of Chap. 11)

$$\xi = 2\sqrt{\pi}\left(\frac{\gamma\sigma}{C_1}\right) = 2\sqrt{2\pi}\left(\frac{\sqrt{\gamma D}}{C_1}\right), \tag{12.48}$$

which is inversely proportional to C_1. The dimensionless ratio in the parentheses in Eq. (12.48) compares a time for capture to a time for motion and represents the extent of *imperfectness* of absorption. If ξ vanishes, that imperfection vanishes and one has a perfect absorber; if ξ is large, one has weak capture. While Eq. (12.47) cannot be Laplace-inverted analytically, and necessitates numerical procedures, for perfect absorption a surprising reduction occurs.

12.5.3 Analytic Inversion for Perfect Absorption

For perfect absorption, $C_1 \to \infty$ making $\xi = 0$ vanish, Eq. (12.47) reduces to

$$\widetilde{Q}(\epsilon) = \frac{1}{\epsilon} - \frac{\Gamma\left(\frac{\epsilon+2\gamma}{2\gamma}\right)\Gamma\left(\frac{\epsilon}{2\gamma}\right)}{2\gamma\sqrt{\pi}\,(\gamma\tau_1)^{1/4}\,\Gamma\left(\frac{\epsilon+\gamma}{2\gamma}\right)}e^{\frac{\gamma\tau_1}{2}}W_{-\frac{1}{4},\frac{1}{4}}(\gamma\tau_1), \tag{12.49}$$

and inversion back to the time domain is readily possible:

$$Q(t) = 1 - \left(\frac{e^{2\gamma t}-1}{\pi^2\gamma\tau_1}\right)^{\frac{1}{4}}e^{\frac{-\gamma\tau_1}{e^{2\gamma t}-1}}W_{-\frac{1}{4},\frac{1}{4}}\left(\frac{\gamma\tau_1}{e^{2\gamma t}-1}\right). \tag{12.50}$$

Here we use well-known scaling and shift rules along with the result of Roberts and Kaufman (1966) that the Laplace transform of

$$e^{-\frac{a}{2}}\left(1-e^{-t}\right)^{-\mu}e^{\frac{-a/2}{(e^t-1)}}W_{\mu,\nu}\left(\frac{a}{e^t-1}\right)$$

is

$$\frac{\Gamma\left(\epsilon+1/2+\nu\right)\Gamma\left(\epsilon+1/2-\nu\right)}{\Gamma\left(\epsilon+1-\mu\right)}W_{-\epsilon,\nu}(a).$$

We see that the survival probability in the time domain Eq. (12.50) also involves the Whittaker function with an argument that is itself a function of time. Furthermore, $W(t)$, in the form it appears in Eq. (12.50), can be defined in terms of the complementary error function (Abramowitz and Stegun 1965),

$$W_{-\frac{1}{4},\frac{1}{4}}(z) = \sqrt{\pi}z^{1/4}e^{z/2}\mathrm{erfc}(\sqrt{z}). \tag{12.51}$$

This has the remarkable consequence that, for perfect absorption, we can derive the simple result

$$Q(t) = \mathrm{erf}\left(\frac{x_0/\sigma}{\sqrt{e^{2\gamma t} - 1}}\right) = \mathrm{erf}\sqrt{\frac{\gamma \tau_1}{e^{2\gamma \tau_1 (t/\tau_1)} - 1}}. \tag{12.52}$$

There is much that can be said about Eq. (12.52). The error function behavior ensures that the survival probability does not change much initially but only after a threshold time has elapsed. All time derivatives of $Q(t)$ of finite order vanish at the origin. The threshold time might be taken to signify that the particle has arrived at the trap. After that, the time scale for the evolution of $Q(t)$ is generally $1/\gamma$ but, in the limit that this time becomes infinite (infinitely flat potential, $\gamma \to 0$), the characteristic time becomes τ_1. One sees here transparently the transition from potential-induced motion to the trap to diffusive motion. In that diffusive limit (no potential), Eq. (12.52) reduces to the well-known result

$$Q(t) = \mathrm{erf}\left(\sqrt{\frac{\tau_1}{2t}}\right)$$

we met earlier in this chapter in Eq. (12.7). Figure 12.2 shows the time dependence of the survival probability for perfect absorption (infinite C_1) for five values (5, 1, 0.1, 0.01, 0.001) of $\gamma \tau_1$. The curves converge to a limit (curves for the lowest two values of $\gamma \tau_1$ practically coincide) that represents pure diffusive motion with no potential pull.

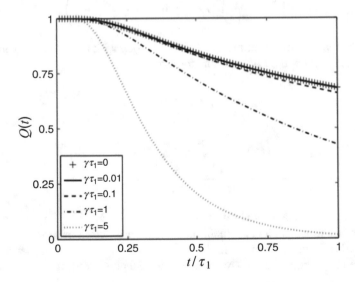

Fig. 12.2 Survival probability for perfect capture at the trap located at center, i.e., in coincidence with the attractive center of the potential. The curves converge to the diffusive limit, Eq. (12.7), (topmost curve) as the potential attraction vanishes, i.e., as $\gamma \to 0$. $Q(t)$ is shown for several $\gamma \tau_1$ values: 5, 1, 0.1, 0.01, 0.001. Reprinted with permission from fig. 1 of Ref. Spendier et al. (2013); copyright (2013) by the American Physical Society

12.5.4 Delocalized Initial Particle Distribution: Superposition

Once the exact solution is known for a point initial condition $P(x_0) = \delta(x - x_0)$ for particle placement, one can solve the problem for any initial condition by summing the results. Thus, provided one has the perfect absorption case, the principle of superposition yields

$$Q(t) = \int_{-\infty}^{\infty} dx_0 \, P(x_0) \mathrm{erf} \left(\frac{x_0/\sigma}{\sqrt{e^{2\gamma t} - 1}} \right). \tag{12.53}$$

The survival probability in Eq. (12.53) for this central-trap perfect-absorber system may be viewed as a transform of the initial probability distribution $P(x_0)$ of the particle. In a number of useful situations the initial particle distribution is non vanishing only on one side of the potential center. Then the lower limit in the integration of Eq. (12.53) becomes 0 and a situation akin to the Laplace transform occurs. The error function takes the place of the exponential in the Laplace transform, and the quantity $(\sigma \sqrt{e^{2\gamma t} - 1})^{-1}$ plays the role of the transform variable ϵ.

We display two useful consequences of this transform. For an initial exponential distribution $P(x_0) = (1/d) \exp(-x_0/d)$ only on one side, i.e., for $x_0 > 0$ (and vanishing $P(x_0)$ elsewhere), with characteristic distance d, the survival probability is

$$Q(t) = e^{\zeta^2(t)} \mathrm{erfc} \left(\zeta(t) \right) \tag{12.54}$$

where $\zeta(t) = (\sigma/2d)\sqrt{e^{2\gamma t} - 1}$. For an initial Rayleigh distribution $P(x_0) = (x_0/d^2) exp[-x_0^2/(2d^2)]$ for $x_0 > 0$ (and vanishing $P(x_0)$ elsewhere), we get

$$Q(t) = \left[1 + \left(\sigma^2/2d^2 \right) (e^{2\gamma t} - 1) \right]^{-1/2}. \tag{12.55}$$

The first of the distributions, often called the *random* or Poisson distribution, arises often and can describe, for instance, the initial placement of coalescing signaling receptor clusters in immune mast cells (Spendier et al. 2010). The second distribution is a biased Poisson distribution which also occurs in several physical systems. We have mentioned both of them because the first concentrates the initial placement of the particle near the attractive center while the second shifts it away by a finite amount. We have used d to denote the average value $\int x_0 P(x_0) dx_0$ in both cases.

Figure 12.3 shows the two cases of the survival probability for the two initial particle distributions. In both of them we see that the $Q(t)$ curves converge to the pure diffusive limit (top line). The characteristic time τ_d in the units of which t is plotted in these curves equals $d^2/2D$, i.e., is the time the particle would take to traverse as a random walker the characteristic distance d for each of the distributions.

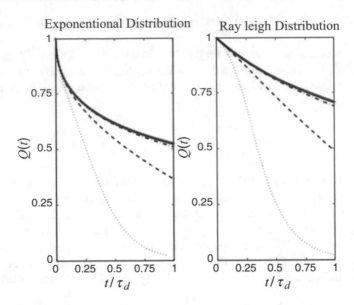

Fig. 12.3 Survival probability for exponential (left panel) and Rayleigh (right panel) initial particle distributions as given in Eqs. (12.54), (12.55). In each panel, the curves correspond to $(d/\sigma)^2 = 5$, 1, 0.1, 0.01, and 0.001, from the bottom to the top curve, respectively. Here $\tau_d = d^2/2D$ and d is the characteristic distance of the initial distribution. The near-origin behavior of Q(t) is substantially different from that in Fig. 12.2. See text. Reprinted with permission from fig. 2 of Ref. Spendier et al. (2013); copyright (2013) by the American Physical Society

A noteworthy feature of Eqs. (12.54) and (12.55), and of Fig. 12.3, is the loss of the reverse Arrhenius behavior near the origin (derivatives of all finite orders vanishing at the origin) brought about by superposition of contributions from multiple initial locations of the particles: $Q(t)$ curves, while *totally flat* as $t \to 0$ in Fig. 12.2, change through superposition to non-drastic variation near the origin in Fig. 12.3. The mathematical mechanism for this conversion is the removal of the isolated essential singularity by integration. It is similar to the one encountered in the temperature (T) dependence of the specific heat of insulators. It is well-known (see any text in solid state physics) that, through a superposition of activated Einstein contributions, each of which fails to describe the correct near-origin temperature behavior, the Debye theory succeeds in predicting the correct, dimension-driven T^3 dependence. In our present problem, the near-origin *time*-dependence of $Q(t)$ plays the role of the near-origin *temperature*-dependence of the specific heat.

Equation (12.47) for arbitrary capture rate, the demonstration of the analytic reduction to the perfect absorber result Eq. (12.52), and the superposition results Eqs. (12.53), (12.54), and (12.55) are our main results in this subsection. We have found that a passing mention of the perfect-absorber central-trap localized result for the centrally placed trap, Eq. (12.52), has appeared in a previous analysis (Bagchi et al. 1983) of the effect of viscosity on electronic relaxation in solution. Perfect-absorber studies for *many* traps have also appeared in papers that, while not directly related to the present analysis, are interesting in their own right (Yuste and Acedo 2001).

12.5.5 Numerical Inversion for Noninfinite Capture

Numerical Laplace inversion of Eq. (12.47) becomes necessary for finite C_1, equivalently for non-vanishing ξ. We use standard inversion routines for this purpose.

Spendier et al. (2013) have explored the survival probability for various values of the capture rate C_1, equivalently of the dimensionless parameter ξ. We have uncovered no surprises. A stronger capture rate makes $Q(t)$ decrease faster as expected. We have not found it instructive to display the resultant figures. We emphasize, however, that our procedure can produce the evolution of the survival probability for arbitrary capture.

12.5.6 Analysis for Arbitrary Relative Locations of Trap and Attractive Center

Situations in which the attractive center, the trap, and the initial placement of the moving particle are at *arbitrary* locations with respect to one another, are rich in their outcome. This is expected. For instance, one might argue that, if the initial location of the particle lies in between the potential center and the trap, the pull provided by the confining potential would tend to act counter to the phenomenon of trapping and that the potential would thus hinder trapping and enhance survival. Yet, since at equilibrium, the particle in the trap-less situation would tend to occupy an extent around the potential center given by the Smoluchowski width, one might expect survival, when the trap is present, to depend on whether the distance of the trap from the potential center is disparate with respect to the Smoluchowski width. Which effect wins over the other in a given set of circumstances? These interesting situations are difficult, or even generally impossible, to study via analytic solutions. To investigate them, the straightforward way is to use a numerical program that starts with an equation such as Eq. (12.42), substitute in it the Laplace transforms of the propagators evaluated *numerically* from Eqs. (12.40) and (12.41), and perform the numerical Laplace inversion by standard methods to produce the final time-dependent survival probability. Spendier et al. (2013) pursued this program systematically.

Symmetrical Placement of Trap and Particle

Let us first consider the case of no potential ($\gamma = 0$), the trap placed at $x_r = L/2$ and the initial location of the particle at $x_0 = -L/2$ so that the distance between the two is L. The survival probability $Q(t)$ is given by the well-known expression (12.7) valid for a diffusion rather than a Smoluchowski equation, with τ_1 replaced by $\tau_L = L^2/2D$. Let us now introduce a potential with its attractive center precisely midway

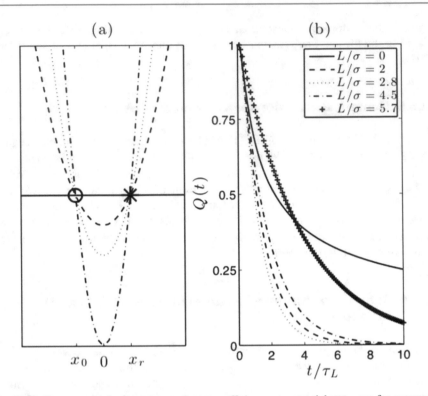

Fig. 12.4 Non-monotonic dependence of capture efficiency on potential steepness for symmetrical placement of trap and particle. Left panel represents the situation visually. Right panel shows the non-monotonic effect as the decay of $Q(t)$ is enhanced by increasing the potential steepness but then hindered on further increase. Curves are labeled by L/σ, the ratio of the distance between trap and initial location of particle to the Smoluchowski width. Four of the traces in the right panel (for $L/\sigma = 0, 2, 2.5$ and 4.5) correspond respectively to the potential curves in the left panel. Reprinted with permission from fig. 3 of Ref. Spendier et al. (2013); copyright (2013) by the American Physical Society

between the trap and the particle (the potential center is at 0), see Fig. 12.4a, and examine the time dependence of the survival probability as the potential steepness measured by γ, or more conveniently the dimensionless ratio L/σ, is varied. We display the results in Fig. 12.4b.

Starting with the pure diffusive case $L/\sigma = 0$, for which σ is infinite and the survival probability is given by Eq. (12.7), we see that increase of potential steepness, equivalently of L/σ, has a remarkable non-monotonic effect. Small increase makes capture more efficient but beyond a certain value it has the opposite effect. Why does this happen? The presence of a potential surely makes the particle move faster, at least initially, towards the trap as it travels to the attractive center. However, past the attractive center, the motion towards the trap is uphill and therefore *hindered* by the potential. The introduction of the potential thus has both a favorable and an unfavorable effect on capture.

There is an approximate but instructive way to think about what is happening by comparing where the trap lies in relation to the Smoluchowski width. The potential pull tends to bring the probability density at the trap location to its equilibrium value in the absence of the capture. That the dependence of this value (indeed the value at any location which is not the potential center) on the steepness of the potential is non-monotonic becomes clear after a moment's reflection.

12.5.7 One-Sided Placement of Trap and Particle

Let us now place the trap at the center and the particle placed initially uphill at some distance L. This means $x_r = 0$ and $x_0 = L$. Nothing particularly interesting emerges as a potential is introduced (with attractive center at the origin, as earlier) and its steepness is varied: steeper potentials make trapping easier. On the other hand, if we reverse the positions of the trap and the particle, so that $x_r = L$ and $x_0 = 0$, interesting non-monotonic behavior is encountered again with variation in potential steepness. These two cases of uphill particle and uphill trap (respectively) are shown in Fig. 12.5, τ_L being, as in Fig. 12.4, the time taken by the particle to

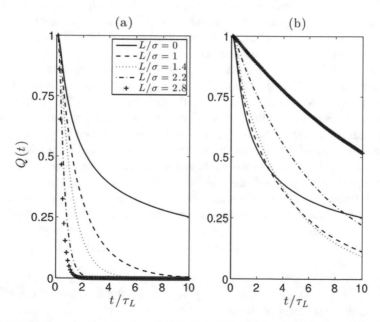

Fig. 12.5 Difference in the behavior of $Q(t)$ for uphill and downhill placement of trap with respect to initial particle location. The trap is at the potential center in the left panel but uphill in the right panel (initial particle location being at the potential center in this case). The effect of increasing the potential steepness is monotonic in the former but displays interesting features as in Fig. 12.4 in the latter. Reprinted with permission from fig. 4 of Ref. Spendier et al. (2013); copyright (2013) by the American Physical Society

traverse the distance from its initial placement to the trap under purely diffusive condition.

Why is there no symmetry in the effects of the placement of the stationary trap and the moving particle? After all, survival probability depends merely on their meeting. The answer is obvious. A particle placed uphill with the trap at the potential center is always helped by the potential steepness to get faster to the trap. The curves in the left panel of Fig. 12.5 therefore show more capture as the potential steepness increases.

These take-aways that I have described from only a part of the analysis of Spendier et al. (2013) should be sufficiently intriguing for the reader to go to the original publication for a number of other discussions I have not touched upon here. They include additional methods of investigation such as time-dependent effective transfer rates and elucidation of the physical significance of some of the tendencies we have seen on their basis.

12.6   Brief Remarks on a Theory of Coalescence

To end the discussion of reaction diffusion phenomena analyzed in this chapter with the help of the defect technique developed in the continuum, here are some brief remarks on a reverse application of the defect technique that Spendier and I carried out to address coalescence of signal receptor clusters that she and her collaborators observed in her experiments on mast cells.

To represent the actual experiments carried out on the aggregation phenomenon, we addressed the following simplified problem in 2-dimensional space. A certain amount of material was thought to be divided into parts. The first part was considered to be made into a disk of uniform density and radius $R(0)$ and placed with its center at the origin. The rest of the material was considered to be distributed throughout the 2-dimensional space external to the disk as point particles undergoing diffusion with diffusion constant D. The model assumed that when the point particles touch the edge of the disk they may be absorbed into the disk with finite probability. When absorbed, they undergo a quick rearrangement process whose details are excluded from consideration but whose consequence is to introduce an overall delay in the capture process expressed by a capture memory. The rearrangement process results in an increase of the radius of the disk, the density being kept constant.

The primary ingredient of our analysis is that we treat the aggregation as a trapping problem whose consequence is to change the location of the trapping entity, the circumference of the disk. The two simplifying assumptions crucial to the calculation were that the shape of the disk remains circular and that the particles are absorbed into the central disk only, no other aggregation locations being formed. We then calculated $R(t)$ as a result of the absorption of the incoming material and succeeded in producing theoretical curves for the rise of the radius from its initial to its final value as observed. However, an interesting observation forced upon us in the analysis was that an ordinary capture process without the rearrangement process was unable to explain the experiment. This was so because the observed rise curve of the cluster size was not mono-exponential but was characterized by a two-time-constant nature. This necessitated the assumption of the capture memory representing the rearrangement process. We were able to extract reasonable values for the system parameters by applying the theory to Kathrin's experiments (Spendier 2020). I presented the work at an international conference in Kyoto[5] and its detail was included in Kathrin's dissertation (Spendier 2012). We are in the process of embellishing the theory with a number of additional features. I have therefore refrained from providing details of the theory.

12.7 Chapter 12 in Summary

Developed primarily for motion on discrete lattices representative of molecular crystals described in Chap. 11, the defect technique was extended to the continuum in this chapter and analyzed in novel ways for simple 1-dimensional systems, the motion without defects being described by the diffusion equation. The first focus was on taking the continuum limit of discrete lattice results. The second was to obtain and catalog 1-dimensional results for perfect and imperfect absorption. The third was to do the same for higher dimensional systems with high symmetry, explicit results being obtained in each of these cases in terms of known special functions such as the error function and Bessel functions. The basic equation was next generalized to one in which, before the introduction of the defect, the moving particles have a tendency to be attracted to a center, necessitating, therefore, the replacement of the diffusion equation by the Smoluchowski equation. Novel results were displayed in this scenario including an effect involving non-monotonicity relative to the strength of the attraction. Finally, a brief discussion was given of how to build a theory of the coalescence of signaling receptor clusters in immune cells by reversing the defect technique.

---

[5]As a keynote lecture on "Theoretical framework for the description of signal receptor cluster aggregation in cells" at the Fourteenth International Membrane Research Forum, Kyoto, Japan, on March 16, 2013.

Memory Functions from Static Disorder: Effective Medium Theory

The purpose of this chapter is to extend the formalism of memory functions to systems with static, rather than dynamic, disorder and make thereby a small but practical contribution to the venerable theory known in the literature as the "effective medium approximation". Practiced by many investigators in fields outside physics, including civil engineering, the general philosophy of that theoretical approach may be found in the work of Bruggeman (1935), Kirkpatrick (1973), Odagaki and Lax (1981), Haus et al. (1982), Haus and Kehr (1983), Haus and Kehr (1987), McCall et al. (1991) and Dyre and Schrøder (2000), as also in the studies by Parris and collaborators (Parris 1987, 1989; Parris and Kenkre 2005; Candia et al. 2007; Parris et al. 2008; Kalay et al. 2008) and the relatively recent book by Ping Sheng in the specific context of the scattering of light waves (Sheng 2006). There is no doubt in my mind that this selection of references merely skims the surface of the vast collection of sources on this popular subject: my choice has been dictated primarily by my own familiarity, and, in some instances, by clarity of exposition of the original authors.

It has long been my impression that there is a bit of magical flavor associated with this theory. The intention of what follows below is in part to examine this theoretical tool thoroughly, make, to the extent possible, a modest attempt to give it a slightly firmer footing, at least in my own mind, adding to an understanding of its domain of validity; and in part to teach its methodology by working out an explicit example. The example will be taken up in Chap. 14. The plan in both chapters is to avoid mere statements of philosophy of which there is no dearth in the literature.

Before undertaking the analysis of the effective medium theory (EMT) in the next section, it might be useful to mention a few *other* research approaches to the subject of static disorder that do not involve the EMT. They are not part of the development in these two chapters. While we will not have the space available to dwell on them in any detail whatsoever, it is suggested that the reader familiarize herself with the work of Gochanour et al. (1979), Machta (1981), Zwanzig (1982),

© The Author(s), under exclusive license to Springer Nature Switzerland AG 2021
V. M. (Nitant) Kenkre, *Memory Functions, Projection Operators, and the Defect Technique*, Lecture Notes in Physics 982,
https://doi.org/10.1007/978-3-030-68667-3_13

Wong and Kenkre (1982), Parris (1986), Bookout and Parris (1993), Dunlap et al. (1996, 1999), Novikov et al. (1998), Parris et al. (2001b) and Kenkre et al. (1998a).

13.1   General Features of Effective Medium Theory

Atoms or molecules comprising a *real* crystal do not sit at precisely the positions that the theorist who idealizes the system would like them to occupy so that she can use mathematical tools based on translational invariance. There are always interstitials, or broken bonds, or other similar irregularities. How should we treat such systems that have in them non-negligible elements of static disorder? One answer to this question is the effective medium theory in which the given spatially disordered system is replaced by a spatially *ordered* counterpart, to which Fourier transforms and related techniques may be applied, paying the price for this simplification by introducing time non-locality in the equations of motion which were originally time-local. What this means is that we introduce *memory functions* into the governing equations of motion to represent the disorder. That such a subterfuge is possible, and bonafide, should be obvious: a discussion why will be provided at the end of the present chapter. The emphasis in the book will be on explicit practical prescriptions (Kenkre et al. 2009) to calculate the memory functions from the description given about the static disorder. It will turn out that the Defect Technique introduced in Chaps. 11 and 12 will be crucial to constructing such prescriptions.

13.1.1 Explicit Prescription

We consider a system in arbitrary number of dimensions in which a particle at a lattice point m (which can be a vector in general) moves to other sites n via a Master equation such as Eq. (2.2) (no time non-locality) but with the understanding that F_{mn} do *not* obey translational invariance. For simplicity, let us focus on a one-dimensional system with nearest-neighbor interactions but bond disorder. The starting equation in traditional notation for a one-dimensional chain is

$$\frac{dP_m(t)}{dt} = F_{m+1}[P_{m+1}(t) - P_m(t)] + F_m[P_{m-1}(t) - P_m(t)]. \qquad (13.1)$$

The probability flow between the sites at $m-1$ and at m is associated with the transition rate F_m (described by a single index m). The disorder is expressed via the fact the F's are not constant throughout the system but vary probabilistically via a distribution $\rho(f)$. What this means is that a transition rate F_m can have any positive value f with probability density $\rho(f)$ normalized such that $\int_0^\infty \rho(f)df = 1$. No correlations exist in the actualization of rates at different locations. Clearly, the discrete Fourier transform technique is of little use for the diagonalization of this problem because of a lack of translational invariance.

The replacement of the given disordered time-local system by an ordered system with memory proceeds by writing in place of Eq. (13.1), a GME

$$\frac{dP_m(t)}{dt} = \int_0^t ds\, \mathcal{F}(t-s)[P_{m+1}(s) + P_{m-1}(s) - 2P_m(s)], \tag{13.2}$$

which is translationally invariant and describes elemental transfer interactions that are nearest neighbor as in the original (disordered) problem.

This, in essence, is the effective medium theory (EMT).

The crucial step to learn here in what smacks a bit of a recipe for the witches' brew is how to convert the probability distribution $\rho(f)$ into the memory function $\mathcal{F}(t)$. To that end, the reader is strongly invited to work out a simple exercise based on what our Chap. 11 has taught. Consider a particle moving on a 1-dimensional infinite chain with the same nearest neighbor rate F everywhere (thus a translationally invariant system) except between two adjacent defective points, a site at r and a site at $r + 1$. Between these sites in both directions, let the rate be given by $f \neq F$. Calculate $P_r - P_{r+1}$ for an initial occupation at the origin which, for simplicity, we assume to be r. This is a simple exercise in the application of the defect technique. You will obviously obtain an expression in the Laplace domain containing F and f. Now repeat the same exercise by taking $f = F$ so that there is no defect. Call these two the first and second parts of your exercise, respectively.

For each f that you pick from the distribution $\rho(f)$ in the first part of the exercise, you will obviously have a different value of $P_r - P_{r+1}$. Take the average of these values obtained in the first part of the exercise with the given distribution of disorder, $\rho(f)$, and require that the average equals $P_r - P_{r+1}$ calculated in the second part of the exercise in which $f = F$. You will see that the F you have been working with is a quantity in the Laplace domain, i.e., dependent on the Laplace variable. In other words, you have a memory function in the sense of the requirement of a convolution in the time domain. Let us call it $\mathcal{F}(t)$ in the time domain and $\tilde{\mathcal{F}}(\epsilon)$ in the Laplace domain. The aim of the exercise is, as a result of the equality of the results of the two parts of the exercise you have undertaken, to arrive at

$$\int_0^\infty df \frac{\rho(f)}{1 + 2[f - \tilde{\mathcal{F}}(\epsilon)][\tilde{\Pi}_0(\epsilon) - \tilde{\Pi}_1(\epsilon)]} = 1. \tag{13.3}$$

The above equation is an implicit equation for the Laplace transform of the memory $\mathcal{F}(t)$ involving the given probability distribution function $\rho(f)$, and the propagators Π_0 and Π_1 of the original ordered system: the probability of remaining on the site initially occupied is Π_0 whereas the probability of occupation of the adjacent site is Π_1. It is, in principle, a precise prescription to get $\mathcal{F}(t)$ from $\rho(f)$. Let us postpone questions about how magical the flavor of the recipe is to the end of the chapter.

Equation (13.3) that you will have derived as an exercise in applications of the defect technique has appeared in the independent work of numerous authors in the literature. Thus you will find it as Eq. (22) of (Dyre and Schrøder 2000), or Eq. (5.4) of (Kirkpatrick 1973), or Eq. (7) of (Haus et al. 1982), or Eq. (3.17) of (Odagaki and Lax 1981) or Eq. (38) of (McCall et al. 1991).[1] All these authors and others have made valuable contributions to effective medium theory (EMT).

Let us now put this *explicit prescription* that converts disorder information in $\rho(f)$ into the effective medium memory function, into a simple and convenient form (Kenkre et al. 2009)

$$\int_0^\infty df\, \frac{\rho(f)}{f+\xi} = \frac{1}{\tilde{\mathcal{F}}+\xi}. \qquad (13.4)$$

While Eqs. (13.3) and (13.4) are equivalent to each other, the form of the latter will be found to be particularly useful in what follows.

All that is necessary for the conversion of (13.3) into (13.4) is a simple manipulation based on the Laplace transform of the relationship,

$$\frac{d\Pi_0(t)}{dt} = 2\int_0^t ds \mathcal{F}(t-s)\,[\Pi_1(t) - \Pi_0(t)],$$

which is obvious on realizing that the propagators Π_m obey Eq. (13.2), and invoking the spatial symmetry of the propagators. Kenkre et al. (2009) rewrote the above result as

$$\int_0^\infty df\, \frac{\rho(f)}{f + \tilde{\mathcal{F}}(\epsilon)\left[\frac{\epsilon\tilde{\Pi}_0(\epsilon)}{1-\epsilon\tilde{\Pi}_0(\epsilon)}\right]} = \frac{1}{\tilde{\mathcal{F}}(\epsilon)}\left[1 - \epsilon\tilde{\Pi}_0(\epsilon)\right], \qquad (13.5)$$

and then, by introducing a quantity ξ, stated their prescription. The quantity $\xi(\epsilon, \tilde{\mathcal{F}})$ is a function of both ϵ and of $\tilde{\mathcal{F}}(\epsilon)$ since the self-propagator $\tilde{\Pi}_0$ depends explicitly on $\tilde{\mathcal{F}}(\epsilon)$ as well as on ϵ. Generally,

$$\xi = \tilde{\mathcal{F}}(\epsilon)\left[\frac{\epsilon\tilde{\Pi}_0(\epsilon)}{1 - \epsilon\tilde{\Pi}_0(\epsilon)}\right]. \qquad (13.6)$$

For the infinite 1-d chain with nearest neighbor rates, given that, in this case of nearest neighbor rates in a chain, $\epsilon\tilde{\Pi}_0$ equals $[1+4\tilde{\mathcal{F}}(\epsilon)/\epsilon]^{-1/2}$, one has the specific expression

$$\xi = \frac{\epsilon}{4}\left(1 + \sqrt{1 + \frac{4\tilde{\mathcal{F}}(\epsilon)}{\epsilon}}\right). \qquad (13.7)$$

---

[1]The correctness of your solution of the exercise should be, thus, verified at least through the mighty power of democracy. No reason to worry, as some say, that it is teetering these days on the brink of a precipice.

As we will see below, this restatement (13.4) of the basic EMT equation (13.3) allows us to obtain a number of our results in a straightforward fashion. With very few exceptions in the literature, the result (13.3) is used in the long-time limit and therefore involves the Markoffian replacement of $\mathcal{F}(t)$ by $\delta(t)[\int_0^\infty \mathcal{F}(s)ds]$. This is equivalent to the $\epsilon \to 0$ limit. Note that Eq. (13.4) smoothly reduces to the well known result (Zwanzig 1982) that the effective transfer rate $F_{eff} = \int_0^\infty \mathcal{F}(s)ds = \tilde{\mathcal{F}}(\epsilon \to 0)$ equals the harmonic mean of the disordered f's:

$$\frac{1}{F_{eff}} = \frac{1}{\tilde{\mathcal{F}}(0)} = \int_0^\infty df \, \frac{\rho(f)}{f}. \tag{13.8}$$

That Eq. (13.8) is blessed with real physics content should be appreciated on asking what one might expect the effective rate F_{eff} to be for motion on a chain for which there may be a broken bond. That means a $\rho(f)$ that is non-zero for $f = 0$. The obvious result, that F_{eff} vanishes for this situation rather than equaling the arithmetic average of f, i.e., $\int df \, f\rho(f)$, is contained in the prescription.

While the basic equation (13.3) has been known to numerous workers in effective medium theory, our contribution will be to go a step beyond by deriving some general features of the EMT memory in section, describing our extensions of the theory for times that are not asymptotic, studying the validity of the results vis-a-vis exact solutions (obtained numerically). We will also analyze the perhaps surprising emergence of spatially long range memories and study finite size effects. A step by step demonstration of how to use the EMT in new problems will also be part of our study, and occupy us in the next chapter.

The first of the results of our present investigation is the reformulation implicit in Eq. (13.4) interpreted as a *transform* of the distribution function $\rho(f)$ (disorder information) into the effective medium quantity $\mathcal{F}(t)$ (temporal memory). Specifically, we can regard Eq. (13.4) as related to a double Laplace transform. One applies the *direct* Laplace transform twice: first to $\rho(f)$, with a dummy variable y as the Laplace variable, to obtain $g(y)$, and then to $g(y)$ with ξ as the Laplace variable to obtain $h(\xi)$:

$$g(y) = \int_0^\infty \rho(f)e^{-yf}df; \quad h(\xi) = \int_0^\infty g(y)e^{-\xi y}dy.$$

The prescription for extracting the memory $\mathcal{F}(t)$ in the EMT equation (13.2) from the disorder distribution $\rho(f)$ consists, thus, of computing the double transform $h(\xi)$ of the disorder distribution, equivalently performing the integral on the left side of (13.4), and inverting into the time domain the memory transform $\tilde{\mathcal{F}}(\epsilon)$ after solving for it from the implicit equation

$$h(\xi) = \frac{1}{\tilde{\mathcal{F}}(\epsilon) + \xi(\epsilon, \tilde{\mathcal{F}})}. \tag{13.9}$$

One has, thus, a practical prescription to obtain the t dependence of the memory $\mathcal{F}(t)$ from the disorder $\rho(f)$.[2]

13.1.2 Nature of the EMT Memory and the Exponential Approximation

If you apply the prescription of Eq. (13.4) to various distribution functions $\rho(f)$ explicitly to obtain $\widetilde{\mathcal{F}}(\epsilon)$, you discover that the results share a number of common features. These features become apparent on inverting the transform to obtain $\mathcal{F}(t)$, the memory function in the time domain, and can be understood simply as we argue below. They also suggest a simple approximation to express the EMT memory in terms of arithmetic and harmonic averages of the transfer rates.

After performing the Laplace inversion, we typically find that the EMT memory $\mathcal{F}(t)$ consists of two pieces, a δ-function at the origin of time ($t = 0$) and a part that is negative but finite. Consider the actual system evolution equation (13.1) on the one hand, and the representative EMT equation (13.2) on the other, both for an initial occupation of only the site m. Let us first evaluate the first time derivative of $P_m(t)$ at the initial time. The respective results are

$$\left[\frac{d P_m(t)}{dt}\right]_{t=0} = -(F_m + F_{m+1}) \tag{13.10}$$

for the actual Master equation, and

$$\left[\frac{d P_m(t)}{dt}\right]_{t=0} = \int_{0-}^{0+} ds \mathcal{F}(t-s)[P_{m+1}(s) + P_{m-1}(s) - 2P_m(s)] \tag{13.11}$$

for the representative EMT equation. A configuration average over the distribution $\rho(f)$ converts the right hand side of Eq. (13.10) into $-2\langle f \rangle = -2 \int df \rho(f) f$. It is impossible for Eq. (13.11) to yield a non-zero result (because of the limits of integration) unless $\mathcal{F}(t)$ contains a δ-function at the origin. As Eqs. (13.10) and (13.11) must yield results that equal each other, we deduce that the form of the EMT memory function is

$$\mathcal{F}(t) = \langle f \rangle \delta(t) - Q(t). \tag{13.12}$$

The origin of the δ-function at $t = 0$ is clear from the above analysis. That the additional part must have a time integral for all time which is negative follows from the general result (13.8) that the integral over all time of $\mathcal{F}(t)$ is the harmonic mean, i.e., the reciprocal of $\langle 1/f \rangle = \int df \rho(f)/f$. The harmonic mean is always smaller

---

[2]The usefulness of the form of the basic equation we have presented, Eq. (13.4), should be already clear by comparison to the well-known asymptotic result for the effective rate Eq. (13.8).

than (or equal to) the arithmetic mean $\langle f \rangle$, as proved briefly but explicitly in Kenkre et al. (2009). The interested reader might want to take up this proof as an exercise and consult footnote 36 in the original publication as the solution to check against.

We also note that the integral of $Q(t)$ over all time is now completely determined as the difference between the arithmetic and harmonic averages of the transition rate with the given probability distribution:

$$\int_0^\infty Q(t)dt = \langle f \rangle - \frac{1}{\langle 1/f \rangle}. \tag{13.13}$$

Additional information can be obtained in this exact manner about the memory function, for instance, the initial value of $Q(t)$. Differentiation of Eq. (13.1) with respect to time yields the initial *second* time derivative

$$\left[\frac{d^2 P_m(t)}{dt^2} \right]_{t=0} = 2 \left(F_{m+1}^2 + F_m^2 + F_{m+1} F_m \right). \tag{13.14}$$

Similarly, differentiation of the EMT generalized master equation (13.2) yields, after a configuration average,

$$\left[\frac{d^2 P_m(t)}{dt^2} \right]_{t=0} = 6\langle f \rangle^2 + 2Q(0). \tag{13.15}$$

Carrying out the configuration average of the former result, which gives $4\langle f^2 \rangle + 2\langle f \rangle^2$, and equating the two values of the second time derivatives at the initial time, we can evaluate $Q(0)$ *exactly* for any distribution function as

$$Q(0) = 2 \left[\langle f^2 \rangle - \langle f \rangle^2 \right] = 2 \left[\left(\int df \rho(f) f^2 \right) - \left(\int df \rho(f) f \right)^2 \right]. \tag{13.16}$$

It is also straightforward to continue in this manner with further differentiations to obtain exact initial values of higher derivatives of $Q(t)$. For instance, in terms of the A-matrix appearing in Eq. (4.29), we can evaluate the initial value of the rth derivative of P_m via

$$\left[\frac{d^r P_m}{dt^r} \right]_0 = (-1)^r (A^r)_{mm}$$

and proceed as shown above with configuration averages. For our nearest neighbor rate system we have $A_{mn} = -F_{m+1}\delta_{m,n+1} - F_m\delta_{m,n-1} + (F_{m+1} + F_m)\delta_{m,n}$.

Let us use these straightforward considerations to develop a simple analytical approximation to the memory cast in the form of a difference of a term proportional to a delta function and another to an exponential. Let us call this the 'exponential' approximation to the EMT memory and denote it by $\mathcal{F}_a(t)$. For any given distribu-

tion of the rates, it is specified in terms of the arithmetic and harmonic averages of
the rates as

$$\mathcal{F}_a(t) = \langle f \rangle \delta(t) - 2\left[\langle f^2 \rangle - \langle f \rangle^2\right] e^{-t\left(\frac{2(\langle f^2 \rangle - \langle f \rangle^2)}{\langle f \rangle - (\langle 1/f \rangle)^{-1}}\right)}. \tag{13.17}$$

The subscript a clarifies that the memory is approximate. While the precise shape
of the actual memory function may not be captured by our approximation (13.17),
examples we have worked out make clear that the approximate memory can be
remarkably good. Indeed, it is represented by the dotted line in Fig. 13.1 for an
example of a specified distribution that will be treated, and remarked on, in the
next section. The near-coincidence of the approximate result with the exact EMT
memory (solid line) is quite remarkable!

The general behavior of the time dependence of the memory function consisting
of a decay (infinitely fast for our system here) to negative values and then a rise
which is slower is typical in many systems having nothing to do with disorder. It
is usually encountered in studies of the velocity autocorrelation $\langle v(t)v \rangle$ which is,
needless to say, very closely related (in our case simply proportional) to the memory
function. The small time behavior represents initial transfer at a higher rate; the
subsequent behavior is affected by disorder or imperfections in the system as they

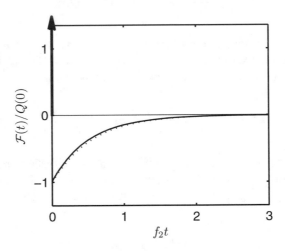

Fig. 13.1 Typical shape of the effective medium memory function $\mathcal{F}(t)$ showing the delta
function of strength $\langle f \rangle$ at the origin and the negative piece $Q(t)$. The time integral from 0 to
∞ of the memory $\mathcal{F}(t)$ is $1/\langle 1/f \rangle$. While the shape of the memory is typical, solid lines do depict
the exact (numerical) memory as calculated for a specific case treated in Sect. 13.1.3, the double-
delta distribution (13.18). For this plot, the concentration of the smaller of the rates is 0.9 and the
ratio of the transition rates is 10. Also plotted is the exponential approximation (dotted line) given
by our formula (13.17). It is surprising how close the agreement is, given the coarse nature of
the approximation. Time is plotted in units of the reciprocal of one of the two possible rates that
can occur randomly. Reprinted with permission from fig. 3 of ref. Kenkre et al. (2009); copyright
(2009) by the American Physical Society

are encountered in the motion. Indeed, the velocity autocorrelation for a random walker completely confined to a finite space exhibits this very behavior, the overall integral of $\langle v(t)v \rangle$ for all time being precisely zero because of the confinement: the mean square displacement saturates in this case (see, e.g., Sheltraw and Kenkre (1996) for a nuclear magnetic resonance context.)

13.1.3 Some Specific Distributions for Transition Rates

Let us now inspect EMT results for a few specific distributions $\rho(f)$.

A natural distribution to consider is $\rho(f) = \sum_{i=1}^{M} \alpha_i \delta(f - f_i)$, the multi-delta distribution, wherein the nearest-neighbor transition rates may take one of M values f_i each with a weight α_i, with $\sum_{i=1}^{M} \alpha_i = 1$. Let us focus on the case $M = 2$, so that

$$\rho(f) = \alpha\delta(f - f_1) + (1 - \alpha)\delta(f - f_2). \tag{13.18}$$

There are here only two admissible values to the transfer rate, f_1 and f_2, and they appear with respective probabilities α and $1 - \alpha$. The arithmetic mean of the rates is clearly $\langle f \rangle = \alpha f_1 + (1 - \alpha)f_2$. Also, it follows that

$$\frac{1}{\langle 1/f \rangle} = \frac{f_1 f_2}{\alpha f_2 + (1 - \alpha)f_1}, \quad \langle f^2 \rangle = \alpha f_1^2 + (1 - \alpha)f_2^2. \tag{13.19}$$

The distribution $\rho(f)$ for this case is shown as the two arrows in Fig. 13.2.

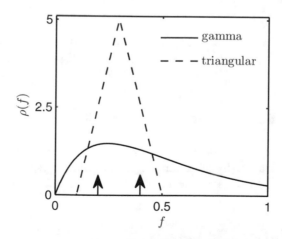

Fig. 13.2 Examples of probability distributions $\rho(f)$. The two arrows represent the double-delta distribution with equal weight $\alpha = 1 - \alpha = 0.5$. A gamma distribution with $n = 1$ and $\gamma = 4$ and a triangular distribution with $f_0 = 0.3$ and $b = 0.2$ are depicted by the solid and dashed lines respectively. Units of f in the plot are arbitrary and the same as those of f_0 and b for the triangular distribution, and reciprocal to those of γ for the gamma distribution. Reprinted with permission from fig. 2 of ref. Kenkre et al. (2009); copyright (2009) by the American Physical Society

Another distribution to consider is the gamma distribution (related closely to the Poisson distribution):

$$\rho(f) = \frac{\gamma^{n+1}}{\Gamma(n+1)} f^n e^{-\gamma f}.$$ (13.20)

The arithmetic mean is $\langle f \rangle = (n+1)/\gamma$. Furthermore,

$$\frac{1}{\langle 1/f \rangle} = \frac{n}{\gamma}, \quad \langle f^2 \rangle = \frac{(n+1)(n+2)}{\gamma^2}.$$ (13.21)

A plot of $\rho(f)$ itself is displayed for the particular case of $n = 1$ and $\gamma = 4$ in Fig. 13.2.

The third case we consider is the triangular distribution given by

$$\rho(f) = \begin{cases} (f - f_0 + f_b)/f_b^2 & f_0 - f_b \le f \le f_0, \\ (-f + f_0 + f_b)/f_b^2 & f_0 \le f \le f_0 + f_b, \\ 0 & \text{elsewhere.} \end{cases}$$ (13.22)

The minimum possible rate is $f_0 - f_b$ and the maximum possible rate is $f_0 + f_b$. The distribution rises linearly from the minimum value with slope $1/f_b^2$ until it attains the value $1/f_b$ at $f = f_0$ and then descends with the same magnitude of the slope down to the maximum value. The meaning of f_0 is that it is the value of f at the apex (and hence the mean of the distribution), and f_b is half the length of the base of the triangle. The distribution is shown in Fig. 13.2 for $f_0 = 0.3$ and $f_b = 0.2$. It leads to $\langle f \rangle = f_0$, and to

$$\langle f^2 \rangle = f_0^2 + \frac{f_b^2}{6}, \quad \frac{1}{\langle 1/f \rangle} = \frac{f_b}{\ln\left(1 + \frac{2f_b}{f_0 - f_b}\right) + \frac{f_0}{f_b} \ln\left(1 - \frac{f_b^2}{f_0^2}\right)}.$$ (13.23)

13.1.4 Evaluation of Memories for Given Distributions

The approximation to the memory given by the formula (13.17) is easily evaluated for the three distributions by substituting in the formula the respective values of $\langle f \rangle$, $\langle f^2 \rangle$ and $1/\langle 1/f \rangle$. As one example, note that for the gamma distribution it is given by

$$\mathcal{F}_a(t) = \left(\frac{n+1}{\gamma}\right) \delta(t) - 2\left(\frac{n+1}{\gamma^2}\right) e^{-\frac{2(n+1)t}{\gamma}}.$$ (13.24)

The memory function $\tilde{\mathcal{F}}(t)$, whether derived from (13.4) or the simpler (13.17), can be used immediately to calculate other, more directly observable, quantities.

A useful quantity is the (dimensionless) mean square displacement $\langle m^2 \rangle = \sum_m m^2 P_m(t)$ for initial localization at the origin. It is simply twice the double time integral of the memory:

$$\langle m^2 \rangle = \sum_m m^2 P_m(t) = 2 \int_0^t ds \int_0^s \mathcal{F}(y)dy.$$

The time-dependent diffusion coefficient $D(t)$, a quantity often used in transport theory to describe the instantaneous state of motion, may be defined as one half the product of the square of the intersite distance a and the time derivative of the mean square displacement. It is proportional to a single time integral of the memory function:

$$D(t) = \frac{a^2}{2}\left(\frac{d\langle m^2 \rangle}{dt}\right) = a^2 \int_0^s \mathcal{F}(s)ds.$$

These have exact expressions in terms of $\langle f \rangle$ and $Q(t)$ appearing in Eq. (13.12). If we use our simple exponential approximation for $Q(t)$, they become

$$\frac{\langle m^2 \rangle}{2} = \frac{t}{\langle 1/f \rangle} + \frac{\left(\langle f \rangle - \langle 1/f \rangle^{-1}\right)^2}{2(\langle f^2 \rangle - \langle f \rangle^2)}\left(1 - e^{-t/\tau}\right), \tag{13.25}$$

$$\frac{D(t)}{a^2} = \langle f \rangle - \left(\langle f \rangle - \langle 1/f \rangle^{-1}\right)\left(1 - e^{-t/\tau}\right), \tag{13.26}$$

where the time constant τ is given by

$$\tau = \left(\frac{\langle f \rangle - (\langle 1/f \rangle)^{-1}}{2(\langle f^2 \rangle - \langle f \rangle^2)}\right).$$

It is straightforward to get expressions particular to the distribution functions chosen. As expected, the mean square displacement starts out linearly with slope twice the arithmetic mean of the rates and ends up also linearly with slope twice the harmonic mean of the rates. Correspondingly, the time-dependent diffusion constant decays from a higher to a lower value.

There are a number of ways the above simple analysis can be put to use to extract physical information. For instance, the mean square displacement of a walker initially localized at a single site will first grow linearly but then saturate to a finite value at long times if there are broken bonds in the 1-d infinite system. Broken bonds correspond to a $\rho(f)$ that has a non-zero value at $f = 0$ which means that there are bonds at which the transition rate is zero. In such a case, $1/\langle 1/f \rangle$, the harmonic mean of the rates, and consequently the long time $D(t)$, vanish. Equation (1.12) can then be used to extract the value at which the mean square displacement saturates at long times:

$$\lim_{t \to \infty} \langle m^2 \rangle = \frac{\langle f \rangle^2}{\langle f^2 \rangle - \langle f \rangle^2}. \tag{13.27}$$

This consequence of the exponential approximation (13.17) to the memory is simply a case of the general EMT result

$$\lim_{t \to \infty} \langle m^2 \rangle = -2 \lim_{\epsilon \to 0} \frac{d\tilde{Q}(\epsilon)}{d\epsilon}. \tag{13.28}$$

This may be proved from the Laplace transform of the non-delta part of the memory in Eq. (13.12) via a Taylor expansion:

$$\tilde{\mathcal{F}}(\epsilon) = \langle f \rangle - \tilde{Q}(\epsilon) = \frac{1}{\langle 1/f \rangle} - \epsilon \left[\frac{d\tilde{Q}(\epsilon)}{d\epsilon} \right]_{\epsilon=0} - \frac{\epsilon^2}{2} \left[\frac{d^2\tilde{Q}(\epsilon)}{d\epsilon^2} \right]_{\epsilon=0} \cdots$$

In the presence of broken bonds in 1-d, the harmonic mean of f's vanishes. Since the mean square displacement is twice the double time integral of $\mathcal{F}(t)$, the limit $\epsilon \to 0$ and the use of an Abelian theorem establish Eq. (13.28) quite generally. If $\mathcal{F}(t)$ is expressed via the exponential approximation (13.17), the general result reduces to Eq. (13.27).

Despite what appears as an impressive agreement of the exponential approximation that we see displayed in Fig. 13.1 for a double delta distribution with $\alpha = 0.9$, and $f_2/f_1 = 10$, the approximation generally will not capture the actual decay in time for all distribution functions and may be regarded only as a highly simplified manner of description. For greater accuracy than can be provided by the relatively coarse approximation of Eq. (13.17), it is necessary to return to the prescription of Eq. (13.4), calculate $\tilde{\mathcal{F}}(\epsilon)$ through the solution of the implicit equation, and then invert the transform to obtain the memory.[3]

The calculation of $\tilde{\mathcal{F}}$ from Eq. (13.4) is easy and analytically doable for the double-delta distribution. The definition $\eta = (1 - \alpha) f_1 + \alpha f_2$, leads to the soluble cubic

$$\tilde{\mathcal{F}}^3 - 2\tilde{\mathcal{F}}^2 \left(2\eta^2/\epsilon + f_1 + f_2 \right)$$

$$+\tilde{\mathcal{F}} \left(8\eta f_1 f_2/\epsilon + \left(2 f_1 f_2 + (f_1 + f_2)^2 - \eta^2 \right) \right)$$

$$-4 f_1^2 f_2^2/\epsilon - (2 f_1 f_2 (f_1 + f_2) - 2\eta f_1 f_2) = 0. \tag{13.29}$$

Standard analytic formulae yield the appropriate solution which can then be numerically Laplace-inverted. Similar procedures can be used for the gamma

---

[3] When the EMT memory is calculated in the Laplace domain via our prescription based on Eq. (13.4), the derived quantities $D(t)$ and $\langle m^2 \rangle$ can be obtained very simply in the Laplace domain by dividing $\tilde{\mathcal{F}}(\epsilon)$ by ϵ and ϵ^2 (except for proportionality constants) respectively.

distribution and the triangular distribution. Explicit polynomials do not result for $\widetilde{\mathcal{F}}$ in those cases but the equations can be solved numerically and inverted. Kenkre et al. (2009) have carried out these procedures for these two distributions as well and reported the results after inversions into the time domain.

13.2   Validity of EMT for Short as well as Long Times

Let us now display the results of the predictions of effective medium theory and the numerically obtained exact evolution not only for long times as is usually done, but for short and intermediate times as well. For each distribution, the exact and the full EMT results can both be explicitly calculated. The exact results are obtained via numerical matrix operations. With the exception of single-run studies to be reported further below, the operations are repeated tens of thousands of times, each time using a different realization of the chain. Then all runs are averaged to produce the quantity desired. The effective medium theory prediction for that quantity is also determined via the effective medium memory function both in its full form as given from our Eq. (13.4), and from the analytic approximation, Eq. (13.17).

We first treat the case when a single site is initially fully occupied.

13.2.1 Localized Initial Condition

Kenkre et al. (2009) carried out the comparison graphically. We refer the reader to that demonstration given in that original investigation, the quantity selected being the time-dependent diffusion coefficient normalized to its initial value: $D(t)/D(0)$. All three distributions were considered. The agreement of the effective medium theory with the exact evolution is found to be remarkably good for all cases considered and for all intermediate times as well. The description appears thus excellent for the parameters considered for times that need not be asymptotic.

In order to explore parameter values for which the agreement may *not* be as good, those authors also restricted themselves to the double-delta distribution in (a) of Fig. 13.2, took the two possible rates f_1 and f_2 to occur with equal weight ($\alpha = 0.5$), but varied the ratio: $f_1/f_2 = 0.5, 0.1, 0.01$. Perhaps surprisingly, the EMT was still found to provide a fine description for all times but to deviate more from the exact description as the rates become more disparate. To drive this situation to an extreme where the EMT would serve *worst*, a broken bond system was considered which we show here in (b) of Fig. 13.3. This means we take $f_1 = 0$ and $f_2 \neq 0$ for different values of the concentration α. The large time value of $D(t)$ is now zero and the mean square displacement $\langle m^2 \rangle$ (proportional to the integral of $D(t)$) saturates. Physically, the saturation value measures the size of clusters (separated by broken bonds from other clusters) on which the walker is localized at long times.

Analytic expressions for the saturation value from the full EMT are already available in Eq. (13.28) and from the exponential approximation in Eq. (13.27). Figure 13.3 is an attempt at looking at EMT in the worst possible light by comparing

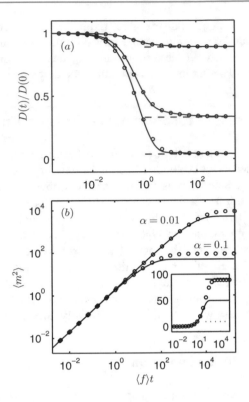

Fig. 13.3 Worst-case scenario comparison of EMT and exact results. In (**a**) we plot the time-dependent diffusion coefficient (normalized to its initial value) as a function of the dimensionless time $\tau = \langle f \rangle t$ for the double-delta distribution function for $\alpha = 0.5$ and, from top to bottom, $f_1/f_2 = 0.5$, $f_1/f_2 = 0.1$, $f_1/f_2 = 0.01$. While good, the agreement gets worse for disparate f's. To explore a regime in which the agreement is bad, in (**b**) we consider two broken-bond systems (the ratio of the f's being zero and therefore extreme) with two different concentrations $\alpha = 0.1, 0.01$ as shown. Plotted is the mean square displacement showing saturation at long times. Here $f_1 = 0$ and $f_2 = 0.2$. Open circles correspond to the exact (numerical) solution obtained by averaging over 20,000 different realizations of the disordered chain which consists of 801 sites. Solid lines are theoretical results from the EMT. The inset shows the $\alpha = 0.01$ case, the ordinate being on a linear scale in units of 10^4. See text for discussion. Reprinted with permission from fig. 5 of ref. Kenkre et al. (2009); copyright (2009) by the American Physical Society

the time evolution of $\langle m^2 \rangle$ predicted by it to that given by exact calculations. We do this for the broken bond case ($f_1 = 0$) for two concentrations α of broken bonds: 0.01 and 0.1 as shown. The main display in Fig. 13.3 shows the two $\langle m^2 \rangle$ curves. To make the discrepancy of the saturation value particularly clear, we show the inset in which the one case $\alpha = 0.01$ is displayed on a semilogarithmic scale. The abscissa is the dimensionless time $\langle f \rangle t$ as in the main figure. The ordinate is $\langle m^2 \rangle$ on a linear scale, the values displayed as 0.5 and 1 being 5000 and 10, 000 (i.e., in units of 10^4). The accumulated values of the mean square displacement, the localization cluster sizes, are $8.71 \dot{x} 10^3$ from the exact calculations but only 5.00×10^3 from the EMT: both are denoted by solid lines in the inset. The corresponding values for the $\alpha = 0.1$ case are 88.6 and 49.5 respectively. The exponential approximation to the EMT is way off as it predicts 99 for the $\alpha = 0.01$ case and 9 for $\alpha = 0.1$. (The latter is denoted by a dotted line in the inset.) This is to be expected from the crudeness of that approximation.

13.2.2 Spatially Extended Initial Condition (Single Runs)

An actual experiment in a real physical situation is performed not on an ensemble but on an individual system. How can EMT, which has at its root an ensemble average, provide a valid description for the experiment? Standard Gibbs-Boltzmann arguments do not help as an answer here because our interest in using EMT is not only for asymptotic times when the system might have completed the mixing process but for all times. One possible answer to this question might lie in the nature of the *initial* condition. If it is *extended in space*, various configurations of transition rates in a random system may be realized even at short times. With this idea in mind we now describe our investigation of extended initial conditions for *single runs*. In particular, we study the agreement of EMT and single-run evolution of the actual system as we vary the spatial extent of the initial condition.

We carry out calculations from exact numerical considerations for systems of 801 sites without changing the configurations of the transition rates once set in accordance with the double-delta distribution, and take initial conditions that are not of the form $P_m(0) = \delta_{m,0}$, but of the extended form $P_m(0) = [1/(2\mu + 1)] \sum_{r=-\mu}^{\mu} \delta_{n,r}$ which represents a *patch* initial condition of spatial extent of $2\mu + 1$ sites. We call this the initial width. The limit $\mu = 0$ gives us back the initial condition we have used in the studies above. We find that larger patches result in smaller deviations of the EMT predictions from the exact results.

The inset of Fig. 13.4 shows values of $D(t)/D(0)$ for a single configuration for two different values of the width (open circles represent $\mu = 50$, crosses represent $\mu = 5$) along side the corresponding prediction of effective medium theory (solid line). The integrated difference between the EMT result and the exact results depicted in that inset (as plotted on a logarithmic time scale) provides a convenient measure of the error. We thus define, for each value of μ, a measure of the relative error, through the expression

$$E_R = \int_{-\infty}^{\infty} \frac{D^{EMT}(s) - D^{EX}(t)}{D^{EX}(s)} ds$$

where $s = \ln(\langle f \rangle t)$. A plot of the (numerically evaluated) relative error as a function of the initial width μ is presented in the main graph in Fig. 13.4, and clearly shows that the relative error decreases monotonically as the patch width increases.

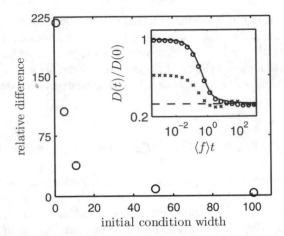

Fig. 13.4 Relative difference between exact results and EMT, or error of the EMT, for a single run. Plotted as large open circles is the error (see text for definition) as a function of the number of sites initially occupied, i.e., the value $2\mu + 1$. No averages are performed. The error is seen to decrease as the initial width increases, allowing the walker to sample different configurations. Inset: Comparison of $D(t)/D(0)$ curves for two different values of the initial width, $\mu = 5$ (crosses) and $\mu = 50$ (open circles), with the EMT (solid line). It is clearly seen that EMT agrees with the simulations for spatially extended initial conditions without averaging. Reprinted with permission from fig. 6 of ref. Kenkre et al. (2009); copyright (2009) by the American Physical Society

13.2.3 Correlation Type Observables

There are, in general, different kinds of observables that can be computed from the solution to the Master equation. Simple observables O are those which associate with each site (state) m, a value O_m that the observable takes when the particle is in that state. The mean value associated with such an observable at any time t can then be written as

$$\langle O(t) \rangle = \sum_m O_m P_m(t) = \sum_{m,n} O_m \Pi_{m,n}(t) P_n(0), \qquad (13.30)$$

where in the second form we have expressed the result in terms of the propagators Π, and the *initial* probability distribution governing the particle's occupation of the possible states of the system. This can be put in the form

$$\langle O(t) \rangle = \sum_n \langle O(t) \rangle_n P_n(0) \qquad (13.31)$$

where

$$\langle O(t) \rangle_n = \sum_m O_m \Pi_{m,n}(t) \qquad (13.32)$$

is the mean value of the observable given that the particle started in state n at time $t = 0$. Simple observables can thus be calculated by incorporating into the averaging process an average over the different possible starting locations of the particle.

Correlation type observables, also of great interest in statistical physics, do not correspond to (simple) observables of this type. Indeed, they *span* two or more different states (or the same state at two different times). An example is $\langle A(t) B(0) \rangle$ given by

$$\sum_{m,n} A_m \Pi_{m,n}(t) B_n P_n(0) = \sum_n \langle A(t) \rangle_n B_n P_n(0).$$
(13.33)

Here A_m and B_m are, respectively, the values of A and B when the particle is in the state m, and

$$\langle A(t) \rangle_n = \sum_m A_m \Pi_{m,n}(t)$$
(13.34)

is the mean value of A at time t if the particle started in state n at $t = 0$. Consider, for instance A^ℓ with components

$$A_m^\ell = \delta_{m,\ell}.$$
(13.35)

It is an indicator observable taking the value 1 if the particle is at site ℓ and the value 0 otherwise. Then the correlation function

$$\langle A^\ell(t) A^{\ell'}(0) \rangle = \Pi_{\ell,\ell'}(t) P_{\ell'}(0)$$
(13.36)

is just the propagator $\pi_{\ell,\ell'}$ weighted by the relative initial probability of finding the particle in the state ℓ'. This shows that the propagators $\pi_{\ell,\ell'}$ themselves can also be considered as observables of the system. Of course, in a specific disordered system, the self-propagator $\pi_{\ell,\ell}(t)$, e.g., will depend on the location of site ℓ in the disordered chain. The effective medium propagator may not, therefore give a good approximation to any given self-propagator $\pi_{\ell,\ell}(t)$ in any single realization of the disordered system. We intuitively expect, however, that self-propagators, averaged over an initial distribution of starting positions on the same chain, will approach that of the effective medium, as the width of the initial distribution of starting sites is increased, i.e., that

$$\lim_{\mu \to \infty} \frac{1}{2\mu + 1} \sum_{\ell=-\mu}^{\mu} \pi_{\ell,\ell} = \Pi_0.$$

This intuition was indeed shown to be verified by calculations carried out and reported in Kenkre et al. (2009).

13.3   Additional Surprises in Effective Medium Theory

There are additional effects that appear as consequences of EMT. You may or may not find them surprising but they are certainly interesting. One is the appearance of (spatially) long range memories in the manner they did in Chap. 4 for systems described by microscopic Hamiltonians. The other is relevant to finite size effects. Let us study them in turn.

13.3.1 Appearance of Spatially Long Range Memories

Both equations, the original one that describes the disordered system, Eq. (13.1), and the EMT counterpart (13.2), are local in time and nearest-neighbor in the character of its transition rates. Given that EMT provides an approximate rather than exact description of the actual dynamics described by Eq. (13.1), one may ask whether the introduction of non-locality in time in the EMT should be accompanied by non-locality in space as well. Spatially long range memories have already appeared for quantum mechanical systems in the last part of Chap. 4. Stated differently, the question we ask is whether the replacement of Eq. (13.1) by Eq. (13.2) with nearest-neighbor transition memories is sufficient or whether the latter should span longer distances. The answer was found in Kenkre et al. (2009) and is explained in this section.

Consider Eq. (13.1) solved for $\tilde{P}_m(\epsilon)$, the Laplace transform of the probability of occupation of the mth site in terms of the matrix A^μ corresponding to the configuration μ (a particular realization of the transition rates f throughout the system). Carrying out the average over the configurations μ one gets a translationally invariant situation:

$$\tilde{P}_m(\epsilon) = \sum_n \left\langle \frac{1}{\epsilon + A^\mu} \right\rangle_{m-n} P_n(0). \tag{13.37}$$

A discrete Fourier transform yields $\tilde{P}^k(\epsilon)/P^k(0)$. Exploiting the general relation between memories and probabilities explained and used in the passage of Eq. (1.24) to Eq. (1.25), we get the *exact* memory function:

$$\tilde{\mathscr{A}}^k = \left\langle \frac{1}{\epsilon + A} \right\rangle^k - \epsilon. \tag{13.38}$$

The superscript μ on the A has not been displayed because the configuration average has been carried out already at this point.

There is no guarantee whatsoever that the k-dependence of $\tilde{\mathscr{A}}^k$ is of the form $(1-\cos k)$. The exact memories need not, therefore, have nearest neighbor character. The nature of the disorder will influence the k-dependence. It is therefore clear that spatially long range memories will naturally develop, in general. The particular disorder distribution $\rho(f)$ will determine their precise form. Contrast this result of the exact procedure with that of the EMT procedure, which necessarily results in the absence of spatially long range memories. This is so because one *assumes* the memories to be nearest-neighbor in character while obtaining them variationally.

In Fig. 13.5 we display the result of the full numerical exact procedure outlined above carried out on a chain of 100 sites, making sure during each run that the value of $P_m(t)$ is negligible (comparable to the precision of the machine used) at the boundaries of the chain. The distribution used is double-delta, the two rates are in the ratio $f_1/f_2 = 0.1$ and the concentration of each is equal to the other. We plot in (a) the Laplace transform of the nearest-neighbor memory obtained from the exact procedure (solid line), $\widetilde{\mathcal{F}}_1(\epsilon)$, as a function of the Laplace variable ϵ, both the abscissa and the ordinate being expressed in units of the average rate $\langle f \rangle$. Also plotted is the result of the EMT procedure (dots) and the dashed line that represents the asymptotic rate $1/\langle 1/f \rangle$. There is hardly any difference in the exact and the EMT result. The *a posteriori* conclusion is that the non-nearest neighbor memories are *much* smaller in magnitude relative to the nearest neighbor $\widetilde{\mathcal{F}}_1(\epsilon)$. This is shown clearly in (b) where the longer range memory transforms, $\widetilde{\mathcal{F}}_n(\epsilon)$, are shown. The scales in the plots in (a) and (b) differ by a little less than 3 orders of magnitude so

Fig. 13.5 Spatially long range memories obtained from exact numerical considerations plotted as a function of the Laplace variable. Units on both axes are of $\langle f \rangle$. Plotted in (**a**) is the exact $\widetilde{\mathcal{F}}_1(\epsilon)$ (solid line), calculated for rings of 100 sites, and the almost identical EMT memory (open circles) along with the asymptotic rate $1/\langle 1/f \rangle$ (dashed line). Plotted in (**b**) on a scale blown up by almost 3 orders of magnitude are the much smaller long range memories $\widetilde{\mathcal{F}}_n(\epsilon)$ in dotted, dashed and solid lines, for $n = 2, 3, 4$, respectively. The distribution is double-delta with $f_1/f_2 = 0.1$ and $\alpha = 0.5$. Reprinted with permission from fig. 9 of ref. Kenkre et al. (2009); copyright (2009) by the American Physical Society

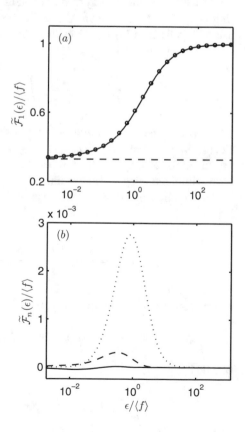

it is indeed clear that the long-range memories are small. It is thus that the EMT can successfully describe the evolution even though it possesses only nearest-neighbor memories. Note that, while $\tilde{\mathcal{F}}_1(\epsilon)$ is sigmoidal in shape, the long-range memories seem to peak for intermediate ϵ and to be negligible for both large and small ϵ.

13.3.2 Finite Size Effects

It appears that effective medium considerations have been used only on infinitely large systems in the past. Presented below are useful EMT results for finite rings of N sites, i.e., chains obeying periodic boundary conditions. The self-propagator for such a system is given in the Laplace domain by

$$\tilde{\pi}_0(\epsilon) = \frac{1}{N} \sum_k \frac{1}{\epsilon + 2\tilde{\mathcal{F}}(\epsilon)(1 - \cos k)} \tag{13.39}$$

where k takes on the values $(2\pi/N)[0, 1, 2, \ldots N - 1]$. In the long time limit, the self propagator $\pi_0(t)$ tends to $1/N$ as one knows both from the explicit limit of Eq. (13.39) or from the physical statement that the probability equalizes over the ring sites. This means via an Abelian theorem that $\epsilon \tilde{\pi}_0(\epsilon) \to 1/N$ as $\epsilon \to 0$. The use of this limit in Eq. (13.6) leads to an important long-time consequence of our general equation (13.4),

$$\frac{1}{F_{eff}} = \frac{N}{N-1} \int_0^\infty df \frac{\rho(f)}{f + F_{eff}(\frac{1}{N-1})}, \tag{13.40}$$

which is an extension to finite systems of the well-known harmonic mean result of Eq. (13.8). Here we have used $F_{eff} = \tilde{\mathcal{F}}(\epsilon \to 0)$ as earlier. Equation (13.40) must be solved for F_{eff} implicitly and becomes explicit only as $N \to \infty$ when the F_{eff} term within the integral disappears.

The implicit equation for the case of the double-delta distribution function of Eq. (13.18),

$$F_{eff} = \frac{N-1}{N} \left[\frac{\alpha}{f_1 + \frac{F_{eff}}{N-1}} + \frac{1-\alpha}{f_2 + \frac{F_{eff}}{N-1}} \right]^{-1},$$

can be converted into a quadratic equation and solved explicitly. With

$$j = f_1(1 - N + N\alpha) + f_2(1 - N\alpha),$$

one has

$$F_{eff} = \frac{j \pm \sqrt{j^2 + 4(N-1)f_1 f_2}}{2}. \tag{13.41}$$

Normally, i.e., when both f_1 and f_2 are non-zero, there is a unique solution as we discard the negative root because F_{eff} must be real.

If one of the two possible rates, e.g. f_1, is zero, i.e., if broken bonds exist in the finite system, an interesting situation arises, *both roots* being of physical interest. The lower root is zero, not negative, in this case. If one varies the concentration α of the broken bonds, a *transcritical bifurcation* occurs as displayed in Fig. 13.6 at the point at which α equals the reciprocal of the number of sites in the ring. As this number increases, the bifurcation point moves towards vanishing concentration. We recover the known result that, for an infinite system, the effective rate is zero for any concentration of broken bonds. Additionally, we get a percolation threshold for finite systems. The two solutions exchange stability at the critical concentration ($\alpha = 1/N$), there being transport throughout the ensemble-averaged system for broken bond concentrations below the critical value.

It is interesting to see how the effective medium nearest neighbor memory function compares with the exact nearest neighbor memory function as $\epsilon \to \infty$ in finite rings. One can obtain the exact memory functions for a ring of N sites by averaging over all possible configurations of the ring. For simplicity, we will consider the double-delta distribution with $\alpha = 1/2$. For rings with $N = 2, 3, 4,$ and 5 sites we have the exact values, $\lim_{\epsilon \to 0} \widetilde{\mathcal{F}}_1^{EX}(\epsilon)$

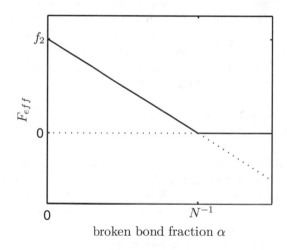

Fig. 13.6 Bifurcation of the effective long time transfer rate for a double delta distribution in a finite system of N sites. Plotted is F_{eff} as a function of the concentration of broken bonds (i.e., bonds with the rate $f_1 = 0$), the rate associated with the remaining fraction $1 - \alpha$ of unbroken bonds being equal to f_2. A transcritical bifurcation occurs when α equals $1/N$. For concentrations higher than this value, the effective rate vanishes but changes linearly with the concentration for lower α. Solid (dotted) lines denote the stable (unstable) solution. Reprinted with permission from fig. 10 of ref. Kenkre et al. (2009); copyright (2009) by the American Physical Society

$$\widetilde{\mathcal{F}}_1^{EX}(0) =$$

N=2: $\quad 2f_2\dfrac{r}{r+1},$

N=3: $\quad 8f_2\dfrac{r(r+2)(2r+1)}{(5r+1)(r+5)(r+1)},$

N=4: $\quad 16f_2\dfrac{r(1+3r)(3+r)(r+1)}{124r(1+r^2)+230r^2+17(1+r^4)},$

N–5: $\quad 16f_2\dfrac{r(3+2r)(2+3r)(1+4r)(4+r)}{(7+3r)(3+7r)(r+1)(7+36r+7r^2)}.$ (13.42)

The effective medium quantity is given by Eq. (13.41) with $\alpha = 1/2$. In order to quantitatively examine how different the exact and effective medium values are, we define a relative difference as

$$\frac{1}{f_2}\left[\frac{F_{eff} - \widetilde{\mathcal{F}}_1^{EX}(0)}{\widetilde{\mathcal{F}}_1^{EX}(0)}\right],$$

and examine it as a function of f_1/f_2 for $N = 3, 4$, and 5. We find that the values predicted by the effective medium theory are slightly different from the exact values. The relative difference between the two decreases as the number of sites in the ring becomes larger and larger. Therefore the effective medium theory predicts the correct values in the limit $\epsilon \to \infty$ when $N \to \infty$, but finite size effects exist otherwise. Note that for finite N, effective medium theory always predicts larger values than those that are calculated exactly.

13.4   General Discussion

Two questions are important to answer in the context of this EMT program of the description of a disordered system. To what extent is the replacement of the spatial disorder by temporal memories possible and meaningful even in principle? And what is the prescription to calculate the memories and effective transfer rates given appropriate information about the disorder in the particular system? Without the first, it is senseless to begin. Without the second, the study is useless.

The first question can be answered quite trivially on a little reflection. Consider the original Master equation on the given disordered system, (13.1) solved. By assigning the solutions for the probabilities, $P_m(t)$, to an appropriate *ordered* lattice, carry out the direct Fourier transform to obtain $P^k(t)$. Put the Laplace transform of the latter into

$$\tilde{\mathcal{A}}_k = \frac{P^k(0)}{\tilde{P}^k(\epsilon)} - \epsilon$$ (13.43)

and Fourier and Laplace invert to get the (translationally invariant) memories $\mathcal{A}_{mn}$, equivalently $\mathcal{W}_{mn}$ appearing in the GME (2.3). The presence of initial conditions in the above prescription means that each possible set of initial conditions would have a corresponding set of memory functions, a situation which is obviously unacceptable for practical purposes. However, in order to turn Eq. (13.43) into a practical prescription for computing memories which is independent of initial conditions, all that is necessary is to carry out an ensemble average over the possible realizations of disorder, compatible with what is known (for instance a distribution function) about the disorder. Such an average makes the system translationally invariant after the average. Then the first term in the right hand side of Eq. (13.43) which is the reciprocal of the Fourier and Laplace transform of the (ensemble averaged) propagator, is independent of initial conditions. The propagators directly lead to the memories.

This is precisely the method devised long ago in Kenkre (1978d) (see also the description in Kenkre and Reineker (1982)) to obtain exact expressions for memory functions analytically for a quantum mechanical (not disordered) system, and explained in Chap. 4.[4] Because Eqs. (13.1) and (13.2) as well as the operation of averaging over configurations are linear, it quite unnecessary to make any assumptions or offer demonstrations to be able to state with certainty that the replacement program is possible. Analyzing the problem from the viewpoint of the application of projection techniques (elucidated in Chap. 2) to the problem, it also becomes clear from Zwanzig's formal theory that a memory will automatically appear in a closed description of *any* quantity that is formally projected from another whose evolution equation is time-local. Here the projection is represented by an ensemble-average over disordered realizations. This too requires no calculation, only a moment's reflection.[5] The initial condition problem, rarely discussed in the disorder context, also makes its appearance in the projection formalism (Kenkre 1978b). It appears as a separate term. In the original derivation as given by Zwanzig (1964), it is removed through the initial random phase or diagonality assumption. In our present disorder context it disappears on carrying out the ensemble average we have mentioned above.

What is really needed in the sense of calculations comes to the second question we have posed above, i.e., the finding of an *explicit* and *practical* prescription that would allow one to go from information about the disorder in the real system to the memories (or pausing time distribution functions) in the replacement problem. Very few instances of such a prescription exist in the literature, a noteworthy attempt being in the early work of Scher and Lax that gave support to the well known theory of Scher and Montroll on transport in xerographic materials. That is the kind of prescription that one needs in developing a usable theory.

---

[4]No ensemble average was involved in that context.

[5]It has always seemed to me surprising that there were investigators (Pollak 1977; Silver et al. 1979) who had to be persuaded with the help of a formal exercise (Klafter and Silbey 1980) that disorder could be treated with memory functions.

The present chapter, and the upcoming Chap. 14, focus on such a program of converting disorder information into temporal memories, The prescription we provide comes in a particularly convenient form that transforms the disorder into explicit memory functions via a double Laplace transform procedure. Our prescription facilitates the extraction of the new results we present in subsequent sections. The spirit of the investigations we present is most akin to, among early attempts that have discussed memory functions in the EMT context, the work of Haus and Kehr (1983, 1987). In justifying the EMT methodology one might argue that the EMT method has a variational flavor. The idea of obtaining "a best description" by equating the EMT consequence of the sought-after memory with the result of averaging over the given probability distribution of disorder certainly would support such an argument. I believe a closer probing of the conceptual foundation of the method would have at least a pedagogical value.

13.5   Chapter 13 in Summary

Whereas the rest of the treatment in the book has dealt with ordered systems such as crystals, and with disorder that is dynamic, i.e. arising from the movement of the constituents of the crystals, this chapter showed how memory functions arise in *statically* disordered systems exemplified by amorphous molecular aggregates. We addressed the concept behind the *effective medium approach*. It was shown how it arises from a judicious and noteworthy application of the defect technique to replace a statically disordered time-local Master equation by an ordered counterpart which is non-local in time. The memory associated with this non-local description was not merely postulated but calculated: an explicit prescription to obtain the memory functions from given probability distributions of the randomness of the disorder was provided and used to examine the extent of validity of the procedure. The typical shape of the EMT memory was determined and an exponential approximation was introduced. Numerical calculations were used to obtain the memory in general as well as in specific cases and a number of effects were discovered with its help. Among others, they included behavior in finite size systems and the appearance of spatially long range memories. A brief conceptual discussion of the method was also provided.

Effective Medium Theory Application to Molecular Movement in Cell Membranes

14

It is hoped that this chapter, in which the *modus operandi* of how to apply the EMT when one encounters a new problem in one's research is demonstrated step by step for a topic of modern relevance, will be of practical use to the reader. Let us start for this purpose by examining the work of Kalay and collaborators on effects of disorder in location and size of fence barriers on molecular motion in cell membranes (Kalay et al. 2008). The background of the area of research is described briefly first; the *ordered* system that serves as the backbone of the problem under consideration is treated next; the specific EMT steps are then described leading to final results. Passing mention is then made of the work of Parris and collaborators on the application of EMT to small world networks (Parris and Kenkre 2005; Candia et al. 2007; Parris et al. 2008).

14.1   Physical Background and Relevance of the Selected System

The biophysics of cell membranes is an important as well as active field of current research, issues of interest being cell shaping and movement (McMahon and Gallop 2005), cell division, signal transduction (Krauss et al. 2003), and molecule trafficking. Observations of the lateral movement of molecules on the surface of the cell have given rise to the idea that the moving (transmembrane) molecules are confined within certain regions of the cell membrane. One possible source of this confinement has been suggested by Kusumi et al. (2005) as being collisions of membrane molecules protruding into the cytoplasm with the cytoskeleton (Howard et al. 2001). The model views the molecules as moving freely, their motion being hampered as they traverse adjacent compartments. As the actin filament that forms the compartment boundary dissociates due to thermal fluctuations, the moving molecule is envisaged as overcoming the barrier potential and hopping to the adjacent compartment.

© The Author(s), under exclusive license to Springer Nature Switzerland AG 2021
V. M. (Nitant) Kenkre, *Memory Functions, Projection Operators, and the Defect Technique*, Lecture Notes in Physics 982,
https://doi.org/10.1007/978-3-030-68667-3_14

Experimentalists interested in comparing their observations with theoretical analysis had been voicing their frustration at the scarcity of available expressions to be put in direct comparison to the experiment (Kusumi et al. 2005). Static disorder is a key feature of this system. It appeared to us, therefore, of some value to undertake an effective medium analysis of the phenomena. We had available to us calculations on an ordered counterpart of a model system that could be studied with EMT. This was given in Kenkre et al. (2008). Let us take that as our starting point here and calculate by substantial modifications of that analysis the consequences of disorder on the effective diffusion constant and the time dependence of the molecular mean square displacement as well.

14.2   The Ordered System that Serves as the Springboard

In the model analyzed in Kenkre et al. (2008), we represent the molecule as a random walker in a 1-dimensional infinite chain of sites. The molecule, whose probability of occupation of the m-th site of the chain at time t is $P_m(t)$, hops via nearest neighbor transfer rates. The rate is F within a compartment and has a lower value f at the interface of compartments where there is a barrier hindering the molecular motion. There are $H+1$ sites, equivalently H nearest neighbor bonds, within a compartment; for simplicity, H is taken to be even, with the site 0 at the center of one of the compartments. Specifically, the Master equation

$$\frac{dP_m}{dt} = F\left[P_{m+1} + P_{m-1} - 2P_m\right]$$

$$-(F - f) \sum_r{}' [P_{r+1} - P_r]\left(\delta_{m,r} - \delta_{m,r+1}\right) \tag{14.1}$$

describes that ordered system. The primed summation goes over sites $r = H/2 + (H + 1)l$ which lie to the left of each barrier, l taking all integer values.[1]

Explicit expressions had been derived in Kenkre et al. (2008) for the time dependence of the mean square displacement of the molecule and for the effective diffusion constant, the compartment size H and the transfer rates f and F being reflected transparently in these calculated quantities. The major element missing

---

[1]Long before we decided to undertake the cell membrane analysis, I had some of the primary calculations for the ordered system ready. This was so because, decades earlier, I had worked them out for an unrelated physical system: Frenkel excitons supposedly rattling in cages whose walls were created by energy mismatch of ordinary and deuterated molecules in tetracholobenzene. I find myself always intrigued by Feynman's well-known remark that "the same equations have the same solutions." It is usually attributed to him as having expressed in Chap. 12 of volume II of his Lectures in Physics. I could make little out of the cage rattling in the Frenkel exciton context but it served us well (my collaborators Luca Giuggioli and Ziya Kalay, and me) when we turned to cell membranes!

from that *periodic barrier* theory was static disorder. Our investigation targeted consequences as a result of realistic randomness in compartment sizes and in barrier heights in the linking of compartments to their neighbors. In other words, we asked how to treat stochastic variations of H and f.

The following is a brief description of how Kalay et al. (2008) attempted to fill this gap. The description will do double duty for us in this chapter by simultaneously illustrating the use of the EMT. The trick will be to replace the ordered system properly structured to incorporate disorder.

Barrier locations for the transmembrane molecular motion in the cellular membrane occur at positions that are by no means regular; the consequent variations are substantial within a system, about one order of magnitude, the compartment sizes sometimes being quoted as lying between 30 nm and 240 nm (Kusumi et al. 2005). The precise situation in which the moving molecules find themselves at the barriers also varies, the result being a variation in the effective transfer rate.

We constructed our effective medium theory considerations in three parts. First, the compartment sizes were taken all equal but the barrier heights, consequently the inter-compartment rates of molecular motion f, were selected from each of several specific distribution functions with given mean, variance, and nature. Next, we took the inter-compartment rates to be constant throughout, but allowed the compartment sizes H to vary. We followed that up by allowing both H and f to be random variables. Figure 14.1 illustrates the three respective cases as (a), (b) and (c).

In each case we calculated physical observables typified by the effective diffusion constant. Whereas the f distributions we considered were arbitrary, the H-distributions analyzed in the next step were not arbitrary but determined by the specific f distributions we took to generate them. The results we obtained

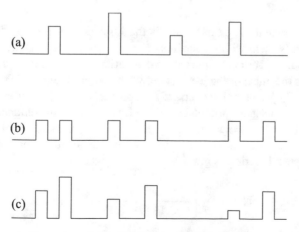

Fig. 14.1 Schematic illustration of the different types of barrier disorder analyzed. In (**a**), the barriers are periodically-spaced, but have energetic heights that are random. In (**b**), the barrier heights are the same throughout, but the distances separating barriers is random. In (**c**), both barrier spacing and barrier heights are independent random variables. Reprinted with permission from fig. 1 of ref. Kalay et al. (2008); copyright (2008) by IOP publishing

through the application of EMT include not only time-independent quantities such as the effective diffusion constant of the molecules but also the underlying memory functions and explicit predicted evolution in time of the mean square displacement. Direct contact was made with the ordered system analysis to learn how to use our EMT technique.

14.2.1 Variation in Barrier Heights

Following the analysis of Kenkre et al. (2008), whose basis is a defect technique calculation for periodically placed imperfections, the probability of occupation of the mth site in the ordered system for arbitrary initial probabilities $P_n(0)$ is expressed as

$$P_m(t) = \sum_n \chi_{m,n}(t) P_n(0)$$

where the transport propagator, which 'propagates' the solution from site n to site m in the presence of the defects, is given in the Laplace domain (ϵ being the Laplace variable and tildes denoting Laplace transforms) in terms of the (translationally invariant) propagators Π_{m-n} in the defect-less system. The explicit expression is

$$\widetilde{\chi}_{m,n} = \widetilde{\Pi}_{m-n} - \frac{F - f}{1 + (F - f)\widetilde{\mu}}$$

$$\times \sum_r{}' (\widetilde{\Pi}_{r-n+1} - \widetilde{\Pi}_{r-n})(\widetilde{\Pi}_{m-r} - \widetilde{\Pi}_{m-r-1}). \tag{14.2}$$

The first term on the right hand side is the Laplace transform of the propagator $\Pi_{m-n}(t)$ of the system without barriers ($F = f$), and is characterized by a single index as a result of complete translational invariance at the site level. The second term describes the effect of the barriers between compartments. It is proportional to the difference $F - f$ and is characterized by a property of the barrierless system, viz., products of the propagator differences in the Laplace domain, summed over barrier locations, and is also characterized by μ, an appropriate summed combination of Π's. This crucial quantity is determined by the locations of defects and for periodically spaced barriers is given by

$$\widetilde{\mu}(\epsilon) = \frac{1}{F} \left[\frac{\tanh{(\xi/2)}}{\tanh{(\xi (H + 1)/2)}} - 1 \right], \tag{14.3}$$

(*) with $\xi = 2\sinh^{-1}(\sqrt{\epsilon/4F})$. The exercise is interesting to do and would serve the prospective reader well in acquiring skills with the defect technique. The solution is given completely in an appendix in Kenkre et al. (2008). To be noted is that,

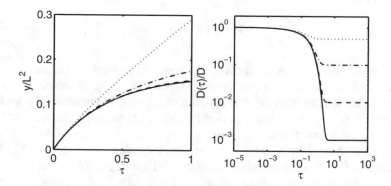

Fig. 14.2 Dependence on the time τ of the mean square displacement and the instantaneous diffusion coefficient for various magnitudes of the confinement effect. Plotted is the normalized mean square displacement, called y in the left panel, and $D(\tau)/D$ in the right panel, averaged over all initial locations in the compartment, for different values of D_{eff}/D: 0.001 (solid line), 0.01 (dashed line), 0.1 (dash-dotted line), 0.5 (dotted line). All quantities considered are made dimensionless appropriately. Reprinted with permission from fig. 2 of ref. Kenkre et al. (2008); copyright (2008) by the American Physical Society

in obtaining this result, one uses the known form of the Laplace transforms of the propagators for the system without barriers,

$$\widetilde{\Pi}_l = \frac{e^{-\xi|l|}}{2F \sinh \xi}. \tag{14.4}$$

Even though the effects of static disorder had not been incorporated in the theory, the analysis in Kenkre et al. (2008) was able to capture a good deal of the observed time dependence in the molecular mean square displacement and the effective diffusion constant. See Fig. 14.2 for a graphical depiction of some of those results. In the left panel y denotes the mean square displacement and the effective *time-dependent* diffusion constant in the right panel is defined as its time derivative. Indeed, details of a successful comparison with experimental observations regarding compartment size can be found in the original publication.

14.3   How to Introduce Static Disorder

Our goal in this section is to analyze the generalization of the system represented by Eq. (14.2) when the intercompartmental transition rates f vary in magnitude throughout the chain and are picked from a distribution function $\rho(f)$. The equation obeyed by the probabilities of occupation is

$$\frac{d P_m}{dt} = F [P_{m+1} + P_{m-1} - 2P_m]$$

$$-\sideset{}{'}\sum_{r}(F - f_r)\left[P_{r+1} - P_r\right]\left(\delta_{m,r} - \delta_{m,r+1}\right). \tag{14.5}$$

Compare Eq. (14.1) which is a form of this equation without the presence of static disorder. While a solution of that equation formed the basis of the study reported in the last section, an exact analytic solution of Eq. (14.5) is practically impossible because the f_r's here vary in a disordered fashion. Therefore, in the spirit of EMT, we replace the actual disordered system with its many f's by an effective medium system characterized by a single quantity (memory function) $\mathcal{F}(t)$ which is time-dependent and to be determined from the distribution $\rho(f)$. The effective medium system is identical to the periodic system represented by Eq. (14.2) except that f is replaced by $\widetilde{\mathcal{F}}(\epsilon)$. This replacement means that the occupation probabilities in the effective medium system obey

$$\frac{dP_m(t)}{dt} = F\left[P_{m+1}(t) + P_{m-1}(t) - 2P_m(t)\right]$$

$$-\int_0^t dt'\left[F\delta(t - t') - \mathcal{F}(t - t')\right]$$

$$\times \sideset{}{'}\sum_{r}\left[P_{r+1}(t') - P_r(t')\right]\left(\delta_{m,r} - \delta_{m,r+1}\right) \tag{14.6}$$

For long-time considerations, one replaces $\mathcal{F}(t)$ by its Markoffian approximation $\delta(t)\int_0^\infty \mathcal{F}(t')dt'$. Then the formal identity to Eq. (14.1) is exact.

Determination of $\mathcal{F}$ follows the usual effective medium prescription that we have learned in Chap. 13. We consider one defect in the otherwise periodic system (14.6) formed by replacing $\mathcal{F}$ by an f drawn from its probability distribution $\rho(f)$, solving the defect problem exactly in the Laplace domain, averaging the solution over the f's in the distribution, i.e., carrying out an ensemble average of the solutions, and then requiring that the ensemble-averaged solution is equal to the solution of the system without the defect.

The propagator for the effective medium system (14.6) is, in the Laplace domain,

$$\widetilde{\chi}_{m,n} = \widetilde{\Pi}_{m-n} - \frac{F - \widetilde{\mathcal{F}}}{1 + (F - \widetilde{\mathcal{F}})\widetilde{\mu}}$$

$$\times \sideset{}{'}\sum_{r}(\widetilde{\Pi}_{r-n+1} - \widetilde{\Pi}_{r-n})(\widetilde{\Pi}_{m-r} - \widetilde{\Pi}_{m-r-1}) \tag{14.7}$$

which is precisely Eq. (14.2) with the replacement of f by $\widetilde{\mathcal{F}}$. For the defective system made by introducing the rate f drawn from its probability distribution and placing it between the sites s and $s + 1$, the propagator is the sum of the propagator given above and an additional term so that the defective propagator is

$$\widetilde{\chi}_{m,n} + \left[\frac{(f - \widetilde{\mathcal{F}})}{1 + (f - \widetilde{\mathcal{F}})\widetilde{\beta}} \right] (\widetilde{\chi}_{m,s} - \widetilde{\chi}_{m,s+1})(\widetilde{\chi}_{s+1,n} - \widetilde{\chi}_{s,n})$$

where, for notational convenience we have introduced the abbreviation

$$\widetilde{\beta} = -\widetilde{\chi}_{s+1,s} + \widetilde{\chi}_{s+1,s+1} + \widetilde{\chi}_{s,s} - \widetilde{\chi}_{s,s+1}.$$

The second term in the defective propagator describes the modification by the barrier lying between the sites s and $s + 1$. We get a different solution for every ensemble member, the difference being in the value of f. We require the self-consistency condition that the ensemble average over the f's give us simply $\widetilde{\chi}_{m,n}$. This must be true whatever the n in the propagator or whichever barrier s characterizes. Therefore, the ensemble average of the factor in the square brackets in the propagator expression above must vanish. This provides a prescription for obtaining the effective quantity $\widetilde{\mathcal{F}}$ through the solution of the implicit equation

$$\int df \rho(f) \left[\frac{f - \widetilde{\mathcal{F}}}{1 + (f - \widetilde{\mathcal{F}})\widetilde{\beta}} \right] = 0. \tag{14.8}$$

(✱) For variety we have exhibited Eq. (14.8) in the more usual form given in the literature rather than that of (13.4) that we have exploited in Chap. 13. To gain familiarity with the techniques employed here, I suggest that the reader work the equivalence out as an exercise here, and obtain the explicit relation between β and ξ that shows that the Laplace transform of the former is the reciprocal of the sum $\xi + \widetilde{\mathcal{F}}(\epsilon)$ and thereby recover the corresponding relations given in Chap. 13.

The chain details are reflected in $\widetilde{\beta}$ and the randomness of the f's in ρ. There is no f-dependence in $\widetilde{\mathcal{F}}$ and $\widetilde{\beta}$, although $\widetilde{\beta}$ is a function of ϵ as well as of $\widetilde{\mathcal{F}}(\epsilon)$. Because of the lack of dependence of $\widetilde{\beta}$ on ϵ, the solution of Eq. (14.8) yields an explicit ϵ-dependence of the effective quantity $\widetilde{\mathcal{F}}(\epsilon)$ that we seek. Different probability distributions result in different expressions for the effective medium quantity $\widetilde{\mathcal{F}}$.

We can also see that, in the long time approximation, the effective value of the intercompartmental rate is indeed given as the reciprocal of the ensemble average of the reciprocals of individual intercompartmental rates:

$$\frac{1}{f_{eff}} = \frac{1}{\widetilde{\mathcal{F}}(0)} = \int df \frac{\rho(f)}{f}. \tag{14.9}$$

Application of effective medium theory has thus reduced the disordered problem of interest into the ordered effective problem which was completely analyzed in Kenkre et al. (2008). Combining that analysis with Eq. (14.9), we find that the overall effective transfer rate for the diffusion of the molecule (taking into account both the existence of the compartments of size H and the existence of disorder in the rates f) is given by

$$F_{eff} = \frac{\left(\frac{H+1}{H}\right)}{\left[\frac{1}{F} + \frac{1}{H}\int df \frac{\rho(f)}{f}\right]}. \tag{14.10}$$

For large H (for instance if $H >> 1$), Eq. (14.10) simply states that the effective overall transfer rate is the harmonic mean of the intracompartment rate F and the effective intercompartmental rate f_{eff} reduced by the size of the compartments.

14.3.1 Calculation of Time-Independent Effective Parameters

Equation (14.10) is one of our main results for the case in which disorder appears only in the values of the intercompartmental rates: it allows us to translate the randomness of the intercompartmental rates as expressed in the form of the distribution $\rho(f)$ directly into the effective diffusion parameters of the system. We will now consider several different cases of $\rho(f)$ for purposes of illustration.

Two expected results emerge in a straightforward fashion: If $\rho(f)$ has a non-zero value at $f = 0$ so that there is a non-zero probability of having disconnected sites, the effective hopping rate at long times must vanish. This is clear from Eq. (14.10) since $\rho(f)$ can be written as $\rho(f) = \delta(f) + R(f)$ where $R(f)$ is some distribution of f obeyed for all f except for $f = 0$. The integral in the denominator of Eq. 14.10 diverges, giving $F_{eff} = 0$. Similarly, if there is only a single value of the intercompartmental rate f, viz. g: $\rho(f) = \delta(f - g)$, Eq. (14.10) reduces to the corresponding expression in the analysis of the *ordered* system (Kenkre et al. 2008).

For the case when there are two values of the intercompartmental rate appearing with different weights:

$$\rho(f) = A_1\delta(f - f_1) + A_2\delta(f - f_2),$$

where obviously the normalization is $A_1 + A_2 = 1$, the overall effective rate is given by

$$F_{eff} = \frac{H + 1}{\frac{H}{F} + \frac{A_2 f_1 + A_1 f_2}{f_1 f_2}}. \tag{14.11}$$

Note that, if either f_1 or f_2 is zero, F_{eff} vanishes as we have stated above.

It is also of interest to exhibit the continuum limit of our results. Particularly for the problem of molecular motion in cell membranes, such results are more directly applicable than their more general discrete counterparts. The continuum limit means that the lattice constant $a \to 0$, transforming hops among discrete sites on a chain to flow on a continuous line. As is well-known, it is necessary that $f, H \to \infty$ as $1/a$ but $F \to \infty$ as $1/a^2$. With this appropriate limiting behavior, the overall effective diffusion constant is given as

$$\lim_{a \to 0} F_{eff} a^2 = D_{eff} = \frac{D}{1 + \frac{D}{L} \int d\mathcal{D}_f \frac{\rho(\mathcal{D}_f)}{\mathcal{D}_f}} \tag{14.12}$$

where D is the continuum limit of Fa^2, $\mathcal{D}_f$ is the continuum limit of fa, and L, the continuum limit of $(H+1)a$, is the size of the compartment. Factors such as $(H+1)/H$ collapse into 1. The ratio $\mathcal{P} = \mathcal{D}_f/D$ has been occasionally called in the literature, the permeability.

To illustrate the effect of the form of the distribution functions, we now consider several explicit realizations of $\rho(\mathcal{D}_f)$. We evaluate Eq. (14.12) for the three respective cases of a constant distribution in an interval, a biased distribution that peaks at a value related to the spread of the distribution, and a biased distribution that peaks at a value independent of the distribution spread. We use the normalization $\int \rho(\mathcal{D}_f)\, d\mathcal{D}_f = 1$.

Uniform Distribution

If $\rho(\mathcal{D}_f)$ is a non-zero constant in an interval of values of $\mathcal{D}_f$, i.e., for $l < \mathcal{D}_f < u$, and vanishes otherwise, then we have from Eq. (14.12)

$$\frac{D_{eff}}{D} = \left[1 + \frac{D \ln(u/l)}{L(u-l)}\right]^{-1}. \tag{14.13}$$

Rayleigh Distribution

If $\rho(\mathcal{D}_f)$ is a biased Gaussian, called sometimes a Rayleigh distribution:

$$\rho(\mathcal{D}_f) = \frac{\mathcal{D}_f e^{-\mathcal{D}_f^2/2\sigma^2}}{\sigma^2}, \tag{14.14}$$

where the mean is $\sigma\sqrt{\pi/2}$, and the variance $\sigma^2 \left(\frac{4-\pi}{2}\right)$ is proportional to the square of the mean, we have

$$\frac{D_{eff}}{D} = \left[1 + \frac{D\sqrt{\pi/2}}{L\sigma}\right]^{-1}. \tag{14.15}$$

Rice Distribution

To have two independently controllable parameters, one deciding the value at which the distribution peaks, and the other the spread, we consider what is called the Rice distribution:

$$\rho(\mathcal{D}_f) = \frac{\mathcal{D}_f}{\sigma^2} e^{-\frac{(\mathcal{D}_f^2 + v^2)}{2\sigma^2}} I_0\left(\mathcal{D}_f \frac{v}{\sigma^2}\right) \tag{14.16}$$

the two parameters being σ and v. The mean is $\sigma\sqrt{\pi/2}L_{1/2}\left(-v^2/2\sigma^2\right)$, and the variance is $2\sigma^2 + v^2 - \pi\sigma^2/2L^2_{1/2}\left(-v^2/2\sigma^2\right)$. Here,

$$L_{1/2}(x) = e^{x/2}\left[(1-x)I_0(-x/2) - xI_1(-x/2)\right]$$

is the Laguerre Polynomial of fractional order and $I_m(x)$ are modified Bessel Functions of the first kind. We find

$$\frac{D_{eff}}{D} = \left[1 + \frac{D\sqrt{\pi/2}}{L\sigma}e^{-v^2/4\sigma^2}I_0(v^2/4\sigma^2)\right]^{-1}. \qquad (14.17)$$

The distributions are plotted in Fig. 14.3. Note that the Rice distribution appears highly symmetric around its peak value although, like the Rayleigh distribution, it has the value 0 at the value $\mathcal{D}_f = 0$.

Notice that, in every case, the expression for the effective diffusion constant depends in essentially the same manner on the ratio of the system (barrier-less) diffusion constant D to the product of the compartment length and a characteristic $\mathcal{D}_f$ value. Except for numerical factors, the latter product is $(u - l)/\ln(u/l)$ for the uniform distribution, $\sigma/\sqrt{\pi/2}$ for the Rayleigh distribution, and $\sigma/\left[\sqrt{\pi/2}e^{-v^2/4\sigma^2}I_0(v^2/4\sigma^2)\right]$ for the Rice distribution.

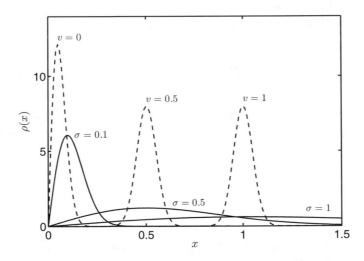

Fig. 14.3 Rayleigh and Rice distributions plotted for various parameter values as shown, represented respectively by solid and dashed lines. For the Rice distribution, σ=0.05 in all cases. Distributions are normalized and x represents the quantity $\mathcal{D}_f$. Reprinted with permission from fig. 2 of ref. Kalay et al. (2008); copyright (2008) by IOP publishing

14.3.2 Determination of the EMT Memory and Calculation of Time-Dependent Quantities

The full exploitation of the convolution in Eq. (14.6) and of the consequent memory effects in the motion has seldom been carried out in the literature. Here are a few exceptions we know: studies in the context of percolative systems in which the long time diffusion constant vanishes at the percolation point (Odagaki and Lax 1981), of the power law tail analysis under the action of a driving field (Parris and Bookout 1996), and prescriptions provided (Kenkre 2001a) for the determination of memory functions for stress distribution in granular compacts. In this subsection we show the results of performing the calculations detailed in Chap. 13 about how to calculate the EMT memory $\mathcal{F}$ in the Laplace domain.

Here we have

$$\tilde{\beta} = 2(\tilde{\Pi}_1 - \tilde{\Pi}_0) - \frac{F - \tilde{\mathcal{F}}}{F^2(1 + (F - \tilde{\mathcal{F}})\tilde{\mu})}\left(1 - 2\epsilon\tilde{\Pi}_0 + \epsilon^2 \sum_r \tilde{\Pi}_{s-r}^2\right),$$

$$(14.18)$$

$$\sum_r' \tilde{\Pi}_{s-r}^2 = \frac{\coth\xi(H+1)}{4F^2\sinh^2\xi}.$$

$$(14.19)$$

Eq. (14.8) then yields:

$$\int df \frac{\rho(f)}{f + (\Gamma - \tilde{\mathcal{F}})} = -\frac{\zeta + \theta((1 + \tilde{\mu}F) - \tilde{\mathcal{F}}\tilde{\mu})}{(1 + \tilde{\mu}F) - \tilde{\mathcal{F}}\tilde{\mu}}.$$

$$(14.20)$$

where

$$\theta = \frac{\coth(\xi/2) - 1}{F} - \zeta,$$

$$\zeta = \frac{1}{\tilde{\mu}F^2}(1 - 2\coth(\xi/2) + \coth^2(\xi/2)\coth\xi(H+1)).$$

$$(14.21)$$

We can now solve for $\tilde{\mathcal{F}}$ for a given $\rho(f)$ by using Eq. (14.20). As a special case of the distribution we use the case when the intercompartmental rate takes on one of two values with different weights:

$$\rho(f) = \alpha\delta(f - f_1) + (1 - \alpha)\delta(f - f_2).$$

$$(14.22)$$

Then Eq. (14.20) gives us, for $\tilde{\mathcal{F}}$,

$$\tilde{\mathcal{F}}^3 + b\tilde{\mathcal{F}}^2 + c\tilde{\mathcal{F}} + d = 0,$$

$$(14.23)$$

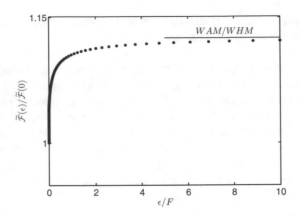

Fig. 14.4 Evaluation of the memory function produced by our effective medium theory. Plotted is the ϵ−dependence of the normalized $\widetilde{\mathcal{F}}$, taking ϵ real for simplicity in display and normalizing it to F. Shown is the case when the distribution of intercompartmental rates f is a sum of two weighted delta-functions: the rates are either f_1 or f_2. Our evaluation shows that $\widetilde{\mathcal{F}}$ equals the (weighted) arithmetic mean (WAM) of the two rates for large ϵ and their (weighted) harmonic mean (WHM) for small ϵ, the latter representing the effective long time intercompartmental rate, as expected. The parameter values chosen for this plot are: $f_1 = 0.1F$, $f_2 = 0.2F$, $H = 10$, $\alpha = 0.5$. Reprinted with permission from fig. 3 of ref. (Kalay et al. 2008); copyright (2008) by IOP publishing

where

$$b = -(f_1 + f_2 - 1/\theta) - \eta/\theta\widetilde{\mu},$$
$$c = f_1 f_2 - [\widetilde{\mu}(f_2 + \alpha(f_1 - f_2)) + (1 + \widetilde{\mu}F)$$
$$- \eta(f_1 + f_2)]/\theta\widetilde{\mu},$$
$$d = [\eta f_1 f_2 - (f_2 + \alpha(f_1 - f_2))(1 + \widetilde{\mu}F)]/\theta\widetilde{\mu},$$
$$\eta = (\zeta + \theta(1 + \widetilde{\mu}F)). \tag{14.24}$$

The numerical solution of the cubic equation yields the memory function explicitly in the Laplace domain. It is plotted in Fig. 14.4. We see that, at $\epsilon = 0$, $\widetilde{\mathcal{F}}$ tends to the value of its Markoffian approximation which is the weighted harmonic mean of the two rates f_1 and f_2. We also see that it tends to the weighted arithmetic mean of the two rates as $\epsilon \rightarrow \infty$. The intermediate behavior corresponds to intermediate times.

14.4   Disorder in Barrier Placement and in Intercompartmental Rates

Kalay et al. (2008) showed a way of analyzing disorder both in barrier placement and in the rates linking the compartments. We study their calculations in turn.

14.4.1 Disorder Only in Barrier Placement

Let us do this in a simplified manner so that the intercompartmental transition rate is either g or F with probabilities α and $1 - \alpha$ respectively. In other words,

$$\rho(f) = \alpha\delta(f - g) + (1 - \alpha)\delta(f - F). \tag{14.25}$$

This means that, starting with exactly the same system as in the ordered problem (involving periodic barriers of the same height g), we replace the barriers randomly and independently with probability $1 - \alpha$ with links that have transfer rates equal to the intracompartmental transfer rate F. Some of the compartments are now merged into one another because the barriers between those compartments are now removed. If a is the lattice constant on the chain, the compartment sizes are now always in multiples of $(H + 1)a$ (the smallest compartment size). In this system, the distance between the consecutive barriers, i.e., the compartment size, is a random variable whose distribution will depend on the parameter α.

The distribution of distances between consecutive barriers is easy to calculate. Consider N points in a discrete linear space. Assign to each point a number s_i which is either 1 or 0. Think of the points in this space as the intercompartmental links in the original problem. In this picture, a point i with $s_i = 1$ will represent a link with transfer rate F and a point with $s_i = 0$ will stand for a link with transfer rate g, which we have been calling a barrier. Equation (14.25) asserts that 0's will occur with probability α and 1's with $1 - \alpha$. If σ is the number of elements in a contiguous sequence of 1's, the distance between two consecutive barriers is simply given by $(\sigma + 1)(H + 1)a$. Note that $\sigma = 0$ corresponds to the case in which the distance between two consecutive barriers is the maximum value $(H + 1)a$. Combining these arguments leads to the conclusion that the number distribution of σ is given by

$$\mathcal{N}(\sigma) = \delta_{\sigma,0} \sum_{j=1}^{N-1} (1 - s_j)(1 - s_{j+1})$$

$$+ (1 - \delta_{\sigma,0}) \sum_{j=1}^{N-\sigma-1} (1 - s_j) \left(\prod_{i=0}^{\sigma-1} s_{j+i+1}\right)(1 - s_{j+\sigma+1}), \tag{14.26}$$

where $\mathcal{N}(\sigma)$ is the number distribution of σ in a *particular realization* of a 1-D chain as described in the beginning of this section. The first and second terms in Eq. (14.26) count the occurrence of compartments of sizes $(H+1)a$ and $(\sigma+1)(H+1)a$ respectively. As s_i's are independently distributed, we can write:

$$\langle s_i \rangle = 1 - \alpha, \tag{14.27}$$

where the angular brackets mean an ensemble average over all realizations of 1-D chains with intercompartmental transition rates sampled from Eq. (14.25). Then the ensemble averaged number distribution is given by

$$\langle \mathcal{N}(\sigma) \rangle = (N - \sigma - 1)\alpha^2 (1 - \alpha)^\sigma, \tag{14.28}$$

and therefore the probability distribution for σ is

$$\langle P(\sigma) \rangle = \frac{\langle \mathcal{N}(\sigma) \rangle}{\sum_{\sigma=0}^{N-1} \langle \mathcal{N}(\sigma) \rangle}. \tag{14.29}$$

On taking the limit $N \to \infty$, the probability distribution for σ is found to be

$$\langle P_{N \to \infty}(\sigma) \rangle = \alpha(1 - \alpha)^\sigma. \tag{14.30}$$

The ensemble averaged compartment size distribution, expressed as a function of the dimensionless compartment size $q = (\sigma + 1)(H + 1)$ is

$$P(q) = P(\sigma, \alpha) = \alpha(1 - \alpha)^{\frac{q}{H+1} - 1}. \tag{14.31}$$

The mean and variance of $P(\sigma, \alpha)$ are

$$\bar{q} = \frac{H + 1}{\alpha}, \tag{14.32}$$

$$(\Delta q)^2 = \overline{(q^2)} - \overline{(q)}^2 = (H + 1)^2 \frac{1 - \alpha}{\alpha^2}. \tag{14.33}$$

The earlier arguments provide us, in light of the distribution we have obtained above, a prescription to calculate the effective long time transfer rate F_{eff} in terms of the mean (dimensionless) compartment size $\bar{q}$ and the rates F and f:

$$F_{eff} = \frac{\bar{q}}{1/f + (\bar{q} - 1)/F}. \tag{14.34}$$

In the continuum limit, obtained by multiplying F_{eff} by a^2 and letting a tend to zero appropriately, we note that H also tends to infinity, producing the limit of $\bar{q}a$ as the mean compartment size Q which has dimensions of length. We get

$$\frac{D_{eff}}{D} = \left[1 + \frac{D}{Q \mathcal{D}_f} \right]^{-1}. \tag{14.35}$$

14.4.2 Disorder Additionally in the Intercompartmental Rates

What effective rates and diffusion constants should we get when *both* heights and places of the barriers are random? The analysis of Kalay et al. (2008) has an answer. Consider

$$\rho(f) = (1 - \alpha)\delta(f - F) + \eta(f, \alpha), \qquad (14.36)$$

where $\eta(f, \alpha)$ is a distribution normalized to α, with the understanding that $\eta(0, \alpha) = 0$. According to Eq. (14.36), and the development in Sect. 14.4.1, a fraction α of the intercompartmental links are barriers whose heights are sampled from the distribution $\eta(f, \alpha)$ and the rest are just intracompartmental links with transition rates F that in turn give rise to the variability in compartment sizes. Note that the statistics of different compartment size distributions do not change even if the barrier heights are not the same. Therefore, the compartment size distribution can still be obtained from Eq. (14.30).

$$F_{eff} = \frac{\frac{H+1}{H}}{\frac{1}{F}\left(\frac{H+1-\alpha}{H}\right) + \frac{1}{H} \int df \frac{\eta(f, \alpha)}{f}}, \qquad (14.37)$$

and the effective diffusion constant in the continuum limit becomes

$$\frac{D_{eff}}{D} = \left[1 + \frac{D}{L} \int d\mathcal{D}_f \frac{\eta(\mathcal{D}_f, \alpha)}{\mathcal{D}_f}\right]^{-1}. \qquad (14.38)$$

Here, $\int_0^\infty dx \eta(x, \alpha) = \alpha$, x being f and $\mathcal{D}_f$ in the two respective equations above. Note that when $\alpha = 0$, so that there are no barriers, $\eta(x, 0)$ vanishes identically as it is a positive function, and the results reduce to $F_{eff} = F$ and $D_{eff} = D$.

14.5 Comparison of the Time Dependence of Effective Diffusion in the Ordered and Disordered Models

It is interesting to compare the time dependence of effective diffusion in the ordered model analyzed in Kenkre et al. (2008) and its disordered version studied in Kalay et al. (2008). In Fig. 14.5 the effective time-dependent diffusion constant defined as the time derivative of the mean square displacement y is plotted as a consequence of the ordered model. The effective diffusion constant tends to start out at a high value and decrease in time as the molecular motion is hampered at the walls. The variation in the time dependence in the several cases considered occurs because each case corresponds to a different location of the starting site within the compartment. The time evolution is strongly dependent on the initial site location (its distance from the wall) at short times but the eventual value of the effective diffusion constant

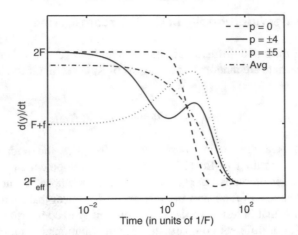

Fig. 14.5 Results of Ordered Model Time dependence of the time derivative of the molecular mean square displacement y plotted as a function of time in units of the reciprocal of the intracompartmental rate F. Effect of initial placement of the random walker at site p within a compartment on the time dependence of the effective transfer rate which, in the continuum limit, would be proportional to the time-dependent effective diffusion coefficient. Each compartment has 11 sites, i.e., $H = 10$, and $f/F = 0.01$. Averaging equally over all initially localized placements results, as shown by the dash-dotted line, in a monotonically decreasing transfer rate. However, interesting structures appear for initial placements at the center of, end of, and elsewhere in the compartment as shown. See text for explanation. Reprinted with permission from fig. 5 of ref. Kenkre et al. (2008); copyright (2008) by the American Physical Society

is always the same as the consequences of walls and barrier-free motion blend together. That initial location is labeled in the plot by p.

We also see an interesting dip and rise phenomenon. While the decrease in effective diffusion and eventual motion at the lowered value is compatible with experiments, we know of no observational evidence of observed dips. Yet in a theoretical work prior to ours, a remark has been made by Powles et al. (1992) about the dip. No explanation has accompanied the remark however in that work, the dip being referred as an "unexplained minimum."

Kenkre et al. (2008) did explain the dip and other features in the following way. The molecule, if placed centrally within the compartment, tends to move initially with the transfer rate or diffusion constant characteristic of the barrier-less system until it meets the barrier. At this point it crosses the barrier more slowly and the effective transfer rate drops. When the molecule has diffused to the next compartment it is outside the immediate influence of the barrier and the effective diffusion is therefore faster. The combined effect of repeated free diffusion and barrier-hindered diffusion eventually brings the effective diffusion constant to its long time value. If, however, the initial placement of the molecule is at the compartment edge, it already begins moving with an effective diffusion constant lower than the free value because of the immediate effect of the barrier. It diffuses faster from then on until other barrier encounters including the one at the other edge of the initial compartment decrease the rate of diffusion. For intermediate initial placements these effects happen one after the other as the molecule encounters first the barrier on one side and then the one on the other side of the compartment.

We now turn to the description of the EMT effects as given in Kalay et al. (2008).

In Fig. 14.6, we display a comparison of the results of our effective medium theory with numerical solutions of the disordered Master equation for various realizations of f-disorder, with H held constant. We see in the plot that the agreement is excellent at sufficiently long times. We considered several different kinds of distribution and found the results to be essentially identical. The results displayed correspond to three Rice distributions with different parameters, ($v = 0.1$, $s = 0.02$), ($v = 0.2$, $s = 0.06$) and ($v = 0.4$, $s = 0.02$), the probability distribution functions being shown in the inset. The same three kinds of curve, dashed, solid and dash-dotted curves respectively, used to display the distributions are used correspondingly in the main figure to show the results of the numerical solution of Eq. (14.5) followed by performing an ensemble average. The specific plots are of the instantaneous transfer rate $F(t)$, normalized to the barrier-less system transfer rate F. This $F(t)$ is one half the time derivative of the (dimensionless) mean square

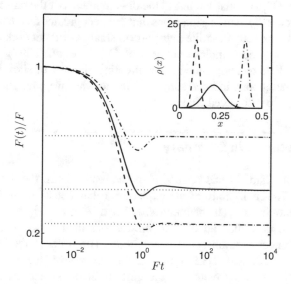

Fig. 14.6 Results of the EMT Model for the time-dependent effective transfer rate $F(t)$ obtained as proportional to the derivative of the molecular mean square displacement displayed in the main figure as a function of time in units of the reciprocal of the intracompartmental rate F. Three different continuous distributions of the intercompartmental rate are considered and shown in the inset. The same line types are used in the inset and the main figure for the three distributions. They are characterized by ($v = 0.1, s = 0.02$), ($v = 0.2, s = 0.06$),, and ($v = 0.4, s = 0.02$) and respectively denoted dashed, solid and dash-dotted curves in both the inset and the main figure. Effective transfer rates of EMT are reached asymptotically in each case. See text for details. Reprinted with permission from fig. 5 of ref. Kalay et al. (2008); copyright (2008) by IOP Publishing

displacement, and should not be confused with the memory $\mathcal{F}(t)$. Two features are visible: dips below the eventual asymptotic values, and the coincidence of the asymptotic values with horizontal dotted lines that represent our effective medium theory. As explained in the ordered model discussion, the dips arise from the fact that the random walker is assumed to start initially at the center of one of the compartments: repeated encounters with walls when the effective transfer rate drops are responsible for the dips. The asymptotic coincidence of the numerical solutions with the effective medium theory provides graphical validation of the latter.[2]

14.6   Application of EMT to Small-World Networks

It is also interesting to examine recent applications of the EMT that Parris has pioneered with his collaborators to explore motion on small world networks, particularly of the Neumann-Watts kind: Parris and Kenkre (2005), Candia et al. (2007), Parris et al. (2008). In those systems standard rings (finite chains with periodic boundary conditions) with nearest neighbor hopping rates for the random walker form the ordered part and additional small world connections make up the disordered part. Of particular interest to the developments of the present paper is the use of EMT to develop memory functions that connect greater than nearest-neighbor pairs. Indeed, to correctly describe transport on small world networks, as well as on the partially disordered complex networks of Candia et al. (2007), it is generally necessary to include memory functions connecting all pairs of sites on the network *except* nearest neighbors, in interesting contrast to what we have shown here to be the case for the 1-d disordered chain.

14.7   Chapter 14 in Summary

The function of Chap. 14 has been to help the reader learn the details of how to use the EMT. The main body of the explanation has centered on the description of molecules moving in cell membranes. After a brief description of the physics of the field, we described an ordered model and then a disordered version of that model to understand observed phenomena. The ordered model consisted of periodically placed defects representing barriers for molecular motion arising from interaction with the cytoskeleton of the cell. It involved a standard application of the defect technique and was successful in explaining experiment. However, addressing static disorder in the barrier heights or barrier placements was beyond its reach. To describe that aspect we considered the disordered model in which we allowed the barrier heights to be random variables, then did the same with

---

[2]Needless to say, if applied, the exponential approximation, Eq. (13.17) given in Chap. 13, does a pretty good job of showing the essence of the evolution of the time-dependence of the effective rate of transfer from the arithmetic average $\langle f \rangle$ initially to the harmonic average $1/\langle 1/f \rangle$ eventually.

barrier placements and finally with both quantities. EMT methods were introduced and explained step by step. They led both to the calculation of time-independent and time-dependent effective quantities of relevance to the observations. Various probability distributions were used as input and the theory was worked out in detail. Mention was also made of the application of EMT to motion on networks. A careful examination of these examples should help the reader in learning the methods of the EMT.

A Mathematical Approach to Non-Physical Defects

15

Surprising as it may sound, we will describe in this chapter the application of the defect technique to three instances where there are no physical defects anywhere in the system. The first will be in the solution of the stochastic Liouville equation (SLE), the second in the description of the mutual annihilation of excitons, and the third in the transmission of infection in epidemics.

15.1 Solution of the Stochastic Liouville Equation with Its Terms viewed as Defects

The job we are undertaking is to find solutions of the simple (Avakian) SLE given by Eq. (2.25) through a method which employs the defect technique in spite of the fact that translational invariance is not broken anywhere in the system we analyze. The trick is to notice that the density matrix is a two-index entity. If we plot the two indices m and n on orthogonal axes, we notice that the evolution equation is translationally invariant everywhere in the index-space except on the diagonal where $m = n$. We have an infinite-size (but systematic) defect in the m-n space although physically there are no defects anywhere.

If we rewrite the SLE (2.25) in the form

$$\frac{d\rho_{mn}}{dt} + \alpha\rho_{mn} = -i(L_V\rho)_{mn} + \alpha\delta_{mn}\rho_{mn}, \tag{15.1}$$

a picture might arise before our mind's eye of an entity moving in the plane defined by the orthogonal m and n axes and simultaneously undergoing decays. One of the decays is constant in space and occurs throughout the two-dimensional index space at rate α while the other actually represents a growth (rather than decay as we have often encountered in trapping problems) at the same rate α but occurring only along the diagonal. We have learned by now to take care of the former by an exponential

© The Author(s), under exclusive license to Springer Nature Switzerland AG 2021
V. M. (Nitant) Kenkre, *Memory Functions, Projection Operators, and the Defect Technique*, Lecture Notes in Physics 982,
https://doi.org/10.1007/978-3-030-68667-3_15

transformation that introduces a simple multiplicative factor into the solutions and the latter by the standard defect technique.

Therefore, calling

$$\eta_{mn}(t) = \sum_{m'n'} \Pi_{m-m',n-n'}(t)\rho_{m'n'}(0)$$

the homogeneous solution (the solution of Eq. 15.1) for vanishing α for whatever given initial conditions $\rho_{m'n'}(0)$, we can write down, as a Green function statement,

$$\tilde{\rho}_{mn}(\epsilon) = \tilde{\eta}_{mn}(\epsilon + \alpha) + \alpha \sum_{m'} \tilde{\Pi}_{m-m',n-m'}(\epsilon + \alpha)\tilde{\rho}_{m'm'}(\epsilon), \qquad (15.2)$$

in the Laplace domain, as well as its consequence for $m = n$:

$$\tilde{\rho}_{mm}(\epsilon) = \tilde{\eta}_{mm}(\epsilon + \alpha) + \alpha \sum_{m'} \tilde{\Pi}_{m-m',m-m'}(\epsilon + \alpha)\tilde{\rho}_{m'm'}(\epsilon), \qquad (15.3)$$

which involves only diagonal elements of the density matrix in the representation of m, n, etc. Solution of Eq. (15.3) is immediately possible through the use of discrete Fourier transforms defined through relations such as

$$\rho^k = \sum_m \rho_{mm} e^{ikm}.$$

The explicit solution in the Fourier-Laplace domain is

$$\tilde{\rho}^k(\epsilon) = \frac{\tilde{\eta}^k(\epsilon)}{1 - \alpha\tilde{\Pi}^k(\epsilon)}. \qquad (15.4)$$

Given that inversion of the discrete Fourier transform is straightforward, we can be assured that we have found the explicit solution of the diagonal elements of the density matrix through the application of the defect technique!

It is also a routine matter to substitute (15.4) in Eq. (15.2) to get the entire density matrix (off-diagonal elements as well) explicitly in the Laplace domain. The task we had undertaken of finding the solution is complete indeed. The coherent chain propagator which is $\tilde{\Pi}_{m-m',n-n'}(\epsilon)$ in the absence of α, is converted in the presence of α to

$$\tilde{\Pi}_{m-m',n-n'}(\epsilon) + \frac{\alpha}{N} \sum_{r,s,k} \left[\frac{e^{ik(s-r)}}{1 - \alpha\tilde{\Pi}^k(\epsilon + \alpha)} \right] \tilde{\Pi}_{m-r,n-r}(\epsilon + \alpha)\tilde{\Pi}_{s-m',s-n'}(\epsilon + \alpha).$$

$$(15.5)$$

This result is exact and explicit. It is explicit in that once one knows the propagators of Eq. (2.25) in the absence of the bath,[1] one can write down solutions of the full Eq. (2.25) for arbitrary initial conditions. The practical usefulness of the above result depends on the simplicity, or lack thereof, of the quadrature problem involved in inversions of the transform.[2] Furthermore, it is not difficult to show that, upon inversion of the discrete Fourier transform, Eq. (15.4) produces the solution for the probability propagator (4.35) which we rewrite here, for convenience:

$$\Pi_m(t) = J_m^2(2Vt)e^{-\alpha t} + \int_0^t du\, \alpha e^{-\alpha(t-u)} J_m^2\left(2V\sqrt{t^2 - u^2}\right).$$

15.1.1   Solution for the Complete SLE with the Additive Hopping Term

As we know from Chap. 4, the Avakian form of the SLE, (15.1), is sometimes found in augmented form in which the so-called hopping terms are added, converting the equation into

$$\frac{d\rho_{mn}}{dt} + \alpha\rho_{mn} = -i(L_V\rho)_{mn} + \delta_{mn}\left[\alpha\rho_{mn} + \sum_r(\gamma_{mr}\rho_{rr} - \gamma_{rm}\rho_{mm})\right]. \quad (15.6)$$

As an additional part of their analysis that we have reproduced above, Kenkre and Brown (1985) also showed how to solve the full SLE of this form through the defect technique. One expresses the γ matrix in terms of the A matrix defined through $A_{mr} = -\gamma_{mr}$ for $m \neq r$ and $A_{mm} = \sum_r \gamma_{rm}$. The immediate consequence is that Eq. (15.2) is replaced by the extended form

$$\widetilde{\rho}_{mn}(\epsilon) = \widetilde{\eta}_{mn}(\epsilon + \alpha) + \sum_{m'}\widetilde{\Pi}_{m-m',n-m'}(\epsilon + \alpha)\left[\alpha\widetilde{\rho}_{m'm'}(\epsilon) + \sum_r A_{m'r}\widetilde{\rho}_{rr}(\epsilon)\right].$$

$$(15.7)$$

The translational invariance of the system considered implies that A_{mr} are functions of $m - r$ and we therefore have the solution for the full SLE in a simple form

$$\widetilde{\rho}^k(\epsilon) = \frac{\widetilde{\eta}^k(\epsilon + \alpha)}{1 - (\alpha - A^k)\widetilde{\Pi}^k(\epsilon + \alpha)}, \quad (15.8)$$

---

[1] We do happen to know these to be $i^{n-m} J_m(2Vt)J_n(2Vt)$ in the time domain from solutions of the amplitude equation.

[2] I undertook the job of finding this defect technique solution of the SLE as a task under the conditions of a challenge from a friend who thought it could not be done because there was no defect in sight.

involving the discrete Fourier transforms represented by the superscripts k. We remind the reader that these transforms are to be calculated by multiplying any quantity that depends on m by e^{ikm} and summing over all m.

15.1.2 Calculation of the Scattering Function Lineshapes

The solutions (15.4) and (15.8) of the simple and augmented SLE's, respectively, were not only found in Kenkre and Brown (1985) but also then used in that publication and others to calculate the scattering function of light interstitials such as hydrogen atoms and muons moving in solids. The investigations included various issues as explained in Chap. 4, including some that focused on the effects of coherence and the consequences of temperature variation. This might sound surprising to the astute reader who might be aware of a well-known shortcoming of the SLE. The assumption of a universal rate α for the decay of all the off-diagonal elements ρ_{mn} of the density matrix forces the SLE to produce equilibrium solutions that do not possess the correct thermal behavior: All off-diagonal elements of ρ in the site representation vanish in equilibrium, and all k-states are occupied with equal probability. This result violates the principle of detailed balance unless $T \rightarrow \infty$, given that each k-state has a different energy. It is clear that one might expect this failing of the SLE to affect the calculated scattering function.[3]

It is possible to devise a way to address the problem of detailed balance. I had worked on a procedure myself to facilitate Rahman-Knox's theory of fluorescence depolarization in stick dimers (Rahman et al. 1979) but it had the limitation of being valid only for a two-site system. A method for optical lineshapes had been devised by Lindenberg and West (1983). For the study of neutron scattering, Brown and I developed an independent procedure by calculating from the SLE a symmetrized correlation function and obtaining from it the actual scattering lineshape by introducing detailed balance factors. We have seen in Chap. 4 that the scattering function is the Fourier transform of the correlation function $I(k, t)$ given by Eq. (4.20). We construct the symmetrized correlation function

$$I^s(k, t) = \frac{1}{2}[I(k, t) + I(-k, -t)], \tag{15.9}$$

and show exactly (Kenkre and Brown 1985) that the scattering function can be displayed exactly through the relation

$$S(k, \omega) = e^{\beta\omega/2}\mathrm{sech}(\beta\omega/2)\frac{1}{2\pi}\int_{-\infty}^{\infty} dt\, I^s(k, t)e^{-i\omega t}. \tag{15.10}$$

where $\beta = 1/k_B T$.

---

[3]It was David Brown who contributed to the article these words "On balance, line shape symmetry is dependent on the detail of the equilibration process." In addition to his fine physics abilities, he was highly appreciated in our research circle for his easy and unsurpassed facility at being a wordsmith.

The general relation (15.10) is valid independently of the dynamics or the approximation inherent in the SLE. Yet the symmetrized correlation function calculated from the SLE contains the specific dynamics particular to the latter. By using this combination method, we were able to derive a scattering function that contains the dynamics characteristic of the SLE as well as the correct detailed balance properties.

To entice the reader to study the original paper (Kenkre and Brown 1985) that happens to have a large number of interesting discussions about detailed balance and related matters, I display in Fig. 15.1, without too detailed a discussion, a plot of the scattering function calculated in that publication showing the effect of varying temperature on the neutron scattering function of light interstitials moving in solids via the simple stochastic Liouville equation. The band energy is $V(k) = 2V \cos ka$ where a is the lattice constant, k the momentum transfer and ω (with $\hbar$ taken as 1) as the energy transfer in the scattering. Different lines in the plot represent different temperatures as explained in the caption. The explicit disappearance of the T-induced shewness in the linesape as T increases to infinity is evident as one passes from curve a to curve d.

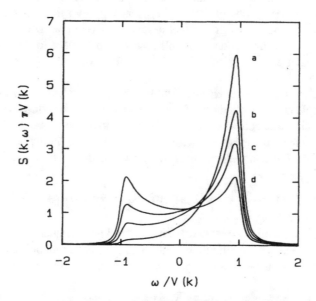

Fig. 15.1 Scattering function $S(k, \omega)$ calculated from the SLE, with the symmetrization prescription applied, as a function of the energy transfer of the neutrons scaled to $V(k)$, see text. The degree of (in)coherence is $\alpha/V(k) = 0.1$ and the momentum transfer is $k = \pi/a$ where a is the lattice constant. The four curves shown, a, b, c, and d are for respective temperatures given by the values 1, 2, 4 and ∞ of the ratio $k_B T/V$. The skewness of the scattering function is visibly lessened as T increases, and disappears at infinite temperature, as expected. Reprinted with permission from fig. 4 of (Kenkre and Brown 1985); copyright (1985) by the American Physical Society

15.2   Excitons as Defects for Excitons: Annihilation Theory

When two Frenkel excitons meet, they may annihilate each other, as a result of the
deexcitation of one participating molecule and the excitation to a higher state of the
other molecule. Used by many experimentalists to study the motion of excitons, this
phenomenon has been well reviewed (Avakian and Merrifield 1968; Swenberg and
Geacintov 1973). There have been approximate treatments such as those based on
diffusion equations with bilinear terms, see, e.g., Eq. (11.36), to represent the actual
process of mutual destruction but there are questions concerning their validity as
were raised, for instance in Kenkre (1981b). The best-known theory constructed
before what will be described below is due to Suna (1970).

For a dilute system of excitons that move with arbitrary coherence but annihilate
on contact with a finite (non-zero but also non-infinite) rate of destruction, an
appropriate procedure to build a theory of the process would be to consider just
a pair of excitons. Construct a space of twice as many dimensions as the crystal,
and study the evolution of $P_{m,n}$, the probability that the first exciton is at site m
and the second at site n. A mathematical (not physical) defect is envisaged to be
throughout the region $m = n$ and, if the system point that represents the exciton
pair reaches that region, the probability $P_{m,n}$ is supposed to leak out at whatever
rate, say b, which describes the mutual destruction. While m, n can be vectors in
the right number of dimensions, we will consider them here to be points and the
evolution of each exciton to be 1-dimensional (therefore that of the system point,
2-dimensional).

Such a theory was given in Kenkre (1980) and will be reviewed briefly here as our
second example of the application of the defect technique for non-physical defects.

For simplicity, first consider the excitons to be moving on a 1-dimensional chain
incoherently via nearest-neighbor transfer rates F and annihilating on contact at rate
b when they reach the same location. The joint probabilities $P_{m,n}$ would obey

$$\frac{dP_{m,n}}{dt} = F(P_{m+1,n}+P_{m-1,n}+P_{m,n+1}+P_{m,n-1}-4P_{m,n})-\delta_{m,n}2bP_{m,n} \quad (15.11)$$

Unlike approximate treatments that introduce ad hoc bilinear terms, this starting
point for the analysis of annihilation is linear in the joint probabilities. Application
of the defect technique is therefore possible and results in the usual manner in

$$\widetilde{P}_{m,n}(\epsilon) = \widetilde{\eta}_{m,n}(\epsilon) - 2b \sum_r \widetilde{\Pi}_{m-r,n-r}(\epsilon)\widetilde{P}_{r,r}(\epsilon), \quad (15.12)$$

which yields, when one puts $m = n$, the explicit solution for P^k, the discrete Fourier
transform of $P_{m,m}$,

$$\widetilde{P}^k(\epsilon) = \sum_m \widetilde{P}_{m,m}(\epsilon)e^{ikm} = \frac{\widetilde{\eta}^k(\epsilon)}{1 + 2b\widetilde{\Pi}^k(\epsilon)}. \quad (15.13)$$

Given that the propagators of the defect-less problem are known in terms of modified Bessel functions, and that the homogeneous solutions $\eta_{m,n}(t)$ are also easily obtained from the fact that they equal $\sum_{m',n'} \Pi_{m-m',n-n'}(t) P_{m',n'}(0)$, the problem is solved exactly. The result can be used for calculating the observables as explained in detail for the case of sensitized luminescence in Chap. 11. Thus, the luminescence intensity in the presence of a radiative lifetime τ_H is obtained as

$$\tilde{n}_H(\epsilon) = \frac{1}{\epsilon'}\left[1 - 2b\left(\frac{\tilde{\eta}^0(\epsilon')}{1 + 2b\tilde{\Pi}^0(\epsilon')}\right)\right] \tag{15.14}$$

where $\epsilon' = \epsilon + (1/\tau_H)$. Putting $\epsilon = 0$, one can also calculate the quantum yield:

$$Q = \frac{1}{\tau_H}\int_0^\infty dt\, n_H(t) = 1 - \left(\frac{\tilde{\eta}^0(1/\tau_H)}{(1/2b) + \tilde{\Pi}^0(1/\tau_H)}\right). \tag{15.15}$$

It is useful to pause and understand that, in these expressions, η^0 is the initial value of the total joint probability that the two excitons happen to be on top of each other and Π^0 is the total conditional probability that they remain in that state of colocation if they are in that state initially. It might also help the reader to undertake a simple exercise consisting of the following items. Show for this example of nearest-neighbor transfer rates on a chain that $\sum_m \Pi_{m,m+s}(t) = e^{-4Ft} I_s(4Ft)$. Calculate in the Laplace domain the luminescence intensity and the yield for three separate initial conditions: (i) that each exciton occupies initially a single site, the two being separated by a distance d, (ii) that the localized condition holds with $d = 0$ so both excitons are on top of each other initially, and (iii) each exciton occupies initially a delocalized state with equal occupation on all sites. Partial answers to the exercise are explicitly given in Kenkre (1980).

The reader will find in that publication a number of other details of interest including analysis valid for higher dimensions and arbitrary coherence which are simply treated by replacing the 1-dimensional incoherent propagator Π_0 by the appropriate-dimensional partially coherent propagator. A general expression for the annihilation rate γ', see, e.g., Eq. (11.38), obtained by dividing the annihilation constant γ by a unit-cell volume v (in 3-dimensions)[4] is

$$\gamma' = \frac{1}{\frac{1}{2b} + \tilde{\Pi}_0\left(\frac{1}{\tau_H}\right)}$$

---

[4]Clearly, one replaces v by a unit-cell area or length if considering a 2-dimensional or 1-dimensional system, respectively.

with both quantities in the denominator referring an exciton pair: thus the destruction rate is $2b$ and the transfer rate in the self-propagator expression is twice the rate for the single exciton.

Effects of coherence, dimensionality, and the relative magnitudes of the motion and destruction rates have been discussed in detail in Kenkre (1980).[5] As an example of dimensionality effects for incoherent motion in which the motion is slow enough to determine the annihilation constant completely, let us examine the annihilation rate. It is given by

$$\gamma' = \frac{\sqrt{1 + 4F\tau_H}}{\tau_H} \tag{15.16a}$$

$$\gamma' = \left[\frac{1 + 4F\tau_H}{2\tau_H/\pi} \right] \left[\mathbf{K} \left(\frac{4F\tau_H}{1 + 4F\tau_H} \right) \right]^{-1} \tag{15.16b}$$

$$\gamma' = 2F \left[I \left(0, 0, 0; 1; 1 + \frac{1}{6F\tau_H} \right) \right]^{-1} \tag{15.16c}$$

in 1, 2, and 3 dimensions, respectively.

In Eqs. (15.16), $\mathbf{K}$ is the complete elliptic integral of the first kind and

$$I(a, b, c; \alpha; \beta) = \int_0^\infty dt \, e^{-(2+\alpha)\beta t} I_a(t) I_b(t) I_c(\alpha t)$$

is a function that has been defined and tabulated in the context of lattice dynamics by Maradudin et al. (1960).

15.3   Transmission of Infection in the Spread of Epidemics

The third and final example we will dwell on in our application of the defect technique to non-physical defects is an analysis of the transmission of infection from a rodent to another rodent when they meet as random walkers moving on the terrain. This came up during the investigation that Sugaya and I carried out in studies of the spread of the Hantavirus epidemic (Kenkre and Sugaya 2014; Sugaya and Kenkre 2018). The situation is much like in our study of the annihilation of excitons but with an important difference. Whereas the excitons diffuse freely, the rodents move

---

[5]Demet Gülen, who had worked on some of her research with me initially, and later completed her dissertation under the supervision of Bob Knox, found an error of assumption regarding the validity of the chain rule that I had made in a small part of my calculations in that article. I have always found it a matter of celebration when a student finds, and points out, an error in my calculations. Every instructor should. It is a sign of the student successfully educating herself. Although most of my publication on exciton annihilation remained unscathed by the error, Demet, Bob, Paul Parris and I published an article (Gülen et al. 1988) correcting the error and taking the opportunity to add a number of new results we had discovered.

in a space-constricted manner, being attracted all the time to their burrows located at separate positions for the two rodents in interaction.

The underlying equation of motion for the rodents is therefore a Smoluchowki equation. We have applied the defect technique to a single particle undergoing such constrained motion in Sect. 12.5. We will extend that scenario here to a two-particle system.

15.3.1 Two-Rodent Model and Method of Analysis

Our analysis follows the guidelines of the problem of mutual annihilation of a pair of excitons described above. We start with a pair of animals, one initially infected and the other initially uninfected, respectively denoted by 1 and 2, performing random walks around respective attractive centers at R_1 and R_2, with a diffusion constant D. Their motion is thus 'tethered' in that each rodent is always attracted towards a "home", separate for each individual. As the two perform their random walks, they may meet. If they do, the uninfected individual may thereby get infected. On contact, this infection process occurs at a rate proportional to C_d where d denotes the dimensionality of the space. The details of a reaction diffusion process, in which the motion obeys a diffusion equation in the continuum, have been described in Chap. 12.

The central quantity that serves as the focus of our calculation is the joint probability density $P(r_1, r_2, t)$ that the infected rodent is at r_1 and the uninfected rodent (often called susceptible following ecological usage) is at r_2. When the susceptible rodent gets (definitely) infected, one can assert that $P(r_1, r_2, t)$ vanishes. The infection problem becomes, therefore, formally similar to the problem of the annihilation of a pair of Frenkel excitons that we have analyzed above, following (Kenkre 1980).

The present problem is, however, considerably more complex than the exciton problem. In the latter, the random walkers move freely. Here they perform a constrained walk. i.e., a walk under the influence of a potential. The most natural form of the potential to assume is a quadratic one.[6] Kenkre and Sugaya (2014) were thus led to consider a capture problem in a space of twice the number of dimensions as the space in which each walker moves in an attractive quadratic potential of steepness γ around its individual home or burrow. Generally in s-dimensions, we start with,

$$\frac{\partial P}{\partial t} = \nabla_1 \cdot [\gamma (r_1 - R_1) P] + \nabla_2 \cdot [\gamma (r_2 - R_2) P]$$

$$+ D \left(\nabla_1^2 + \nabla_2^2 \right) P - \delta(r_1 - r_2) C_d P. \quad (15.17)$$

---

[6]See, e.g., a mathematically more convenient alternative that is available as a result of the evaluation of the propagators in both Laplace and time domains given for a linear (rather than a quadratic) restoring potential by Chase et al. (2016).

In terms of $\Pi(r_1, r_1^0, r_2, r_2^0, t)$, the propagator for the homogeneous problem, the solution in the absence of the infection rate for any initial placement of the two animals given by $P(r_1, r_2, 0)$, would be

$$\eta(r_1, r_2, t) = \int_{-\infty}^{\infty} \int_{-\infty}^{\infty} d^s r_1^0 d^s r_2^0 \, \Pi(r_1, r_1^0, r_2, r_2^0, t) P(r_1^0, r_2^0, 0). \tag{15.18}$$

When infection is present, we write, as a consequence of the linearity of the equations,

$$P(r_1, r_2, t) = \eta(r_1, r_2, t)$$
$$- C \int_0^t dt' \int_{-\infty}^{\infty} d^s r_1' \, \Pi(r_1, r_1', r_2, r_1', t - t') P(r_1', r_1', t'). \tag{15.19}$$

The situation is ripe for the application of the defect technique. Laplace transforming Eq. (15.19), setting $r_1 = r_2$, and integrating over r_1 in the appropriate space of s dimensions, we find

$$\int_{-\infty}^{\infty} d^s r_1 \, \tilde{P}(r_1, r_1, \epsilon) = \int_{-\infty}^{\infty} d^s r_1 \, \tilde{\eta}(r_1, r_1, \epsilon)$$
$$- C \int_{-\infty}^{\infty} d^s r_1' \int_{-\infty}^{\infty} d^s r_1 \, \tilde{\Pi}(r_1, r_1', r_1, r_1', \epsilon) \tilde{P}(r_1', r_1', \epsilon), \tag{15.20}$$

where ϵ is the Laplace variable and tildes denote Laplace transforms. Motivated by the ν-function analysis introduced in Kenkre (1982), Kenkre and Parris (1983) and explained in Chap. 12, and assisted by the observation that the integral of $\tilde{\Pi}(r_1, r_1', r_1, r_1', \epsilon)$ over the entire domain of r_1 (i.e., all space) appearing in Eq. (15.20) is independent of r_1', we introduce the symbol $\tilde{\nu}(\epsilon)$ to denote that integral,

$$\tilde{\nu}(\epsilon) = \int_{-\infty}^{\infty} d^s r_1 \, \tilde{\Pi}(r_1, r_1', r_1, r_1', \epsilon), \tag{15.21}$$

and succeed in obtaining, in the Laplace domain, an *explicit* solution for the joint probability (density) that the two animals occupy the same position,

$$\int_{-\infty}^{\infty} d^s r_1' \, \tilde{P}(r_1', r_1', \epsilon) = \frac{\tilde{\mu}(\epsilon)}{1 + C\tilde{\nu}(\epsilon)}. \tag{15.22}$$

The expression in Eq. (15.22) contains two quantities that are key to the analysis. The first of these, $\nu(t)$, whose Laplace transform is defined in Eq. (15.21), is the probability (density) that the locations of the two animals coincide (whatever that location) if at a time t earlier their locations also coincided. The second key quantity, $\mu(t)$, whose Laplace transform is

$$
\tilde{\mu}(\epsilon) = \int_{-\infty}^{\infty} d^s r'_1 \ \tilde{\eta}(r'_1, r'_1, \epsilon)
$$

$$
= \int_{-\infty}^{\infty} d^s r'_1 \int_{-\infty}^{\infty} \int_{-\infty}^{\infty} d^s r_1^0 d^s r_2^0 \ \tilde{\Pi}(r'_1, r_1^0, r'_1, r_2^0, \epsilon) P(r_1^0, r_2^0, 0),
$$

$$(15.23)$$

is the probability (density) that the two animals occupy the same location at the present time (whatever that location) if at a time t earlier they occupied locations as per the *given initial condition* of the problem. Both refer to the problem when there is no transmission of infection ($C = 0$). They are integrals (over the s-dimensional space) of the two-particle joint probability density and have the dimensions of reciprocal length raised to s. The rest of the calculation is straightforward. Knowledge of the propagators of the system generally in the presence of constraining potentials gives ν and, in combination with the given initial conditions, yields μ. The two, together with Eq. (15.22), provide all that is necessary to obtain the infection probability and the nuances of its behavior.

15.4 Infection Curve and Its Non-Monotonic Dependence

When a definite infection event occurs, the joint probability density $P(r_1, r_2, t)$ *drops to zero*. The infection probability is, therefore,

$$
I(t) = 1 - \int_{-\infty}^{\infty} \int_{-\infty}^{\infty} d^s r_1 d^s r_2 \ P(r_1, r_2, t), \tag{15.24}
$$

and, from Eq. (15.19), is obtained in the Laplace domain as

$$
\tilde{I}(\epsilon) = \frac{1}{\epsilon} \left[\frac{\tilde{\mu}(\epsilon)}{(1/C) + \tilde{\nu}(\epsilon)} \right]. \tag{15.25}
$$

Further insight requires the evaluation of the key quantities μ and ν, which follows from the form of the propagators appropriate to Eq. (15.17). These are well-known to be Gaussian and to be multiplicative in Cartesian coordinates as one proceeds to higher dimensions. We have encountered them in Eq. (12.40) in Chap. 12 and we have seen that they involve the saturating time $\mathcal{T}(t) = (1/2\gamma)(1 - e^{-2\gamma t})$ that emerges from standard Ornstein-Uhlenbeck arguments (Reichl 2009; Risken 1984) and given already in Eq. (12.41). The 2s-dimensional propagator and the resulting ν and μ functions, the latter for arbitrary initial placement, are

$$\Pi(r_1, r_1^0, r_2, r_2^0, t) = \left(\frac{1}{4\pi DT(t)}\right)^s \prod_{\beta=1}^s e^{-\frac{\left(x_1^\beta - h_1^\beta - (x_1^{0\beta} - h_1^\beta)e^{-\gamma t}\right)^2 + \left(x_2^\beta - h_2^\beta - (x_2^{0\beta} - h_2^\beta)e^{-\gamma t}\right)^2}{4DT(t)}},$$

$$v(t) = \left(\frac{1}{\sqrt{8\pi DT(t)}}\right)^s \prod_{\beta=1}^s e^{-\frac{(1-e^{-\gamma t})^2\left(h_1^\beta - h_2^\beta\right)^2}{8DT(t)}},$$

$$\mu(t) = \left(\frac{1}{\sqrt{8\pi DT(t)}}\right)^s \prod_{\beta=1}^s e^{-\frac{\left(h_1^\beta - h_2^\beta + \left((x_1^{0\beta} - h_1^\beta) - (x_2^{0\beta} - h_2^\beta)\right)e^{-\gamma t}\right)^2}{8DT(t)}}, \qquad (15.26)$$

where the label β runs from 1 to s, and the initial position and home range center of the susceptible animal have the respective x-components $x_2^{0\beta}$ and h_2^β. The rest of the notation is obvious.

15.4.1 Simplification for 1-Dimensional Motion

We were able to understand a number of nuances of what happens in this problem in our simpler 1-dimensional study (Kenkre and Sugaya 2014). We will describe only that study in this book and refer the reader to the investigation of arbitrary dimensions carried out in Sugaya and Kenkre (2018) and also reviewed in a recent book (Kenkre and Giuggioli 2020).

For the motion of two 1-dimensional walkers ($s = 1$), we do not need the index β and, if we make the natural assumption that the animals are located initially at their own respective centers, the quantities $v(t)$, $\mu(t)$, which are closely related to Smoluchowski propagators connecting the two home range centers, are given by

$$v(t) = \frac{e^{-\frac{H^2}{8DT(t)}}(1-e^{-\gamma t})^2}{\sqrt{8\pi DT(t)}}, \qquad (15.27a)$$

$$\mu(t) = \frac{e^{-\frac{H^2}{8DT(t)}}}{\sqrt{8\pi DT(t)}}. \qquad (15.27b)$$

They equal each other for large times but begin quite differently at the initial time: $\mu(0)$ vanishes while $v(0)$ is infinite. Here, $H = h_1 - h_2$ is the distance between the two home-range centers.

The infection curve $I(t)$, defined in Eq. (15.24), is now obtained by calculating the Laplace transforms of Eq. (15.27), substituting them in Eq. (15.25), and inverting the transform. Kenkre and Sugaya (2014) performed the inversion numerically.[7]

---

[7]In order to be sure of the procedure, we also undertook the task of verifying results by direct numerical solution of the partial differential equation (15.17).

Our calculated $I(t)$ for initial location of the animals at their home range centers, and for an assumed contact rate parameter C_1 equal to 0.3 in units of $2D/H$, is displayed in Fig. 15.2 as a function of t scaled to τ_H, for various steepness values of the confining potential. Here τ_H is no radiative lifetime, of course. Rather, it is the time required for either animal to traverse diffusively the distance between the two home centers. It equals $H^2/2D$. We attach the suffix 1 to C to emphasize that this result is 1-dimensional.

A cursory glance at Fig. 15.2 may not reveal a striking feature that is apparent in the plots. We know that $\sigma = \sqrt{2D/\gamma}$ is the width of the steady-state distribution of the Smoluchowski walker in $1d$. We keep D and the inter-center distance H constant, and increment γ, thereby changing σ. The case of no confining potential corresponds to the thick solid curve ($H/\sigma = 0$). We gradually increase the steepness of the confinement potential, giving the latter parameter the respective values 0.6 (thin solid line), 1.0 (dotted), 1.64 (dot-dashed) and 2.12 (dashed). Generally, as time proceeds, $I(t)$ rises from 0 and saturates to 1. Infection may be said to occur faster as the confining potential becomes steeper *but only for relatively small values* of γ. Further increases make the infection proceed more slowly. Vertical arrows

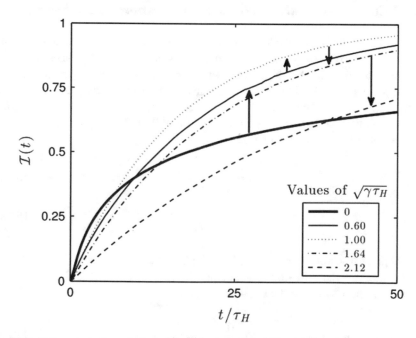

Fig. 15.2 Non-monotonic variation of the infection curve $I(t)$ with change in γ, the steepness of the potential confining the animals to their home ranges. The value shown for each line in the legend is of $\sqrt{\gamma\tau_H}$. This quantity equals H/σ, the ratio of the inter-center distance to the steady-state Smoluchowski width. Time is scaled to τ_H; C_1 scaled to $2D/H$ equals 0.3. Starting with the unconfined case ($\gamma=0$), an increase in γ makes infection more effective for small values of γ *but less effective* for larger values. Reprinted from Fig. 1 of (Kenkre and Sugaya 2014); copyright (2014) with permission from Springer Nature

between curves show this march graphically. Reversal in their direction even as the steepness is increases surely marks an interesting phenomenon.

This non-monotonic behavior is obviously worth noticing. It is also observed if the diffusion constant of the animals is varied keeping the potential steepness constant. It arises from the interplay of three quantities, the diffusion constant D, the steepness γ and the inter-center distance H which here is also the distance between the initial locations of the animals. For a given value of H, changes in D or γ exhibit the phenomenon. Varying H does not: maximum transmission occurs when $H = 0$, i.e., when the animals do not have to move to find each other for the infection to be propagated. The key parameter is $\gamma \tau_H = H^2 \gamma / 2D$ which is nothing other than $(H/\sigma)^2$: for a given H, optimum transmission of infection occurs when the parameter equals 1, particularly in the contact-limited case. More generally, the critical value is different from 1. By the term contact-limited we mean the case that C is much smaller than the corresponding motion parameter. In such a case, the contact process of infection when the animals meet, rather than the motion, determines the overall infection event (see Eq. 15.25).

15.4.2 Case of No Confinement to the Neighborhood of Burrows

What do we expect in the transmission of infection in case the attraction to burrows is not particularly strong for the rodents we study? Typically, such attraction is known to be weak for itinerant juvenile members of some rodent species. It is important to ask, therefore, what our model calculation predicts for zero confinement, i.e., for the simpler case of free diffusion. In that case, an *explicit* analytic solution is possible. With $\gamma \rightarrow 0$, $\nu(t)$ and $\mu(t)$ in $1d$ are simple propagators of the diffusion equation,

$$\nu(t) = \frac{1}{\sqrt{8\pi Dt}}, \tag{15.28a}$$

$$\mu(t) = \frac{1}{\sqrt{8\pi Dt}} e^{-\frac{H^2}{8Dt}}. \tag{15.28b}$$

Their Laplace transforms are known. With the introduction of a time $\theta = 8D/(\pi C_1^2)$ that incorporates the diffusion constant and the contact parameter C_1, we have for the infection probability in the Laplace domain,

$$\tilde{I}(\epsilon) = \frac{1}{\epsilon} \left(\frac{e^{-\sqrt{\epsilon \tau_H}}}{1 + \sqrt{\epsilon \theta}} \right). \tag{15.29}$$

It is not difficult to invert this Laplace transform explicitly. One obtains the analytic time domain result

$$\mathcal{I}(t) = \text{erfc}\left(\sqrt{\frac{\tau_H}{4t}}\right) - e^{\left(\sqrt{\frac{\tau_H}{4t}} + \frac{t}{\theta}\right)} \text{erfc}\left(\sqrt{\frac{\tau_H}{4t}} + \sqrt{\frac{t}{\theta}}\right). \tag{15.30}$$

This is surely a novel result in the epidemic literature. However, it makes contact with the results of several authors in totally unrelated reaction diffusion contexts (Carslaw and Jaeger 1959; Abramson and Wio, 1995; Redner 2001; Spendier and Kenkre 2013). The further simplification of an infinite contact rate (motion limit), leading to a vanishing θ, yields the simple diffusion result that the infection curve is given by a complementary error function of argument $\sqrt{\tau_H/4t}$. The time dependence of Eq. (15.30) is depicted as the thick solid line $\gamma = 0$ in Fig. 15.2.[8]

15.4.3 An Effective Rate of Infection

The foregoing analysis, while exact for dilute systems, is not immediately applicable to dense systems because they contain numerous (rather than one) interacting pairs. The dynamics, and even identity, of the pairs, evolve in time. Required is an approximate *kinetic equation theory* applicable to such situations, along the lines of earlier analysis (Kenkre 2003). For use in theories of that ilk, we can extract from the above single-pair analysis what could be termed an effective infection rate. The spirit of this extraction is the same as in the calculation of a Fermi Golden Rule rate for describing transitions in a complex quantum system. The detailed procedure is, however, different.

First, we notice that Eq. (15.25) for the infection probability in the Laplace domain can be cast in the form

$$\tilde{\mathcal{I}}(\epsilon) = \frac{\tilde{\alpha}(\epsilon)}{\epsilon\,(\epsilon + \tilde{\alpha}(\epsilon))},$$

where we will term the quantity $\tilde{\alpha}(\epsilon)$ an infection *memory*. It is given in the Laplace domain by

$$\tilde{\alpha}(\epsilon) = \frac{\epsilon\,\tilde{\mu}(\epsilon)}{(1/C) + \tilde{\nu}(\epsilon) - \tilde{\mu}(\epsilon)}. \tag{15.31}$$

If the Markoffian approximation were to be made on this infection memory, i.e., if $\tilde{\alpha}(\epsilon)$ were to be replaced by an ϵ-independent constant α, the infection probability in the time domain, $\mathcal{I}(t)$, would be simply an exponentially rising function $1 - e^{-\alpha t}$. It is clear that this function has, in essence, the typical shape seen in Fig. 15.2.

---

[8]If you feel Eq. (15.30) looks oddly familiar to you, you should skip back to Chap. 12 of this book and look at Eqs. (12.8) and (12.7).

Obviously, the extraction of a single infection rate from the full dynamics of the infection probability is provided by taking the Markoffian approximation of the infection memory $\tilde{\alpha}(\epsilon)$ through the limit $\epsilon \to 0$.

$$\alpha \equiv \lim_{\epsilon \to 0} \tilde{\alpha}(\epsilon) = \frac{\mu(\infty)}{(1/C) + (1/M)}. \tag{15.32}$$

Here, $\mu(\infty)$ is the limit of $\mu(t)$ as $t \to \infty$, and we have introduced a *motion parameter* M as the reciprocal of $\int_0^\infty dt\,[\nu(t) - \mu(t)]$ in the precise spirit of the discussions in Chap. 11. An Abelian theorem has been used in the second equality in Eq. (15.32) to express α in terms of quantities in the time domain. The effective rate now appears as the product of the probability in the steady state that the two walkers occupy the same position, independently of the initial condition (essentially the numerator), and a combined rate involving the contact parameter and a motion parameter (essentially the reciprocal of the denominator). Thus, α equals simply $C\mu(\infty)$ in the contact-limited case, i.e., when $C << M$. In the opposite limit $M << C$, infection is governed by the motion and α is $M\mu(\infty)$.

This limiting behavior for extreme relative values of the contact and motion parameters is clear in the left panel of Fig. 15.3. The motion parameter M describes an accumulated integral of the difference between the two probability densities explained above of the two walkers coinciding in location. The non-monotonicity effect is displayed in the right panel of Fig. 15.3 where the infection rate α rises, peaks, and drops as the potential steepness is varied.

Equation (15.26) allows the evaluation of $\mu(\infty)$ in Eq. (15.32) for arbitrary dimensions s as being $\left[(1/\sigma\sqrt{2\pi})e^{-H^2/2\sigma^2}\right]^s$ where $\sigma = \sqrt{2D/\gamma}$ is the width of the steady-state distribution in $1d$. Calculating M involves the evaluation of an improper integral which is convergent in 1-dimension but presents the standard difficulties that arise in reaction diffusion problems in dimensions higher than 1 if reaction is taken to occur at points (regions of vanishing dimension) as we have done here. This problem has been discussed further by Satomi Sugaya in her dissertation.

In other contexts, it is also of interest to include the consequences of the introduction of an additional decay into the system. Such a decay may arise from radiative lifetimes as explained for excitons in molecular crystals earlier (Kenkre and Reineker 1982), from finite lifetimes τ of the infected animals as they may die from natural death or from predator attack, or from finite lifetime of the infection itself. The latter may be caused by the animals recovering from being infective. In such cases one takes the limit $\epsilon \to 1/\tau$ rather than $\epsilon \to 0$, and Eq. (15.32) is replaced by

$$\alpha\tau = \frac{\int_0^\infty dt\,e^{-t/\tau}\mu(t)}{(1/C) + \int_0^\infty dt\,e^{-t/\tau}[\nu(t) - \mu(t)]}. \tag{15.33}$$

In case a finite lifetime is absent in the given problem, it is perfectly natural to introduce it as a probe time associated with measurement.

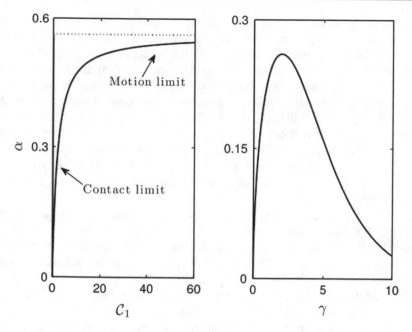

Fig. 15.3 Dependence of the effective infection rate α from Eq. (15.32) on the contact rate C_1 (scaled to $2D/H$) in the left panel and on the confining potential steepness γ in the right panel. Both α and γ are scaled to $1/\tau_H$. The left panel shows that α is linear in the contact rate for small values of the latter but saturates to the motion-limited value (0.56 in this example) for large values. The right panel shows the non-monotonicity effect on infection: as confinement steepness γ increases, the infection rate α rises to a peak and decreases for larger γ. For the right panel, C_1 in units of $2D/H$ is 15. Reprinted from Fig. 2 of (Kenkre and Sugaya 2014); copyright (2014) with permission from Springer Nature

15.4.4 Concluding Remarks

The calculation we have presented is precise for the limited model considered and is valid for movement both with and without spatial constraints imposed on the moving animals. The spatial constraint would represent the existence of home ranges. In the presence of spatial constraints, the analysis has uncovered a remarkable phenomenon: the infection efficiency is non-monotonic when the steepness of the confining potential, or the animal diffusion constant, is varied. A similar simpler phenomenon occurs in reaction diffusion scenarios for *trapping* considerations under a confining potential, as was shown in Spendier et al. (2013) and discussed in this book in Chap. 12.

In our present context, each of the two quantities, the steepness of the confining potential and the animal diffusion constant, has a critical value on both sides of which infection becomes inefficient. An understanding of the curious effect we observe can be achieved at various levels. The effect involves three quantities, the distance H between the centers of the home ranges, the diffusion constant D, and the potential steepness γ. Combined into a single parameter $\sqrt{H^2\gamma/2D}$, which equals H/σ, the quantities signal inefficient transmission of infection when variations in D or γ make the parameter differ from its optimum value. In the contact-limited case, the optimum value is 1 and corresponds to the static statement that the width of the steady-state distribution of the Smoluchowski equation equals the distance between the home centers; or to the dynamic statement that the time taken by the walker to traverse the inter-home distance H diffusively equals the time $1/\gamma$ characteristic of free motion of the walker to the center under the action of the potential. Away from the contact limit, the optimum value changes from 1 because of contributions from what has been explained as the motion parameter M (see earlier text). Thus, in the right panel of Fig. 15.3, for the particular variable values we have assumed, it happens to equal 1.97.

In extending this investigation to higher-dimensional systems, Sugaya and Kenkre (2018) found a number of additional new results and gained novel insights. It would take us too far away from the stated aims of this book, however, to initiate discussions of those matters. Let me refer the reader to that original publication as well as the recent book by Kenkre and Giuggioli (2020) for that purpose.

The (quadratic) confinement potential we have considered in Eq. (15.17), and consequently throughout the analysis, has been selected for two reasons. The first is the simplicity of a linear restoring force it represents. The second is the analytic tractability it provides. Explicit expressions for the propagators can also be obtained for linear (rather than quadratic) and box potentials (Chase 2016). They too result in the behavior we have discussed. Generally the analysis we have given carries over, in its qualitative conclusions, for any confinement potential.

15.5   Chapter 15 in Summary

While the defects studied in Chaps. 11 and 12 were physical objects that actually break the translational invariance of the host, the treatment in this chapter is of formal defects in perfectly symmetrical systems in which there are no *physical* defects. The defects analyzed arose from mathematical terms in the governing equations. The method was applied to the problem of calculating the scattering function in a quantum mechanical system described by the so-called stochastic Liouville equation obeyed by the density matrix. Next, the problem for which the defect technique application was originally extended in this fashion for non-physical defects, the mutual annihilation of Frenkel excitons moving on a lattice, was explained. Annihilation rates and constants computed from the theory were displayed for various dimensions and it was indicated how coherence effects in the motion are incorporated. Finally, for our third example of non-physical defects, related methods were applied on the basis of a Smoluchowski equation to the study of the transmission of infection in the spread of epidemics such as the Hantavirus and the time-dependence of the infection curve was obtained.

Concluding Remarks

16

There are several ways of doing science. I think of two as I look out of my window first, and then turn my sight inward inside the room. Out there I see an agile man balanced precariously on a lift crane, intense concentration in his eyes as he applies his exceptional expertise to his job of helping erect an edifice. As I turn my eyes inward, I spot a kitten playing with a ball of wool on the floor. The apparent care-free activity[1] as the ball is rolled and thrown in every which direction also involves intense concentration but, prominently, what looks to me like pure joy with little responsibility. Many of my colleagues, including my teachers and my students, bless them, do science in the manner of the man. My own way has been, for most part, the other one. Not really a matter of choice for me. I have merely followed the path of least resistance. If you, the reader, understand the distinction, you might appreciate this book a tad better than if you don't.

This ending chapter has primarily two functions. The first is to summarize what we dwelt on; not piece-meal as we have done at the end of each chapter but in a unified manner so that the reader is aware of the logical connection, at least the relation, between the various pursuits described. The second is to mention a few topics that are relevant but did not naturally fit in the flow of the story in the book. Let me begin with the first.[2]

---

[1] My colleague who is much more familiar with kittens than I am warns me that the activity of the kitten with the ball of wool involves the visualization of mice it would love to pounce on with an intensity not a whit less than that of the man I see through my window. My response to my colleague is that the fact that he is, very much, one of those intense scientists perhaps supports what I am attempting to communicate.

[2] A few words about the style of writing which might remind you of the British-born philosopher Alan Watts (if you are into his peculiar books and talks) who, I remember hearing in the 70s on the radio, would chuckle heartily at his own jokes. I think I use my words to make life more livable for myself. For instance, the reader might have noticed in an earlier chapter a footnote about the "mighty power of democracy." Were it not for such quietly innocent outlets, I would

© The Author(s), under exclusive license to Springer Nature Switzerland AG 2021
V. M. (Nitant) Kenkre, *Memory Functions, Projection Operators, and the Defect Technique*, Lecture Notes in Physics 982,
https://doi.org/10.1007/978-3-030-68667-3_16

16.1   What Was Covered

We began our conversation (you, the reader, have had ample evidence that this book has been written in the spirit of a conversation) with the observation that extremes of movement we see around us are oscillations on the one hand and decays on the other. We found that memory functions are an excellent tool to examine and ultimately unify these two extremes and the whole range in between. We cited a number of instances in the first chapter, first classical and then quantum mechanical. The reader was promised hands-on exercises sprinkled throughout the book.

That memories are not mere artificial constructs but appear naturally during the passage from underlying equations of evolution at the microscopic level to macroscopic phenomena was made clear in the second chapter. Our second research device, projections, also emerged during that passage and we spent some time and effort familiarizing ourselves with the technical aspects of that tool.

The need for, as well as the method of, injecting coarse-graining into projections occupied us in the third chapter. We made our acquaintance with the field of energy transfer in that chapter, learned about some puzzles and problems in that area of research, and showed how our memory tool renders excellent help in resolving them.

We examined more of the energy transfer or exciton transport field in the fourth chapter where we saw the relations that the memory formalism as a tool bears to other theories such as the stochastic Liouville equation and to other physics constructs such as correlation functions and scattering functions. We saw connections to pausing time distributions of random walkers, conduction of charges, and neutron scattering of interstitials. We familiarized ourselves with the computational device of the intermediate coherence propagator obtained through explicit solution of the GME.

Questions of exciton coherence were addressed in the fifth chapter by developing a theory of transient grating and Ronchi ruling experiments. These being methods of observation as direct as one could wish for, we shaped, and applied, our theory of generalized master equations in close proximity to experiment in that chapter. We even produced a definitive chart about exciton coherence expressed in the form of a concrete table of extracted parameter values.

Then came a similar analysis close to experiment where there was a connection to memory functions that made their presence felt through velocity correlation functions but not for exciton transport, rather for the motion of photo-injected charge carriers in the aromatic hydrocarbon crystals. The temperature dependence of their mobility is what we sought to explain with the help of our GME theory in this sixth chapter. The discussion was presented as a continual dialog between theory and experiment.

---

die of intellectual suffocation every single time I hear of a president's antics or a senator's idea of consistency.

Under the guise of generalizing into the coherent domain the Montroll-Shuler equation for vibrational relaxation of a molecule embedded in a reservoir, in the seventh chapter, we addressed the passage from microscopic to macroscopic levels of description of systems inherently non-degenerate in energy. We answered some questions that might arise regarding the simultaneous processes of decoherence and population equilibration in a quantum mechanical system. We were able to reveal some new behavior in this venerable and old area of research involving non-monotonic dependence of relaxation observables on temperature.

A return to the projection tool was undertaken in the eighth chapter where we learned how projections could be diversified to do much more than diagonalize or coarse-grain. We defined and used the technique to calculate the electrical resistivity of an arbitrary system, to perform integrations in classical systems, to deal with problems in the study of nuclear magnetic resonance microscopy, and for other matters as well.

Through an analogical leap where time is replaced by depth, we analyzed stress compaction in granular compacts in the ninth chapter. Through the so-called $t - z$ transformation and eigenvalue analysis, we understood stress distributions in compacts via the perhaps unexpected analogy.

In spite of the obvious linearity of the Zwanzig projection tool, we found a way of using it to analyze nonlinear problems in the tenth chapter. The first of the investigations was to the nonlinear Schrödinger equation which is a patently nonlinear entity carrying, as it does, a cubic nonlinearity in the amplitude. The other nonlinear investigations were in the ecological field.

The third primary tool in the triad on which the book focuses, the defect technique, was introduced in the eleventh chapter. First, a number of procedures, old and novel, were introduced to treat systems in which perturbations destroy translational invariance of the unperturbed system via small and large changes. Applications were then described in the close context of experiments in molecular crystals. The exquisitely important issue of whether motion does or does not limit an overall reaction-diffusion process was investigated. This chapter suggested all was not well in practices followed in the field of pure molecular crystals.

The elucidation of additional questions and answers that are naturally addressed with the defect technique occupied us in the twelfth chapter where a survey was given of a number of results, some of them new and others poorly known, and an application of the technique to solve problems of aggregation was described in the context of the clustering of signal receptors in immunology. What happens when the defect technique is used in conjunction with an equation like Smoluchowki's wherein the moving particle is tethered to a home was also worked out.

Memory functions in statically disordered systems was the subject matter of the thirteenth and fourteenth chapters. The defect technique know-how gathered in the previous two chapters was used to develop and investigate in detail the theory of the effective medium formalism in the first of these two chapters while a tutorial on how to apply the formalism with attention to individual steps was the subject of the second.

The use of the defect technique in a system in which there are no physical defects at all, only mathematical ones in the formalism, was the content of the fifteenth chapter. The stochastic Liouville equation was solved through the methodology developed, mutual annihilation of moving excitons was addressed, and the transmission of infection in epidemics was investigated.

That brought us to this final chapter.

16.2 What Was Not Covered

During the course of the writing, many additional topics suggested themselves but seemed not completely appropriate to include at the places at which they arose in the mind. Some of them seem important enough to mention here in this final chapter. They are touched on below.

16.2.1 The Many Faces of the Word "Coherence"

I have seen few words in physics research confused by their users as much as coherence in the context of excitons. In part this has arisen from misconceptions, in part from varied legitimate choices by investigators that have little to do with one another. Because I have seen such Babel's tower miscommunication happen at conferences and in the literature, I venture into a bit of clarification but not without some trepidation.[3]

The meaning of the word that I have used in this book, as well as in conducting coherence research on Frenkel excitons, should be completely clear from the development in Chap. 1 and the evolution through the next four chapters ending with charts of the amount of coherence of triplet and singlet excitons in the aromatic hydrocarbon crystals constructed from observations in Chap. 5. Coherence in our usage stands for oscillations, when the oscillations pass into decays we think of the process as decoherence, and when we say that an exciton in a certain system at a certain temperature has such and such degree of coherence, we refer to this oscillation aspect and measure the degree with the help of a tangible ratio such as of the mean free path to the lattice constant.[4]

First, let me make the probably unnecessary comment that exciton coherence has nothing to do with the so-called *coherent states* that one finds attributed to Sudarshan, Klauder, and Glauber. Those coherent states have been called "displaced

---

[3]Emotions run high in this subject, I seem to recall.

[4]Hence, we have regarded a harmonic oscillator with small damping as coherent and so too propagation on a line via the wave equation with little scattering, whether quantum mechanical or classical. By the same token, a highly overdamped harmonic oscillator is incoherent, as is the wave equation with so much scattering that we are near the diffusion limit, or a quantum mechanical particle that gets scattered at every lattice site in its travels.

oscillator states" by some condensed matter physicists and "minimum uncertainty states" by others. The interested reader will find them treated in detail in a book such as Klauder and Skagerstam (1985).

Second, it is best not to mix exciton coherence as analyzed by countless experimentalists and theorists in the area of molecular crystals and aggregates with the issue of quantum mechanical decoherence. This latter issue has been treated widely in the literature and here in this book in Chap. 7. In this book it has naturally arisen allied to our extension of the Montroll-Shuler vibrational relaxation Master equation into the quantum domain by deriving a GME. In a general manner, this use of the word coherence is related to the belief that, if we encounter an isolated system in weak interaction with a reservoir (i.e., open to the universe), we postulate not only its probabilities of the occupation of its Hamiltonian eigenstates to be in Boltzmann ratios (if we are in a canonical situation) but also, *in addition*, its off-diagonal density matrix elements in the representation of those Hamiltonian eigenstates to be vanishing. The process whereby the system goes into that random phase state is called (quantum) decoherence.

This has led some exciton investigators to identify a coherent exciton as one in which this process, required by thermalization in quantum mechanics, has not been completed yet. It appears that such a discussion is part of two relatively recent publications in photosynthesis (Engel et al. 2007; Cao et al. 2020).

Our own viewpoint in the 1970s and 1980s, equally shared by experimentalists and theorists at least in the field of exciton motion in molecular crystals and aggregates, was that if a particle moved with little scattering, its motion being therefore largely ballistic or wave-like, it was said to be moving coherently. If the particle did not maintain its velocity, in magnitude or direction, but was repeatedly scattered, the limit was said to be incoherent. Very important to realize was that this is *independent* of whether the description is classical or quantum mechanical.[5] The same picture of coherence applies to both. In Chap. 1 I have given an ample number of examples illustrating this situation.

It is also important not to confuse matters such as spin coherence to be mixed in with transport coherence; the latter is what the above discussion is about. The charts of triplet and singlet exciton coherence provided in Chap. 5 extracted in collaboration with Schmid from Ronchi ruling and transient grating experiments reported by others should make crystal clear what some of us mean by coherence.

16.2.2 How I Began with the Energy Transfer Problem

Although I have had loads of fun applying the three devices that form the focus of the book in a diversity of fields, coherence issues in energy transfer are where it all

---

[5]Just as I want to think of a material as being red or green or hard or soft, independently of whether I look at it quantum mechanically or classically, I seek a definition of the exciton as being coherent or incoherent independently of the brand of mechanics I use to approach the matter.

began for me. That a memory formalism could make a substantial contribution to the resolution of coherence issues was quite clear to me from the outset. However, the first attempt I made to implement the idea was so simple that it bordered on the simplistic. I based it on the telegrapher's equation and coherence signatures to be found in the mean square displacement of the exciton. It was correctly, and promptly, bounced out by the experts.[6] Bob Knox, and sometimes Raoul Kopelman, provided kind encouragement. However, harsh criticism from colleagues such as Bob Pearlstein, Katja Lindenberg, and Bob Silbey, all of whom had already made respectable and worthwhile contributions to the field, was prompt and effective. They certainly had the right, perhaps even the duty, to curb, and thereby educate, an overenthusiastic tyro. They justifiably pointed out that the telegrapher's equation was too classical in flavor and did not have the correct dispersion to be able to describe the motion of excitons. My telegrapher's attempt suffered an early demise. It was frustrating to attempt to convince the experts that the primary point was about how the memories naturally combined coherent and incoherent motion, not the details of dispersion. A powerful weapon that was made available to me by Bob Knox was the English translation he had made of the Förster paper.[7] The generalization of Förster's theory of excitation transfer including of his spectral prescription that Bob and I then undertook together, started the respectable version of the GME theory of exciton transport.

16.2.3 Direct Versus Indirect Experiments for Determining Motion

I think it was Harris and Zwemer (1978) who classified Frenkel exciton experiments into the categories: optical lineshape, magnetic resonance, and direct migration. Thinking of this topic I have always thought that, if you want to know how fast your horse runs, you might want to start him at one spot and measure with a stopwatch the time he takes to reach a point a fixed known distance apart. This would be a direct experiment. There is no guarantee whatsoever that such experiments would be simple to implement but they do possess conceptual simplicity. By contrast, tying clearly viewed objects to the horse's mane and measuring their trajectories

---

[6]My unfortunate paper was kept for eighteen months on their desks by J. Chem. Phys., but never rejected.

[7]Having had a formalistic training, and little exposure to experiment in physics, I had a great time learning from two of Bob's favorite pieces of advice. One was that a theoretical physicist should focus on observations. The other that it was futile to leave to others the application of one's mathematical creations to experiment, because no one else would be likely to undertake the application. That second piece of advice as well as the chiding thrown into the literature by Powell and Soos (1975), that theorists and experimentalists *had to* intensify their mutual dialogue, shaped my activities for some time, for better or for worse. I must narrate in this regard Bob Silbey's humorous remark. Apropos the Powell-Soos comment, and perhaps in response to my over-enthusiastic attempts to engage experimentalists, he went on to warn, "It is not enough for a theorist to talk to an experimentalist; it is necessary to explain the experiment."

as the horse gallops might be called an indirect experiment. The first two categories mentioned above inclusive of optically detected magnetic resonance observations and ESR lineshape measurements might be said to belong, in this sense, to the class of indirect experiments. Not that they are inferior in any sense. Indeed, they are often extremely clever. However, the analogy might have some usefulness.

In this connection the most direct experiments might be considered to be sensitized luminescence and a closely allied relative, mutual annhilation. Although simple to think about, we have seen in Chap. 11 that in many systems they seem to be *not* gentle experiments. By this is meant that they introduce new features into the system that one cannot neglect, forcing us to pay undue attention to the probe process, e.g., capture.

Warning signs, such as the large discrepancies in the interpreted values of the singlet diffusion constant in anthracene discussed in Chap. 11, might disabuse us in this context and compel us to turn to the experiments discussed in Chap. 5. These, as we saw, involved transient grating or Ronchi ruling procedures. Do the initial condition discussions in that chapter raise red flags and stop us from determining the amount and nature of exciton transport in molecular crystals? Some have felt so but I do not. My reasons are made clear in the detailed analysis in Chap. 5. It is my belief that this is an important issue worthy of careful scrutiny by future investigators so that these matters are not left hanging.[8]

16.2.4 Extensions by Hand of Approximated Equations of Motion

The form and structure of equations of motion that we use for various purposes in physics are determined by the physical considerations that underlie them. Solving the equations so we can find out what they predict is not always easy. Often approximations have to be made before solutions can be obtained. The approximations might involve slicing off a part of the equations, or replacing a part by another part that is more tractable. There is no guarantee that the approximated version maintains some general feature of the solution that may be expected. Consider, for instance, the memory function in the GME for an infinite chain. If approximated by a constant, it predicts probabilities that can go negative. Negative probabilities are not only absurd but offensive as well: they play upon our emotions. Remarkably, in this particular example, the mean square displacement is given accurately by the truncated equation even as it predicts negative probabilities.

When we notice such a misdemeanor on the part of the approximated equation, we are often driven to apply a correction by inventing, essentially in an *ad hoc*

---

[8]My biggest dissatisfaction with the way our community of investigators pursues science is that we are often too eager to chase the next shining object and think little of leaving issues unresolved. A wise colleague once explained to me that he focuses his energies on inventing the next novel experimental idea because finishing matters has a much lower chance of getting him funded. I cringed at the explanation even while I sympathized with my colleague's words but then I have always had difficulty coming to terms with reality.

manner an extension or change in the equation. An example is the SLE for a two-state system that Talat Rahman and Bob Knox had to use for their theory of the depolarization of fluorescence in stick dimers. As it stood, the SLE forced the off-diagonal elements of the density matrix ρ in the site representation to vanish at long times, a result that was patently inaccurate. I provided assistance by adding terms, literally by hand, that would make them go to their correct values. Now we had an equation that no longer possessed the *particular* incorrect feature that we had got rid of. But were we certain that exercise had not introduced some other undesirable feature? Certainly not.

Approximation is an art. Some of the work involved has to be done by feel or on faith. If approximations always included introducing an infinite series of powers of a small number, one might have a situation in which higher order terms could be dropped. However in actual instances of research, convergence of the approximated entities is seldom guaranteed.

The cajoling of SLE's that we did for the fluorescence depolarization for stick dimers can be found in Rahman et al. (1979). A similar exercise carried out further for the calculation of neutron scattering lineshapes, also for a dimer, can be found in Brown and Kenkre (1987). A non-trivial similar modification of a truncated equation that we have described to a slightly larger extent (specifically in Chap. 15) was based on the symmetrization of the scattering correlation function to solve detailed balance woes in the SLE for the infinite chain (Kenkre and Brown 1985).

I have drawn attention to these approximation issues because they are ubiquitous in physics and need to be kept in mind. They cannot be avoided as undesirable practices unless one wishes to work only on toy models or always involve numerical procedures that may not remain transparent as they increase in complexity. There should be no question in the reader's mind that, as stated above, approximation in physics is always fraught with danger.[9] All one needs to do to be reminded of the perils is to recall the repeated random phase approximation that the great Pauli made in order to obtain the Master equation from the microscopic von Neumann equation and how van Hove showed (a quarter of a century later) that it led to a stationary or unchanging density matrix, completely ousting Pauli's derivation!

16.2.5 Substitutional Models and Momentum-Space Description

All descriptions of exciton trapping and its effects set forth in Chap. 11 have been based on the simple picture of the defective behavior as consisting of the leaking out of the exciton probability from a host site. That is typically referred to as the sink model. Other (sometimes more elaborate) ways of treating capture are both possible and have been carried out. Let us take a brief look at some of them.

---

[9]If you want to emphasize the positive side of this matter, you could say to yourself, as I have done above, that it is, indeed, an art.

The first elaboration is to include detrapping. The defective term in an equation such as Eq. (11.3) is written as $-\delta_{m,r}[cP_m - c'P_\theta]$, where $P_\theta(t)$ is the probability that the trap θ is occupied, and augmented by an equation of evolution for the latter,

$$\frac{dP_\theta}{dt} + \frac{P_\theta}{\tau_G} = cP_r - c'P_\theta. \tag{16.1}$$

Here, τ_G is the (e.g., radiative) lifetime of the excitation in the trap and, while we have assumed here for simplicity that a guest site communicates with a single host site and not other guest sites, a generalization is possible easily.

One of the places to see the development of this model laid out in detail is in the book by Kenkre and Reineker (1982) and another in the original paper by Kenkre and Wong (1981). In the latter publication, yet another model is presented to describe the more commonly occurring case of substitutional traps. No capture rates c and c' are introduced, and (memory functions or) transfer rates F_{mn} are generally modified when m or n is a trap site to incorporate the capture phenomenon. Temperature effects associated with detrapping are easily introduced (assuming the energy difference between trap and host sites are not enormously larger than $k_B T$) into the analysis and guidelines can be obtained for deciding when the simpler sink model may (or may not) be used as a representation of the more physical substitutional-trap model.

It is also possible to develop a momentum-space theory of sensitized luminescence in parallel with the defect technique considerations described so far in the book. Inspired by the approach of Fayer and Harris (1974) along these lines, I developed an analysis (Kenkre 1978c, 1979) based on linearized Boltzmann equations which was extended in collaboration with Yiu-man Wong (Wong and Kenkre 1979). The starting point is a coupled equation for the probabilities of occupation, f_k of the k-space state of the host, and f_θ of the trap:

$$\frac{df_k}{dt} + \left(\alpha_k + \frac{1}{\tau_H}\right) f_k = \alpha_k^\dagger f_\theta + \sum_{k'} (Q_{kk'}f_{k'} - Q_{k'k}f_k), \tag{16.2a}$$

$$\frac{df_\theta}{dt} + \left(\frac{1}{\tau_G} + \sum_k \alpha_k^\dagger\right) f_\theta = \sum_k \alpha_k f_k. \tag{16.2b}$$

Here $Q_{kk'}$ is the scattering rate from state k' to state k arising from exciton-phonon or exciton-imperfection interactions, α_k is the trapping rate from state k, and $\alpha_k^\dagger$ is the detrapping rate to state k from the trap. No trap-trap interactions are included and the k-dependence of the host lifetime is suppressed. The relaxation time approximation[10] is used for the solution of the equation of motion, model calculations give the trapping and detrapping rates, and the host excitation probability is obtained as

$$n_H(t) = \sum_k f_k(t).$$

While one meets with some fresh technical tools in the process, there are no surprises in the calculated quantities and everything proceeds as expected. The original publication should be consulted for details.

16.2.6 Long-Range Capture in Sensitized Luminescence

In a footnote in Sect. 11.7, during the course of the discussion of whether sensitized luminescence (and associated experiments such as those on exciton annihilation) is an appropriate probe of exciton motion in molecular crystals, I mentioned that the possibility does exist in some systems that long-range reaction processes may be at work (footnote 19 in that chapter to be precise.) Similar suggestions and even rate and 'sphere of influence' calculations have appeared early on (Babenko et al. 1971a,b; Rahman and Knox 1973).

Although the mainstream thinking has been based on motion and reaction processes being sequential in a reaction-diffusion scenario, this possibility suggests that investigations be undertaken to restructure the way one reasons in this context. This is not easy to do. When the parts of a reaction-diffusion phenomenon are not sequential, calculations and even visualization become at once much more complicated. Nevertheless, I would like to bring to the interested reader's attention a theoretical analysis undertaken by Parris and me as early as in 1986 along these lines which applies for any number of dimensions.

For simplicity, it is assumed in Parris and Kenkre (1986) that, in a 1-dimensional chain, the origin is a trap-influenced site and excitons disappear at rate c when they land, not *at that site* but in a neighborhood of that site up to a distance of r sites in either direction. In the Laplace domain, the probability in that trap-influenced region is given by (the notation $\epsilon' = \epsilon + 1/\tau_H$ is used here)

$$\sum_{s=-r}^{r} \widetilde{P}_s(\epsilon) = \sum_{s=-r}^{r} \widetilde{\eta}_s(\epsilon') - c \sum_{s=-r}^{r} \widetilde{\zeta}_s(\epsilon') \widetilde{P}_s(\epsilon)$$

in which we define $\zeta_s(t) = \sum_{j=-r}^{r} \Pi_{s-j}(t)$. If we now invoke the sense of the approximation involved in the ν-function formalism developed in Chap. 12 but apply it here to define an average ζ, we can arrive at a low-concentration expression

---

[10]For a relatively recent extension of the reach of this age-old workhorse of practical transport theory see Giuggioli et al. (2006b). It allows the relaxation time approximation to describe the effect of the initial distribution on relaxation times through a variational formula.

(as in the Kenkre-Wong assumption) at the guest luminescence intensity

$$\tilde{n}_G(\epsilon) = \frac{\rho(2r+1)}{\epsilon'(\epsilon + 1/\tau_G)\left[1/c + \tilde{\zeta}(\epsilon')\right]}.$$

This leads to an explicit expression for the important quantity, the guest yield in the manner shown in Chap. 11. Parris and Kenkre (1986) were able to generalize this simple theory to arbitrary dimensionality and study the effect of dimensionality provided the transport is diffusive. One of their successes was an examination of the applicability of the Smoluchowski coagulation rate expressions to sensitized luminescence in the presence of long-range capture. The Smoluchowski prescription in arbitrary number of dimensions was not only recovered for large capture rate, but for the case when the rate is not too large, explicit correction factors to that prescription that depend on the parameters such as the diffusion constant were provided. I definitely recommend to the reader to examine this publication if interested in this difficult problem.

16.2.7 Time-Dependent Transfer Rates: An Economical Ruse or a Poor Practice?

The literature on energy transfer in molecular crystals and aggregates is riddled with *time-dependent* energy transfer rates and all sorts of interpretations have been forcibly attached to the concept. What is right and what is wrong here, what is useful and what is tautological, deserves at least a bit of comment.

If energy transfer is simplistically described by two coupled rate equations for the numbers n_H and n_G of host and guest excitons, respectively,

$$\frac{dn_H}{dt} + \frac{n_H}{\tau_H} = -kn_H, \tag{16.3a}$$

$$\frac{dn_G}{dt} + \frac{n_G}{\tau_G} = kn_H, \tag{16.3b}$$

the expectation is that the guest luminescence intensity rises and decays as a sum of two exponentials, and the host intensity decays exponentially with a decay constant that is a sum of the reciprocal of the radiative lifetime and the "energy transfer rate" k.[11] Time-resolved spectroscopy introduced into this field by Power and Kepler, and reviewed in Powell and Soos (1975), resulted in observations that could not be explained by the simplistic description of (16.3). It was natural to brandish a time-dependent $k(t)$ defined through

---

[11] The assumption here is, of course, that only the host is excited initially.

$$k(t) = -\frac{1}{n_H(t)}\frac{dn_H(t)}{dt} - \frac{1}{\tau_H} \tag{16.4}$$

and perhaps tempting to fall into the trap[12] of claiming that this rate should be related to what was sometimes referred to (a little bit with awe) as the Chandrasekhar-Smoluchowski expression

$$k(t) = A + \frac{B}{\sqrt{t}},$$

where A and B are constants. Much was done (see, e.g., Powell and Soos (1975)) to fit this expression to observations and deduce quantities such as exciton diffusion constants from the results. There were even statements in the literature that a time-independent $k(t)$ means that the exciton diffusion model is appropriate and that time dependence in the rate signals coherence. Some investigators challenged experimental $k(t)$ reports,[13] some others challenged theoretical derivations of $k(t)$.

Volumes could be written on this subject.[14] However, suffice it to state that the host and the guest have internal structure. There is motion of excitons that is going on from location to location. There is the process of capture by the guest that occurs. There should be no question that, theoretically, the energy transfer could be quite complex requiring more than a single exponential or even other functional dependences for its description. The existence of a time dependence in the energy transfer rate should have nothing to do with the underlying process being diffusive. Exciton diffusion can result in generally a time-dependent $k(t)$. So can exciton waves or almost any other kind of exciton propagation.

It should be equally clear that many features of the process including simply relative magnitudes of various parameters could mask all of this complexity and make the transfer appear mono-exponential. There is no reason to expect that decisive information of any useful kind would emerge from the time dependence of such a coarse-grained quantity as the transfer rate.

On the other hand, it can be sometimes of some value to expect and define an energy transfer memory $k_m(t)$ simply because there are all of these processes going on inside. It would appear in a description in which the last term in the simplistic Eq. (16.3a) would be replaced by $-\int_0^t dt'\, k_m(t - t')n_H(t')$. The situation we have just alluded to, wherein nonexponential transfer is masked by various factors, can be then said to correspond simply to the Markoffian approximation $k_m(t) = k\delta(t)$

---

[12]Given that we have almost reached the end of the book, I hope the reader will be kind enough to pardon the pun.

[13]To me, the most fascinating piece of information was two reports from completely different experimental groups presented at the IX International Molecular Crystals Symposium in 1980, held in Kleinwalsertal, Austria, that closer scrutiny revealed no time-dependence in the energy transfer rate whatsoever.

[14]Indeed, the interested reader will find an entire section on the subject occupying several pages of the book on exciton dynamics coauthored by Kenkre and Reineker (1982).

being appropriate to the description. However, one must remain alert against any tendency to attach one's favorite kind of exciton motion as an interpretation to the memory.

16.2.8 Exact Analysis for Smoluchowski-Like Equations

Discussions of entities that are attracted towards a "home" while they are executing random walks and undergoing a reaction process, whether in a general, abstract, sense or for a specific phenomenon such as the transmission of infection in an epidemic, have appeared in Chaps. 12 and 15, respectively. To represent the attraction, those discussions have employed the Smoluchowski equation which is based on a quadratic (equivalently, harmonic) potential. The reader should be aware of an alternative that is available as a result of analytic work performed by Matt Chase.

The attractive force that we encountered in Chaps. 12 and 15 is linear in the distance from the center of attraction, thus increasing in magnitude with distance. The alternative that I want to draw the reader's attention to is that of a constant force (but of course always directed towards the center). Whereas the standard treatment is based on Eq. (12.39) in the absence of the capture rate C_1, the constant force equation that Chase and I investigated is

$$\frac{\partial P(x,t)}{\partial t} = \Gamma \frac{\partial}{\partial x}\left(\frac{|x|}{x} P(x,t)\right) + D\frac{\partial^2 P(x,t)}{\partial x^2}, \tag{16.5}$$

where Γ is the strength of the potential with units of velocity.

The propagator of the linear force equation was as given by Eq. (12.40). That for the constant force equation is

$$\Pi(x, x_0, t) = \frac{1}{\sqrt{4\pi Dt}} e^{-\frac{(x-x_0)^2+\Gamma^2t^2}{4Dt}} e^{-\frac{|x|-|x_0|}{2\ell}} + \frac{e^{-\frac{|x|}{\ell}}}{4\ell}\left(1 - \mathrm{erf}\left(\frac{|x|+|x_0|-\Gamma t}{\sqrt{4Dt}}\right)\right). \tag{16.6}$$

The advantage of Chase's analysis over that of the standard linear force case is that not only can one write down the propagator explicitly in the time domain as in the linear force case, but an explicit expression for its Laplace transform is also available:

$$\widetilde{\Pi}(x, x_0, \epsilon) = \frac{e^{-\frac{|x|-|x_0|}{2\ell}}}{\Gamma\sqrt{1+\frac{4\ell\epsilon}{\Gamma}}}\left[e^{-\sqrt{1+\frac{4\ell\epsilon}{\Gamma}}\frac{|x-x_0|}{2\ell}} + \frac{e^{-\sqrt{1+\frac{4\ell\epsilon}{\Gamma}}\frac{|x|+|x_0|}{2\ell}}}{\sqrt{1+\frac{4\ell\epsilon}{\Gamma}}-1}\right]. \tag{16.7}$$

This is not true of the standard linear force case and necessitates complex numerical work to be carried out for the solution of the trapping problem as was explained in Chap. 12 or for the solution of the infection transmission problem as was explained in Chap. 15. We found that the time-domain propagator had been previously reported by Touchette et al. (2010) in their study of dry friction but Chase's Laplace transform propagator, or his method of its determination, was not known in the previous literature. In collaboration with Spendier, we applied this alternate way of describing attraction to a center to the analysis of matters of interest to immunology (Chase et al. 2016).

16.2.9 The Initial Term in GMEs

A problem in fundamental quantum statistical mechanics that has received very little attention since the 1960s when it first appeared on the horizon, has to do with the exploration of the initial term in generalized master equations. This is the third term in the right hand side of Eq. (2.7) that arises from the Zwanzig analysis. Specifically, in the projected equation whose left hand side is $\partial \mathcal{P}\rho(t)/\partial t$, it is given by

$$\mathcal{I}(t) = -i\mathcal{P}Le^{-it(1-\mathcal{P})L}(1 - \mathcal{P})\rho(0). \tag{16.8}$$

Generally, this initial term has been put equal to zero, by assumption or approximation, in countless investigations associated with GME's.[15] In Zwanzig's own derivation of the Master equation it vanishes because of the initial random phase assumption. What does its omission exactly mean and what happens when it cannot be put equal to zero?

This question has been lightly commented on in a discussion following Eq. (2.12) at the end of Sect. 2.2 and was asked at an elementary level in Kenkre (1978b). By expanding the exponential in (16.8) in an infinite series and reordering the placement of L, and resumming the series into an exponential again, it was shown that the term is also exactly given by the expression

$$\mathcal{I}(t) = -i\mathcal{P}e^{-itL(1-\mathcal{P})}L(1 - \mathcal{P})\rho(0). \tag{16.9}$$

The advance made is that, whereas Eq. (16.8) ensures the vanishing of the initial term for all time if $(1 - \mathcal{P})\rho(0) = 0$, the new result, Eq. (16.9), guarantees such vanishing even if that condition is not satisfied provided $L(1 - \mathcal{P})\rho(0) = 0$ is satisfied. Even in cases in which the initial density matrix is not completely diagonal, if it is such that its off-diagonal part commutes with the system Hamiltonian, we are assured of the vanishing of the initial term for all t.

The derivation in Kenkre (1978b) shows, thus, that not only will $\mathcal{I}(t)$ vanish for an initially localized state in the crystal but for an initially totally delocalized state as well. The importance of this result is that it leads us to explore initial occupation first of pairs of states, and then of groups of states. For instance, Kenkre (1978b)

---

[15]The one exception is the work of Mori (1965) in which the term plays the central role of the random force in an equation of the Langevin type.

gives explicit calculations for initial occupation of any two states localized at sites m and n, and any two delocalized states with quasimomenta k and q in the band. Expressions are made available there for $\mathcal{I}(t)$ for all time for such pair states.

The reader may recall from Chap. 5 how that early investigation led Kenkre and Tsironis (1985) and Kenkre and Schmid (1987) to show that chances are excellent that initial conditions do not throw a monkey wrench into the conclusions drawn by Kenkre and Schmid (1985) about coherence of singlets in anthracene as interpreted from observations of Fayer and collaborators. It is also possible that the analysis lends support to conclusions drawn by Tyminski et al. (1984) and by Morgan et al. (1986) about coherence in exciton motion in their experiments on inorganic materials where they touched upon our transient grating theory.

What is even more important than these relatively elementary studies is to continue these investigations into the evolution of $\mathcal{I}(t)$ and its consequences in situations that *include reservoirs*. The initial term must be analyzed when the initial density matrix is not simply a product of a reservoir ρ and a system ρ but one in which correlations exist across the system and the reservoir. These are difficult problems to tackle.

16.2.10 Competing Perturbation Schemes

If a bright graduate student were to approach me with a request that I suggest a topic for research, I would probably mention competing approximation schemes. I have in mind three specific questions. One has not been discussed in this book and has to do with semi-classical representations of quantum evolution with and without the nonlinear Schrödinger equation.[16] The other two have appeared in this book in Chaps. 4 and 8, respectively.

The polaron problem has been discussed in Chap. 4. After the dressing transformation is carried out, should we take into account the intersite interaction (that has been called "horizontal" in that discussion) exactly by going into Bloch states and should we do a final perturbation expansion in the left-over interaction? Or should we remain true to our initial assumption that site-localized states form the natural representation as a consequence of the magnitudes of the different parts of the Hamiltonian and do the perturbation calculation in that representation? The fact that the latter procedure appears to have explained naphthalene mobility data (as discussed in Chap. 6) while the former method seems to have failed to do so cannot be weighted too strongly given that such comparison involves so many additional approximations for a realistic system. What is required is a controlled system for which numerical solutions can be obtained along with the approximations.

The upstairs-downstairs methods that compete for the calculation of correlation functions, mentioned in Chap. 8, are a second candidate. Does the $\lambda^2 t$ method

---

[16]The interested reader should see the ending discussion in Kenkre et al. (1996) if motivation is desired.

capture better the correlation function or evaluate better a transport coefficient such as electrical resistivity? Or is the truncated projection method better equipped for the calculation? Does it depend on the nature of the scattering or the spatial distribution of scatterers? One might calculate the physical quantities exactly for a model system of a linear chain on which a charge moves with site energy fluctuations as scattering agents. This exact calculation could be performed either numerically or via the various defect technique procedures we have developed. The two methods of approximation could then be compared *vis-a-vis* the solutions. My hope is that a set of criteria might emerge to decide where one approximation method is preferable to the other. If you are a conscientious reader who decides to bite, please do let me know what transpires.

16.2.11 Experiment vs Theory and Falsification of Observations

The days are long gone when we could drop a stone, time its fall, and construct a theory for its motion. Or split light with a prism and theorize about the wavelengths comprising sunlight, all by ourselves. Division of investigators into experimentalists and theorists is inevitable given the complexity of the procedures involved, particularly in experimental technique.

You might think these are truisms of no importance. But I am plagued by practical questions. If I am a theorist, how can I be certain that the data I am attempting to understand really came from observations? The expectation is that an experimentalist repeats the act of observation under varying conditions (that are supposedly unimportant to the outcome) and that there are many laboratory groups that observe the same processes and find essentially the same results. But as the complexity of the experiments and the expense behind the necessary equipment spirals upwards, is such an expectation realistic? I ask again: how can I be certain that the data I am attempting to understand came really from observations?

Normally, we do not ask this question for the same reasons that we do not obsess with what went on in the kitchen of a restaurant we visit to order food. A bit of faith that all is okay with the world around us, that things are as they seem, or as we are told, is essential to living life without constant chaos. Yet I must remind you that around 2000, scientists engaged in research in organic materials in particular, and condensed matter physics in general, were traumatized by reports of falsification by experimentalists.[17] My colleagues and I who were merrily constructing theories for the saturation, the up-turn with temperature variations, and such other fancy occurrences in carrier mobility, had to stop our efforts precipitously. I advised a

---

[17]The term 'falsification' is used in this subsection not in the sense common to discussions in the philosophy of science where it means the experimental demonstration that a certain theory is inapplicable. I use the word here to represent, instead, the *false* reporting of observations that did not take place. This is the alarming story of Schön, Bell Labs, and pentacene. If interested, the reader should find out the details from the web or other sources.

student of mine who had already completed in part a Ph.D. thesis on transport in organics to change totally his direction of research and work, instead, in ecological science.

So the question remains in my mind, how to proceed. The situation is not any better outside the practice of science. In the last four years I have learnt that nothing should be taken for its face value and that there is something that is called "alternative facts", a concept that remains totally opaque to me. By mentioning these matters in a book meant at least in part for young researchers starting in theoretical physics, do I run the risk of turning them off the pursuit of science? Perhaps. But I have greater confidence in their own abilities to reason than those who would hide these goings-on from them.

16.3 Parting Comments

It is a great pleasure to end the book by expressing my sincere thanks to several of my colleagues who have spent large chunks of their valuable time by actually reading preliminary versions of my writing and making suggestions. Most notably John Andersen, but also Anastasia Ierides, Ziya Kalay, Bob Knox, Paul Parris, Kathrin Spendier, Mukesh Tiwari and Yiu-man Wong.

The research I have done in physics, I was told recently by a colleague who wants not to be named, appears to be divided into four parts. One is characterized by my attachment to memory functions, projection operators and the defect technique as narrated in this book. Another deals with the spread of epidemics, particularly Hantavirus, and other ecological matters. I have written a book on that subject, in collaboration with Luca Giuggioli; it is being published by Cambridge University Press at the end of 2020 with the title "Theory of the Spread of Epidemics and Movement Ecology of Animals." A third part of my work is about the interplay of quantum mechanics with nonlinearity and I have begun to collect it in a book I have been asked to write.

But then there is a fourth part about miscellaneous topics such as the study of solid friction, of dynamic localization, of microwave interactions with ceramics, and of scanning tunneling microscopy to mention a few. I found it tempting but practically impossible to knead them into the present book given that the title had been established. All those fascinating topics, if advancing age permits, I also want to describe in a collected fashion. I mention this because a few of my students (to all of whom I have dedicated this book) might frown on my omission here of *their* topics. Those will appear in that miscellaneous collection, I assure them, if only I can continue to think and write. This re-inspection of investigations we have had such fun performing over the years is certainly helping my own education in addition to giving me a great deal of enjoyment.

16.4   Chapter 16 in Summary

Presenting a summary of the summary chapter of the book surely smacks of *ouroboros*, the famous snake swallowing its own tail. But why not? In the present chapter, detailed remarks were made, in a summary mode, about what the book teaches in terms of techniques, methods and phenomena. The discussion also addressed several topics that are closely related to the subject matter of the book, which do not fall within the purview of any chapter in the book. They included the many faces of the word "coherence"; how my involvement in the energy transfer problem began; direct versus indirect experiments for determining motion; the method of extending equations of motion by hand. They elaborated upon substitutional models of trapping and momentum-space description of sensitized luminescence. They spoke of long-range capture in sensitized luminescence as well as attacking the use of time dependent transfer rates as a questionable practice. They passingly presented exact analysis for Smoluchowski-like equations. They revived the analysis of the initial term in the Zwanzig treatment of GMEs and the unsolved problem of competing perturbation schemes that need work. And they even touched upon the usually avoided topic of the rare practice of falsification of observations and generally the relationship of experiment to theory in the actual manner physics is carried out.

Bibliography

Abramowitz, M., & Stegun, I. A. (1965). *Handbook of mathematical functions with formulas, graphs, and mathematical tables*. New York: Dover.

Abramson, G., Bishop, A. R., & Kenkre, V. M. (2001). Effects of transport memory and nonlinear damping in a generalized Fisher's equation. *Physical Review E, 64*(6), 066615.

Abramson, G., & Wio, H. S. (1995). Time behaviour for diffusion in the presence of static imperfect traps. *Chaos, Solitons & Fractals, 6*, 1–5.

Adelman, S. A., Muralidhar, R., & Stote, R. H. (1991). Time correlation function approach to vibrational energy relaxation in liquids: Revised results for monatomic solvents and a comparison with the isolated binary collision model. *The Journal of Chemical Physics, 95*(4), 2738–2751.

Agarwal, G. S., & Harshawardhan, W. (1994). Realization of trapping in a two-level system with frequency-modulated fields. *Physical Review A, 50*(6), R4465.

Agranovich, V. M., & Hochstrasser, R. M. (1983). *Spectroscopy and excitation dynamics of condensed molecular systems* (Vol. 4). Amsterdam: North Holland.

Andersen, J. D., Duke, C. B., & Kenkre, V. M. (1983). Injected electrons in naphthalene: Band motion at low temperatures. *Physical Review Letters, 51*(24), 2202.

Andersen, J. D., Duke, C. B., & Kenkre, V. M. (1984). Application of the Silbey-Munn theory to interpret the temperature dependence of the mobilities of injected electrons in naphthalene. *Chemical Physics Letters, 110*(5), 504–507.

Andersen, H. C., Oppenheim, I., Shuler, K. E., & Weiss, G. H. (1964). Exact conditions for the preservation of a canonical distribution in Markovian relaxation processes. *Journal of Mathematical Physics, 5*(4), 522–536.

Anderson, P. W. (1997). *Concepts in solids: Lectures on the theory of solids* (Vol. 58). River Edge, NJ: World Scientific.

Argyres, P. N, & Sigel, J. L. (1974). Discussion of a new theory of electrical resistivity. *Physical Review B, 9*(8), 3197.

Aslangul, C., & Kottis, P. (1974). Density operator description of excitons in molecular aggregates: Optical absorption and motion. I. The dimer problem. *Physical Review B, 10*(10), 4364.

Auweter, H., Braun, A., Mayer, U., & Schmid, D. (1979). Dynamics of energy transfer by singlet excitons in naphthalene crystals as studied by time-resolved spectroscopy. *Zeitschrift für Naturforschung A, 34*(6), 761–772.

Auweter, H., Mayer, U., & Schmid, D. (1978). Singlet-exciton energy transfer in naphthalene doped with anthracene following two-photon picosecond excitation: Dependence on dopant concentration. *Zeitschrift für Naturforschung A, 33*(6), 651–657.

Avakian, P., Ern, V., Merrifield, R. E., & Suna, A. (1968). Spectroscopic approach to triplet exciton dynamics in anthracene. *Physical Review, 165*(3), 974.

Avakian, P., & Merrifield, R. E. (1968). Triplet excitons in anthracene crystals? A review. *Molecular Crystals and Liquid Crystals, 5*(1), 37–77.

© The Author(s), under exclusive license to Springer Nature Switzerland AG 2021
V. M. (Nitant) Kenkre, *Memory Functions, Projection Operators, and the Defect Technique*, Lecture Notes in Physics 982,
https://doi.org/10.1007/978-3-030-68667-3

Aydin, İ., Briscoe, B. J., & Şanlitürk, K. Y. (1994). Density distributions during the compaction of alumina powders: A comparison of a computational prediction with experiment. *Computational Materials Science, 3*(1), 55–68.

Aydin, İ., Briscoe, B. J, & Şanlitürk, K. Y. (1996). The internal form of compacted ceramic components: A comparison of a finite element modelling with experiment. *Powder Technology, 89*(3), 239–254.

Babenko, S. D., Benderskii, V. A., Gol'Danskii, V. I., Lavrushko, A. G., & Tychinskii, V. P. (1971a). Annihilation of singlet excited states in anthracene solutions. *Chemical Physics Letters, 8*(6), 598–600.

Babenko, S. D., Benderskii, V. A., Goldanskii, V. I., Lavrushko, A. G., & Tychinskii, V. P. (1971b). Singlet exciton annihilation in anthracene crystals. *Physica Status Solidi (B), 45*(1), 91–97.

Bagchi, B., Fleming, G. R., & Oxtoby, D. W. (1983). Theory of electronic relaxation in solution in the absence of an activation barrier. *The Journal of Chemical Physics, 78*(12), 7375–7385.

Ben-Naim, E., Redner, S., & Weiss, G. H. (1993). Partial absorption and 'virtual' traps. *Journal of Statistical Physics, 71*(1–2), 75–88.

Benderskii, V. A., Kh. Brikenshtein, V., Lavrushko, A. G., & Filippov, P. G. (1978). Non-linear fluorescence quenching in molecular crystals I. Recombination of localized excitons. *Physica Status Solidi (B), 86*(2), 449–458.

Blees, M. H. (1994). The effect of finite duration of gradient pulses on the pulsed-field-gradient NMR method for studying restricted diffusion. *Journal of Magnetic Resonance, Series A, 109*(2), 203–209.

Blythe, R. A., & Bray, A. J. (2003). Survival probability of a diffusing particle in the presence of Poisson-distributed mobile traps. *Physical Review E, 67*(4), 041101.

Bonci, L., Roncaglia, R., West, B. J., & Grigolini, P. (1991). Quantum irreversibility and chaos. *Physical Review Letters, 67*(19), 2593.

Bookout, B. D., & Parris, P. E. (1993). Long-range random walks on energetically disordered lattices. *Physical Review Letters, 71*(1), 16.

Bouchaud, J.-P., Cates, M. E., & Claudin, Ph. (1995). Stress distribution in granular media and nonlinear wave equation. *Journal de physique I, 5*(6), 639–656.

Braun, A., Mayer, U., Auweter, H., Wolf, H. C., & Schmid, D. (1982). Singlet-exciton energy transfer in tetracene-doped anthracene crystals as studied by time-resolved spectroscopy. *Zeitschrift für Naturforschung A, 37*(9), 1013–1023.

Broude, V. L., Vidmont, N. A., & Korshunov, V. V. (1978). Singlet exciton annihilation kinetics in anthracene crystals. *Physica Status Solidi (B), 90*(1), K53–K58.

Brown, D. W., & Kenkre, V. M. (1983). Quasielastic neutron scattering in metal hydrides: Effects of the quantum mechanical motion of interstitial hydrogen atoms. In *Electronic structure and properties of hydrogen in metals* (pp. 177–182). New York: Springer.

Brown, D. W., & Kenkre, V. M. (1985). Coupling of tunneling and hopping transport interactions in neutron scattering lineshapes. *Journal of Physics and Chemistry of Solids, 46*(5), 579–583.

Brown, D. W., & Kenkre, V. M. (1986). Neutron scattering lineshapes for nearly-incoherent transport on non-bravais lattices. *Journal of Physics and Chemistry of Solids, 47*(3), 289–293.

Brown, D. W., & Kenkre, V. M. (1987). Neutron scattering lineshapes for hydrogen trapped near impurities in metals. *Journal of Physics and Chemistry of Solids, 48*(9), 869–876.

Brown, D. W, Lindenberg, K., & Zhao, Y. (1997). Variational energy band theory for polarons: Mapping polaron structure with the global-local method. *The Journal of Chemical Physics, 107*(8), 3179–3195.

Bruggeman, D. A. G. (1935). The prediction of the thermal conductivity of heterogeneous mixtures. *Annals of Physics, 24*, 636–664.

Buff, F. P., & Wilson, D. J. (1960). Some considerations of unimolecular rate theory. *The Journal of Chemical Physics, 32*(3), 677–685.

Bukov, M., D'Alessio, L., & Polkovnikov, A. (2015). Universal high-frequency behavior of periodically driven systems: From dynamical stabilization to Floquet engineering. *Advances in Physics, 64*(2), 139–226.

Callaghan, P. T. (1991). *Principles of NMR microscopy*. Oxford: Clarendon.

Campillo, A. J., Shapiro, S. L., & Swenberg, C. E. (1977). Picosecond measurements of exciton migration in tetracene crystals doped with pentacene. *Chemical Physics Letters, 52*(1), 11–15.

Candia, J., Parris, P. E., & Kenkre, V. M. (2007). Transport properties of random walks on scale-free/regular-lattice hybrid networks. *Journal of Statistical Physics, 129*(2), 323–333.

Cao, J., Cogdell, R. J., Coker, D. F., Duan, H.-G., Hauer, J., Kleinekathöfer, U., et al. (2020). Quantum biology revisited. *Science Advances, 6*(14), eaaz4888.

Carslaw, H. S., & Jaeger, J. C. (1959). *Conduction of heat in solids*. Oxford: Clarendon Press.

Casella, R. C. (1983). Theory of excitation bands of hydrogen in bcc metals and of their observation by neutron scattering. *Physical Review B, 27*(10), 5943.

Chandrasekhar, S. (1943). Stochastic problems in physics and astronomy. *Reviews of Modern Physics, 15*, 1–89.

Chase, M. (2016). *Memory Effects in Brownian Motion, Random Walks Under Confining Potentials, and Relaxation of Quantum Systems*. Ph.D. thesis, University of New Mexico, Albuquerque, NM.

Chase, M., Spendier, K., & Kenkre, V. M. (2016). Analysis of confined random walkers with applications to processes occurring in molecular aggregates and immunological systems. *Journal of Physical Chemistry B, 120*(12), 3072–3080.

Cheng, Y.-C., & Silbey, R. J. (2008). A unified theory for charge-carrier transport in organic crystals. *The Journal of Chemical Physics, 128*(11), 114713.

Chester, G. V., & Thellung, A. (1959). On the electrical conductivity of metals. *Proceedings of the Physical Society, 73*(5), 745.

Clayton, R. K. (1980). *Photosynthesis: Physical mechanisms and chemical patterns*. Cambridge: Cambridge University Press.

Clerc, M. G., Escaff, D., & Kenkre, V. M. (2005). Patterns and localized structures in population dynamics. *Physical Review E, 72*(5), 056217.

Clerc, M. G., Escaff, D., & Kenkre, V. M. (2010). Analytical studies of fronts, colonies, and patterns: Combination of the Allee effect and nonlocal competition interactions. *Physical Review E, 82*(3), 036210.

Conwell, E. M., & Basko, D. M. (2003). Negative differential mobility in pentacene. *Journal of Polymer Science Part B: Polymer Physics, 41*(21), 2595–2600.

Cramer, T., Steinbrecher, T., Koslowski, T., Case, D. A., Biscarini, F., & Zerbetto, F. (2009). Water-induced polaron formation at the pentacene surface: Quantum mechanical molecular mechanics simulations. *Physical Review B, 79*(15), 155316.

Cross, M. C., & Hohenberg, P. C. (1993). Pattern formation outside of equilibrium. *Reviews of Modern Physics, 65*(3), 851.

Cruzeiro-Hansson, L., Christiansen, P. L., & Elgin, J. N. (1988). Comment on "Self-trapping on a dimer: Time-dependent solutions of a discrete nonlinear Schrödinger equation". *Physical Review B, 37*(13), 7896.

Davydov, A. S. (1968). The radiationless transfer of energy of electronic excitation between impurity molecules in crystals. *Physica Status Solidi (B), 30*(1), 357–366.

Davydov, A. S. (1971). *Theory of molecular excitons: Translated from Russian by Stephen B. Dresner*. New York, NY: Plenum Press.

de Gennes, P.-G. (1999). Granular matter: A tentative view. *Reviews of Modern Physics, 71*(2), S374.

Dexter, D. L. (1953). A theory of sensitized luminescence in solids. *The Journal of Chemical Physics, 21*(5), 836–850.

Dexter, D. L., & Knox, R. S. (1965). Excitons. In *Interscience tracts on physics and astronomy* (Vol. 125). New York: Wiley.

Dexter, D. L., Knox, R. S., & Förster, Th. (1969). The radiationless transfer of energy of electronic excitation between impurity molecules in crystals. *Physica Status Solidi (B), 34*(2), K159–K162.

Diestler, D. J. (1976). Vibrational relaxation of molecules in condensed media. In *Radiationless processes in molecules and condensed phases* (pp. 169–238). Cham: Springer.

Dlott, D. D., & Fayer, M. D. (1990). Shocked molecular solids: Vibrational up pumping, defect hot spot formation, and the onset of chemistry. *The Journal of Chemical Physics, 92*(6), 3798–3812.

Dlott, D. D., Fayer, M. D., & Wieting, R. D. (1977). Coherent one-dimensional exciton transport and impurity scattering. *The Journal of Chemical Physics, 67*(8), 3808–3817.

Doering, C. R, & Ben-Avraham, D. (1988). Interparticle distribution functions and rate equations for diffusion-limited reactions. *Physical Review A, 38*(6), 3035.

Doetsch, G. (1971). *Guide to the application of the Laplace and Z-transforms* (240 pp.). New York, NY: Van Nostrand Reingold Co.

Drazer, G., Wio, H. S, & Tsallis, C. (2000). Anomalous diffusion with absorption: Exact time-dependent solutions. *Physical Review E, 61*(2), 1417.

Dresden, M. (1961). Recent developments in the quantum theory of transport and galvanomagnetic phenomena. *Reviews of Modern Physics, 33*(2), 265.

Duke, C. B, & Schein, L. B. (1980). Organic solids: Is energy-band theory enough? *Physics Today, 33*(2), 42–48.

Dunlap, D. H., & Kenkre, V. M. (1986). Dynamic localization of a charged particle moving under the influence of an electric field. *Physical Review B, 34*(6), 3625.

Dunlap, D. H., & Kenkre, V. M. (1988a). Dynamic localization of a particle in an electric field viewed in momentum space: Connection with Bloch oscillations. *Physics Letters A, 127*(8–9), 438–440.

Dunlap, D. H., & Kenkre, V. M. (1988b). Effect of scattering on the dynamic localization of a particle in a time-dependent electric field. *Physical Review B, 37*(12), 6622.

Dunlap, D. H., & Kenkre, V. M. (1993). Disordered polaron transport: A theoretical description of the motion of photoinjected charges in molecularly doped polymers. *Chemical Physics, 178*, 67–75.

Dunlap, D. H., Kenkre, V. M., & Parris, P. E. (1999). What is behind the square root of E? *Journal of Imaging Science and Technology, 43*(5), 437–443.

Dunlap, D. H., Parris, P. E., & Kenkre, V. M. (1996). Charge-dipole model for the universal field dependence of mobilities in molecularly doped polymers. *Physical Review Letters, 77*(3), 542.

Duran, J. (2012). *Sands, powders, and grains: An introduction to the physics of granular materials.* Berlin: Springer Science & Business Media.

Dyre, J. C., & Schrøder, T. B. (2000). Universality of AC conduction in disordered solids. *Reviews of Modern Physics, 72*(3), 873.

Ebeling, W., Schweitzer, F., & Tilch, B. (1999). Active Brownian particles with energy depots modeling animal mobility. *BioSystems, 49*(1), 17–29.

Edwards, S. F., & Oakeshott, R. B. S. (1989). The transmission of stress in an aggregate. *Physica D: Nonlinear Phenomena, 38*(1–3), 88–92.

Efrima, S., & Metiu, H. (1979). The temperature dependence of the electron mobility in molecular crystals. *Chemical Physics Letters, 60*(2), 226–231.

Emch, G. (1964). Coarse-graining in Liouville space + master equation. *Helvetica Physica Acta, 37*(6), 532.

Emch, G. (1965). On Markov character of master equations. *Helvetica Physica Acta, 38*(1), 164.

Engel, G. S., Calhoun, T. R., Read, E. L., Ahn, T.-K., Mančal, T., Cheng, Y.-C., et al. (2007). Evidence for wavelike energy transfer through quantum coherence in photosynthetic systems. *Nature, 446*(7137), 782–786.

Erginsoy, C. (1950). Neutral impurity scattering in semiconductors. *Physical Review, 79*(6), 1013.

Ern, V. (1969). Anisotropy of triplet exciton diffusion in anthracene. *Physical Review Letters, 22*(8), 343.

Ern, V., Avakian, P., & Merrifield, R. E. (1966). Diffusion of triplet excitons in anthracene crystals. *Physical Review, 148*(2), 862.

Ern, V., & Schott, M. (1976). Motion of localized excitations in organic solids. In *Localization and delocalization in quantum chemistry* (pp. 249–284). Berlin: Springer.

Evans, M. W., Grigolini, P., & Parravicini, G. P. (1985). *Advances in chemical physics, memory function approaches to stochastic problems in condensed matter* (Vol. 62). New York: Wiley.

Farlow, S. J. (1993). *Partial differential equations for scientists and engineers*. North Chelmsford, MA: Courier Corporation.

Fayer, M. D., & Harris, C. B. (1974). Coherent energy migration in solids. I. Band-trap equilibria at Boltzmann and non-Boltzmann temperatures. *Physical Review B, 9*(2), 748.

Fisher, R. A. (1937). The wave of advance of advantageous genes. *Annals of Eugenics, 7*(4), 355–369.

Fitchen, D. B. (1968). Zero-phonon transitions. *Physics of color centers* (pp. 293–350). New York, NY: Academic.

Förster, Th. (1948). Intermolecular energy transfer and fluorescence. *Annalen der Physik (Leipzig), 2*, 55–75.

Fort, A., Ern, V., & Kenkre, V. M. (1983). Theory of coherence effects in time-dependent delayed fluorescence. II. Application to two-and three-dimensional crystals. *Chemical Physics, 80*(3), 205–211.

Fuentes, M. A., Kuperman, M. N., & Kenkre, V. M. (2003). Nonlocal interaction effects on pattern formation in population dynamics. *Physical Review Letters, 91*(15), 158104.

Fuentes, M. A., Kuperman, M. N., & Kenkre, V. M. (2004). Analytical considerations in the study of spatial patterns arising from nonlocal interaction effects. *Journal of Physical Chemistry B, 108*(29), 10505–10508.

Gallus, G., & Wolf, H. C. (1966). Direct measurement of the diffusion length of singulet excitons in solid phenanthrene. *Physica Status Solidi (B), 16*(1), 277–280.

Garrity, D. K., & Skinner, J. L. (1985). Exciton dynamics and transient grating experiments. *The Journal of Chemical Physics, 82*(1), 260–269.

Gentry, S. T., & Kopelman, R. (1983). Analog metal-to-nonmetal exciton transition: Effect of isotopic alloying on naphthalene triplet transport. *Physical Review B, 27*(4), 2579.

Giuggioli, L., Abramson, G., Kenkre, V. M., Parmenter, R. R., & Yates, T. L. (2006a). Theory of home range estimation from displacement measurements of animal populations. *Journal of Theoretical Biology, 240*, 126–135.

Giuggioli, L., Andersen, J. D., & Kenkre, V. M. (2003). Mobility theory of intermediate-bandwidth carriers in organic crystals: Scattering by acoustic and optical phonons. *Physical Review B, 67*(4), 045110.

Giuggioli, L., Parris, P. E., & Kenkre, V. M. (2006b). Variational formula for the relaxation time in the Boltzmann equation. *The Journal of Physical Chemistry B, 110*(38), 18921–18924.

Giuggioli, L., Sevilla, F. J., & Kenkre, V. M. (2009). A generalized master equation approach to modelling anomalous transport in animal movement. *Journal of Physics A: Mathematical and Theoretical, 42*, 434004.

Gochanour, C. R., Andersen, H. C., & Fayer, M. D. (1979). Electronic excited state transport in solution. *The Journal of Chemical Physics, 70*(9), 4254–4271.

Goldstein, J. C., & Scully, M. O. (1973). Nonequilibrium properties of an Ising-model ferromagnet. *Physical Review B, 7*(3), 1084.

Gordon, S. A. (1961). In M. Burton, J. S. Kirby-Smith, & J. L. Magee (Eds.), *Comparative effects of radiation* (p. xx+ 426). New York: Wiley (1960)

Grifoni, M., & Hänggi, P. (1998). Driven quantum tunneling. *Physics Reports, 304*(5–6), 229–354.

Großmann, F., & Hänggi, P. (1992). Localization in a driven two-level dynamics. *Europhysics Letters, 18*(7), 571.

Grover, M., & Silbey, R. (1971). Exciton migration in molecular crystals. *The Journal of Chemical Physics, 54*(11), 4843–4851.

Gülen, D. (1988). Determination of the exciton diffusion length by surface quenching experiments. *Journal of Luminescence, 42*(4), 191–195.

Gülen, D., Kenkre, V. M., Knox, R. S., & Parris, P. E. (1988). Effects of transport coherence on the mutual annihilation of excitons. *Physical Review B, 37*(4), 1839.

Haan, S. W., & Zwanzig, R. (1978). Förster migration of electronic excitation between randomly distributed molecules. *The Journal of Chemical Physics, 68*(4), 1879–1883.

Haarer, D., & Castro, G. (1976). Singlet exciton diffusion and exciton quenching in phenanthrene single crystals. *Journal of Luminescence, 12*, 233–238.

Haken, H., & Reineker, P. (1972). The coupled coherent and incoherent motion of excitons and its influence on the line shape of optical absorption. *Zeitschrift für Physik, 249*(3), 253–268.

Haken, H., & Strobl, G. (1973). An exactly solvable model for coherent and incoherent exciton motion. *Zeitschrift für Physik A Hadrons and nuclei, 262*(2), 135–148.

Harris, C. B., & Zwemer, D. A. (1978). Coherent energy transfer in solids. *Annual Review of Physical Chemistry, 29*(1), 473–495.

Harris, R. A., & Silbey, R. (1985). Variational calculation of the tunneling system interacting with a heat bath. II. Dynamics of an asymmetric tunneling system. *The Journal of Chemical Physics, 83*(3), 1069–1074.

Haus, J. W., & Kehr, K. W. (1983). Equivalence between random systems and continuous-time random walk: Literal and associated waiting-time distributions. *Physical Review B, 28*(6), 3573.

Haus, J. W., & Kehr, K. W. (1987). Diffusion in regular and disordered lattices. *Physics Reports, 150*(5–6), 263–406.

Haus, J. W., Kehr, K. W., & Kitahara, K. (1982). Long-time tail effects on particle diffusion in a disordered system. *Physical Review B, 25*(7), 4918.

Herrmann, R. G. (1999). Biogenesis and evolution of photosynthetic (thylakoid) membranes. *Bioscience Reports, 19*(5), 355–365.

Hochstrasser, R. M. (1966). Electronic spectra of organic molecules. *Annual Review of Physical Chemistry, 17*(1), 457–480.

Hochstrasser, R. M., & Prasad, P. N. (1972). Phonon sidebands of electronic transitions in molecular crystals and mixed crystals. *The Journal of Chemical Physics, 56*(6), 2814–2823.

Holstein, T. (1959a). Studies of polaron motion: Part I. The molecular-crystal model. *Annals of Physics, 8*(3), 325–342.

Holstein, T. (1959b). Studies of polaron motion: Part II. The "small" polaron. *Annals of Physics, 8*(3), 343–389.

Howard, J. (2001). *Mechanics of motor proteins and the cytoskeleton*. Sunderland, MA: Jonathon Howard Sinauer Associates.

Huang, K. (1987). *Statistical mechanics* (2nd ed.). New York: Wiley.

Hughes, B. D. (1995). *Random walks and random environments: Random walks* (Vol. 1). Oxford: Clarendon Press.

Ierides, A. A., & Kenkre, V. M. (2018). Reservoir effects on the temperature dependence of the relaxation to equilibrium of three simple quantum systems. *Physica A: Statistical Mechanics and Its Applications, 503*, 9–25.

Jaeger, H. M, Nagel, S. R., & Behringer, R. P. (1996a). Granular solids, liquids, and gases. *Reviews of Modern Physics, 68*(4), 1259.

Jaeger, H. M, Nagel, S. R., & Behringer, R. P. (1996b). The physics of granular materials. *Physics Today, 49*(4), 32–39.

Jaeger, J. C., & Carslaw, H. S. (1959). *Conduction of heat in solids*. Oxford: Clarendon Press.

Jones, A. C., Janecka-styrcz, K., Elliot, D. A., & Williams, J. O. (1981). Singlet exciton energy transfer in tetracene-doped p-terphenyl single crystals. *Chemical Physics Letters, 80*(3), 413–417.

Kadanoff, L. P. (1999). Built upon sand: Theoretical ideas inspired by granular flows. *Reviews of Modern Physics, 71*(1), 435.

Kalay, Z., Parris, P. E., & Kenkre, V. M. (2008). Effects of disorder in location and size of fence barriers on molecular motion in cell membranes. *Journal of Physics: Condensed Matter, 20*(24), 245105.

Kenkre, V. M. (1974). Coupled wave-like and diffusive motion of excitons. *Physics Letters A, 47*, 119–120.

Kenkre, V. M. (1975a). Generalized-master-equation analysis of a ferromagnet model. *Physical Review B, 11*(9), 3406.

Kenkre, V. M. (1975b). Relations among theories of excitation transfer. *Physical Review B, 11*(4), 1741.

Kenkre, V. M. (1975c). Relations among theories of excitation transfer. II. Influence of spectral features on exciton motion. *Physical Review B, 12*(6), 2150.

Kenkre, V. M. (1977a). The generalized master equation and its applications. In Landman, U. (Ed.), *Statistical mechanics and statistical methods in theory and application* (pp. 441–461). New York: Plenum.

Kenkre, V. M. (1977b). Master-equation theory of the effect of vibrational relaxation on intermolecular transfer of electronic excitation. *Physical Review A, 16*, 766–776.

Kenkre, V. M. (1977c). Spatially nonlocal transfer rates in exciton transport arising from local intersite matrix elements. *Physics Letters A, 63*(3), 367–368.

Kenkre, V. M. (1978a). Generalization to spatially extended systems of the relation between stochastic Liouville equations and generalized master equations. *Physics Letters A, 65*(5–6), 391–392.

Kenkre, V. M. (1978b). Generalized master equations under delocalized initial conditions. *Journal of Statistical Physics, 19*(4), 333–340.

Kenkre, V. M. (1978c). Model for trapping rates for sensitized fluorescence in molecular crystals. *Physica Status Solidi (B), 89*(2), 651–654.

Kenkre, V. M. (1978d). Theory of exciton transport in the limit of strong intersite coupling. I. Emergence of long-range transfer rates. *Physical Review B, 18*(8), 4064.

Kenkre, V. M. (1979). Theory of the energy transfer rate in sensitized fluorescence in molecular crystals. *Chemical Physics, 36*(3), 377–382.

Kenkre, V. M. (1980). Theory of exciton annihilation in molecular crystals. *Physical Review B, 22*, 2089–2098.

Kenkre, V. M. (1981a). Simple connection between signals in transient grating experiments and memories in generalized master equations for excitons. *Physics Letters A, 82*(2), 100–102.

Kenkre, V. M. (1981b). Validity of the bilinear rate equation for exciton annihilation and expressions for the annihilation constant. *Zeitschrift für Physik B Condensed Matter, 43*(3), 221–227.

Kenkre, V. M. (1981c). Determination of the exciton diffusion constant from variation of quantum yield with penetration length. *Chemical Physics Letters, 82*(2), 301–304.

Kenkre, V. M. (1982). A theoretical approach to exciton trapping in systems with arbitrary trap concentration. *Chemical Physics Letters, 93*(3), 260–263.

Kenkre, V. M. (2000). Memory formalism for quantum control of dynamic localization. *Journal of Physical Chemistry B, 104*(16), 3960–3966.

Kenkre, V. M. (2001a). Spatial memories and correlation functions in the theory of stress distribution in granular materials. *Granular Matter, 3*(1–2), 23–28.

Kenkre, V. M. (2001b). Theory of stress distribution in granular materials: The memory formalism. In S. Sen & M. L. Hunt (Eds.), *Materials Research Society Conference Proceedings on the Granular State* (Vol. 627, pp. BB6.5.1–8). Cambridge/Warrendale, PA: Cambridge University Press.

Kenkre, V. M. (2002). Finite-bandwidth calculations for charge carrier mobility in organic crystals. *Physics Letters A, 305*(6), 443–447.

Kenkre, V. M. (2003). Memory formalism, nonlinear techniques, and kinetic equation approaches. In V. M. Kenkre, & K. Lindenberg (Eds.), *AIP Conference Proceedings on Modern Challenges in Statistical Mechanics: Patterns, Noise, and the Interplay of Nonlinearity and Complexity* (Vol. 658, pp. 63–103). Melville, NY: American Institute of Physics.

Kenkre, V. M., Andersen, J. D., Dunlap, D. H., & Duke, C. B. (1989). Unified theory of the mobilities of photoinjected electrons in naphthalene. *Physical Review Letters, 62*(10), 1165.

Kenkre, V. M., & Brown, D. W. (1985). Exact solution of the stochastic Liouville equation and application to an evaluation of the neutron scattering function. *Physical Review B, 31*(4), 2479.

Kenkre, V. M., & Campbell, D. K. (1986). Self-trapping on a dimer: time-dependent solutions of a discrete nonlinear Schrödinger equation. *Physical Review B, 34*(7), 4959.

Kenkre, V. M., & Chase, M. (2017). Approach to equilibrium of a quantum system and generalization of the Montroll–Shuler equation for vibrational relaxation of a molecular oscillator. *International Journal of Modern Physics B, 31*(20), 1750244.

Kenkre, V. M., & Dresden, M. (1971). Exact transport parameters for driving forces of arbitrary magnitude. *Physical Review Letters, 27*(1), 9.

Kenkre, V. M., & Dresden, M. (1972). Theory of electrical resistivity. *Physical Review A, 6*(2), 769.

Kenkre, V. M., Endicott, M. R., Glass, S. J., & Hurd, A. J. (1996). A theoretical model for compaction of granular materials. *Journal of the American Ceramic Society, 79*(12), 3045–3054.

Kenkre, V. M, Ern, V., & Fort, A. (1983a). Coherence effects in triplet-exciton transport via time-dependent delayed fluorescence. *Physical Review B, 28*(2), 598.

Kenkre, V. M., Fort, A., & Ern, V. (1983b). Steady-state delayed flourescence signals in Ronchi ruling experiments: Theory of coherence effects. *Chemical Physics Letters, 96*(6), 658–663.

Kenkre, V. M., Fukushima, E., & Sheltraw, D. (1997). Simple solutions of the Torrey–Bloch equations in the NMR study of molecular diffusion. *Journal of Magnetic Resonance, 128*(1), 62–69.

Kenkre, V. M. (Nitant), & Giuggioli, L. (2020). *Theory of the spread of epidemics and movement ecology of animals: An interdisciplinary approach using methodologies of physics and mathematics*. Cambridge: Cambridge University Press.

Kenkre, V. M., Giuggioli, L., & Kalay, Z. (2008). Molecular motion in cell membranes: Analytic study of fence-hindered random walks. *Physical Review E, 77*, 051907.

Kenkre, V. M., & Ierides, A. A. (2018). Vibrational relaxation of a molecule in strong interaction with a reservoir: Nonmonotonic temperature dependence. *Physics Letters A, 382*(22), 1460–1464.

Kenkre, V. M., Kalay, Z., & Parris, P. E. (2009). Extensions of effective-medium theory of transport in disordered systems. *Physical Review E, 79*, 011114.

Kenkre, V. M., & Knox, R. S. (1974a). Generalized-master-equation theory of excitation transfer. *Physical Review B, 9*, 5279–5290.

Kenkre, V. M., & Knox, R. S. (1974b). Theory of fast and slow excitation transfer rates. *Physical Review Letters, 33*(14), 803.

Kenkre, V. M., & Knox, R. S. (1976). Optical spectra and exciton coherence. *Journal of Luminescence, 12*, 187–193.

Kenkre, V. M., Kühne, R., & Reineker, P. (1981). Connection of the velocity autocorrelation function to the mean-square-displacement and to the memory function of generalized master equations. *Zeitschrift für Physik B Condensed Matter, 41*(2), 177–180.

Kenkre, V. M., Kuś, M., Dunlap, D. H., & Parris, P. E. (1998b). Nonlinear field dependence of the mobility of a charge subjected to a superposition of dichotomous stochastic potentials. *Physical Review E, 58*(1), 99.

Kenkre, V. M., & Lindenberg, K. (2003). Modern challenges in statistical mechanics: Patterns, noise, and the interplay of nonlinearity and complexity. In *Conference Proceedings* (Vol. 658). Melville, NY: American Institute of Physics.

Kenkre, V. M., Montroll, E. W., & Shlesinger, M. F. (1973). Generalized master equations for continuous-time random walks. *Journal of Statistical Physics, 9*, 45–50.

Kenkre, V. M., & Parris, P. E. (1983). Exciton trapping and sensitized luminescence: a generalized theory for all trap concentrations. *Physical Review B, 27*(6), 3221.

Kenkre, V. M., & Parris, P. E. (1984). Usefulness of sensitized luminescence as a probe for exciton motion. *Journal of Luminescence, 31*, 612–614.

Kenkre, V. M., & Parris, P. E. (2002a). Mechanism for carrier velocity saturation in pure organic crystals. *Physical Review B, 65*(24), 245106.

Kenkre, V. M., & Parris, P. E. (2002b). Saturation of charge carrier velocity with increasing electric fields: Theoretical investigations for pure organic crystals. *Physical Review B, 65*(20), 205104.

Kenkre, V. M., Parris, P. E., & Phatak, S. M. (1984). Motion and capture of quasiparticles in solids in the presence of cooperative trap interactions. *Physica. A, Statistical Mechanics and Its Applications, 128*(3), 571–588.

Kenkre, V. M., Parris, P. E., & Schmid, D. (1985a). Investigation of the appropriateness of sensitized luminescence to determine exciton motion parameters in pure molecular crystals. *Physical Review B, 32*(8), 4946.

Kenkre, V. M., & Phatak, S. M. (1984). Exact probability propagators for motion with arbitrary degree of transport coherence. *Physics Letters A, 100*(2), 101–104.

Kenkre, V. M., & Rahman, T. S. (1974). Model calculations in the theory of excitation transfer. *Physics Letters A, 50*(3), 170–172.

Kenkre, V. M., & Reineker, P. (1982). Exciton dynamics in molecular crystals and aggregates. In *Springer tracts in modern physics* (Vol. 94). Berlin: Springer.

Kenkre, V. M., & Schmid, D. (1983). Comments on the exciton annihilation constant and the energy transfer rate in naphthalene and anthracene. *Chemical Physics Letters, 94*(6), 603–608.

Kenkre, V. M., & Schmid, D. (1985). Coherence in singlet-exciton motion in anthracene crystals. *Physical Review B, 31*(4), 2430.

Kenkre, V. M., & Schmid, D. (1987). Consequences of initial condition analysis for the interpretation of transient grating observations. *Chemical Physics Letters, 140*(3), 238–242.

Kenkre, V. M., Scott, J. E., Pease, E. A., & Hurd, A. J. (1998a). Nonlocal approach to the analysis of the stress distribution in granular systems. I. Theoretical framework. *Physical Review E, 57*(5), 5841–5849.

Kenkre, V. M., & Seshadri, V. (1977). Time evolution of the average energy of a relaxing molecule. *Physical Review A, 15*(1), 197.

Kenkre, V. M., & Sevilla, F. J. (2006). Analytic considerations in the theory of NMR microscopy. *Physica A: Statistical Mechanics and Its Applications, 371*(1), 139–143.

Kenkre, V. M., & Sevilla, F. J. (2007). Thoughts about anomalous diffusion: Time-dependent coefficients versus memory functions. In T. S. Ali & K. B. Sinha (Eds.), *Contributions to mathematical physics: A tribute to Gerard G. Emch* (pp. 147–160). New Delhi: Hindustani Book Agency.

Kenkre, V. M., & Sugaya, S. (2014). Theory of the transmission of infection in the spread of epidemics: Interacting random walkers with and without confinement. *Bulletin of Mathematical Biology, 76*(12), 3016–3027.

Kenkre, V. M., Tokmakoff, A., & Fayer, M. D. (1994). Theory of vibrational relaxation of polyatomic molecules in liquids. *The Journal of Chemical Physics, 101*(12), 10618–10629.

Kenkre, V. M., & Tsironis, G. P. (1985). Initial condition effects on the time dependence of the signal in transient grating experiments. *Journal of Luminescence, 34*(1–2), 107–116.

Kenkre, V. M., & Tsironis, G. P. (1987). Nonlinear effects in quasielastic neutron scattering: Exact line-shape calculation for a dimer. *Physical Review B, 35*(4), 1473.

Kenkre, V. M., Tsironis, G. P., & Schmid, D. (1985b). Coherence investigations and initial condition effects in transient gratings in molecular crystals. *Journal de Physique, Colloque, 46*(1–2), C7–91–C7–94.

Kenkre, V. M., & Wong, Y. M. (1981). Effect of transport coherence on trapping: Quantum-yield calculations for excitons in molecular crystals. *Physical Review B, 23*(8), 3748.

Kenkre, V. M., & Wong, Y. M. (1980). Theory of exciton migration experiments with imperfectly absorbing end detectors. *Physical Review B, 22*(12), 5716.

Kim, Y., Puhl III, H. L., Chen, E., Taumoefolau, G. H., Nguyen, T. A., Kliger, D. S., et al. (2019). VenusA206 dimers behave coherently at room temperature. *Biophysical Journal, 116*(10), 1918–1930.

Kirkpatrick, S. (1973). Percolation and conduction. *Reviews of Modern Physics, 45*(4), 574.

Klafter, J., & Silbey, R. (1980). Derivation of the continuous-time random-walk equation. *Physical Review Letters, 44*(2), 55.

Klauder, J. R., & Skagerstam, B.-S. (1985). *Coherent states: Applications in physics and mathematical physics*. Singapore: World Scientific.

Knox, R. S. (1963). *Theory of excitons, solid state physics* (Supplement Vol. 5). New York, NY: Academic.

Knox, R. S. (1975). In Govindjee (Ed.) *Bioenergetics of photosynthesis* (pp. 183–221). New York, NY: Academic.

Kohn, W., & Luttinger, J. M. (1957). Quantum theory of electrical transport phenomena. *Physical Review, 108*(3), 590.

Kopelman, R., Monberg, E. M., Ochs, F. W., & Prasad, P. N. (1975). Exciton percolation: Isotopic-mixed 1B2u naphthalene. *Physical Review Letters, 34*(24), 1506.

Krauss, G., Schönbrunner, N., & Cooper, J. (2003). *Biochemistry of signal transduction and regulation* (Vol. 3). Hoboken, NJ: Wiley Online Library.

Kubo, R. (1957). Statistical-mechanical theory of irreversible processes. I. General theory and simple applications to magnetic and conduction problems. *Journal of the Physical Society of Japan, 12*(6), 570–586.

Kühne, R., & Reineker, P. (1979). Exact evaluation of the kernel of the generalized master equation for the coupled coherent and incoherent exciton motion. *Solid State Communications, 29*(3), 279–281.

Kusumi, A., Nakada, C., Ritchie, K., Murase, K., Suzuki, K., Murakoshi, H., et al. (2005). Paradigm shift of the plasma membrane concept from the two-dimensional continuum fluid to the partitioned fluid: High-speed single-molecule tracking of membrane molecules. *Annual Review of Biophysics and Biomolecular Structure, 34*, 351–378.

Lakatos-Lindenberg, K., Hemenger, R. P., & Pearlstein, R. M. (1972). Solutions of master equations and related random walks on quenched linear chains. *Journal of Chemical Physics, 56*(10), 4852–4867.

Landman, U., Montroll, E. W., & Shlesinger, M. F. (1977). Random walks and generalized master equations with internal degrees of freedom. *Proceedings of the National Academy of Sciences USA, 74*(2), 430–433.

Lantelme, F., Turq, P., & Schofield, P. (1979). On the use of memory functions in the study of the dynamical properties of ionic liquids. *The Journal of Chemical Physics, 71*(6), 2507–2513.

Laubereau, A., & Kaiser, W. (1978). Vibrational dynamics of liquids and solids investigated by picosecond light pulses. *Reviews of Modern Physics, 50*, 607–665.

Lawson, C. M., Powell, R. C., & Zwicker, W. K. (1982). Transient grating investigation of exciton diffusion and fluorescence quenching in NdxTa1− xP5O14 crystals. *Physical Review B, 26*(9), 4836.

Lax, M. (1952). The Franck-Condon principle and its application to crystals. *The Journal of Chemical Physics, 20*(11), 1752–1760.

Lee, C. T., Hoopes, M. F., Diehl, J., Gilliland, W., Huxel, G., Leaver, E. V., et al. (2001). Non-local concepts and models in biology. *Journal of Theoretical Biology, 210*(2), 201–219.

Lin, S. H., & Eyring, H. (1974). Stochastic processes in physical chemistry. *Annual Review of Physical Chemistry, 25*(1), 39–77.

Lindenberg, K., & West, B. J. (1983). Exciton line shapes at finite temperatures. *Physical Review Letters, 51*(15), 1370.

Lindner, M., Nir, G., Vivante, A., Young, I. T, & Garini, Y. (2013). Dynamic analysis of a diffusing particle in a trapping potential. *Physical Review E, 87*(2), 022716.

Liu, C.-H., Nagel, S. R., Schecter, D. A., Coppersmith, S. N., Majumdar, S., Narayan, O., et al. (1995). Force fluctuations in bead packs. *Science, 269*(5223), 513–515.

Livesay, D. R., Jambeck, P., Rojnuckarin, A., & Subramaniam, S. (2003). Conservation of electrostatic properties within enzyme families and superfamilies. *Biochemistry, 42*(12), 3464–3473.

Luttinger, J. M., & Kohn, W. (1958). Quantum theory of electrical transport phenomena. II. *Physical Review, 109*(6), 1892.

Machta, J. (1981). Generalized diffusion coefficient in one-dimensional random walks with static disorder. *Physical Review B, 24*(9), 5260.

Macleod, H. M., & Marshall, U. (1977). The determination of density distribution in ceramic compacts using autoradiography. *Powder Technology, 16*(1), 107–122.

Madhukar, A., & Post, W. (1977). Exact solution for the diffusion of a particle in a medium with site diagonal and off-diagonal dynamic disorder. *Physical Review Letters, 39*(22), 1424.

Madison, K. W., Fischer, M. C., Diener, R. B., Niu, Q., & Raizen, M. G. (1998). Dynamical Bloch band suppression in an optical lattice. *Physical Review Letters, 81*(23), 5093.

Mainardi, F. (1997). Fractional calculus, some basic problems in continuum and statistical mechanics. In A. Carpinteri & F. Mainardi (Eds.), *Fractals and fractional calculus in continuum mechanics*. Wien: Springer-Verlag.

Manne, K. K., Hurd, A. J., & Kenkre, V. M. (2000). Nonlinear waves in reaction-diffusion systems: The effect of transport memory. *Physical Review E, 61*(4), 4177.

Mannella, R., Grigolini, P., & West, B. J. (1994). A dynamical approach to fractional Brownian motion. *Fractals, 2*(01), 81–94.

Maradudin, A. A., Montroll, E. W., Weiss, G. H., Herman, R., & Miles, W. H. (1960). Green's functions for monatomic cubic lattices. *Acadèmie Royale de Belgique* (Vol. 5–15). Bruxelles: Palais des Académies

McCall, K. R., Johnson, D. L., & Guyer, R. A. (1991). Magnetization evolution in connected pore systems. *Physical Review B, 44*(14), 7344.

McMahon, H. T., & Gallop, J. L. (2005). Membrane curvature and mechanisms of dynamic cell membrane remodelling. *Nature, 438*(7068), 590–596.

Mehta, A., & Barker, G. C. (1994). The dynamics of sand. *Reports on Progress in Physics, 57*(4), 383.

Meijer, P. H E. (1966). In *Quantum statistical mechanics*. New York/London/Paris: Gordon and Breach.

Merrifield, R. E. (1964). Theory of the vibrational structure of molecular exciton states. *The Journal of Chemical Physics, 40*(2), 445–450.

Metiu, H., Oxtoby, D. W., & Freed, K. F. (1977). Hydrodynamic theory for vibrational relaxation in liquids. *Physical Review A, 15*(1), 361.

Mitra, P. P., & Halperin, B. I. (1995). Effects of finite gradient-pulse widths in pulsed-field-gradient diffusion measurements. *Journal of Magnetic Resonance, Series A, 113*(1), 94–101.

Moessner, R., & Sondhi, S. L. (2017). Equilibration and order in quantum Floquet matter. *Nature Physics, 13*(5), 424–428.

Mogilner, A., & Edelstein-Keshet, L. (1999). A non-local model for a swarm. *Journal of Mathematical Biology, 38*(6), 534–570.

Mokshin, A. V., Yulmetyev, R. M., & Hänggi, P. (2005a). Diffusion processes and memory effects. *New Journal of Physics, 7*(1), 9.

Mokshin, A. V., Yulmetyev, R. M., & Hänggi, P. (2005b). Simple measure of memory for dynamical processes described by a generalized Langevin equation. *Physical Review Letters,* **95**(20), 200601.

Monetti, R., Hurd, A., & Kenkre, V. M. (2001). Simulations for dynamics of granular mixtures in a rotating drum. *Granular Matter, 3*(1–2), 113–116.

Montroll, E. W. (1962). *Fundamental problems in statistical mechanics* (Vol. 230). Compiled by E. G. D. Cohen. Amsterdam: North-Holland Publishing Company.

Montroll, E. W. (1964). Random walks on lattices. In R. Bellman (Ed.), *Stochastic Processes in Mathematical Physics and Engineering. Proceedings of Symposia in Applied Mathematics* (Vol. 16, pp. 193–220) New York, NY: American Mathematical Society.

Montroll, E. W. (1969). Random walks on lattices containing traps. In *Physical Society of Japan Journal Supplement, Proceedings of the International Conference on Statistical Mechanics* (Vol. 26, p. 6).

Montroll, E. W., & Potts, R. B. (1955). Effect of defects on lattice vibrations. *Physical Review, 100*, 525–543.

Montroll, E. W., & Shuler, K. E. (1957). Studies in nonequilibrium rate processes. I. The relaxation of a system of harmonic oscillators. *The Journal of Chemical Physics, 26*(3), 454–464.

Montroll, E. W., & Weiss, G. H. (1965). Random walks on Lattices II. *Journal of Mathematical Physics, 6*(2), 167–181.

Montroll, E. W., & West, B. J. (1979). On an enriched collection of stochastic processes. In E. W. Montroll & J. J. Lebowitz (Eds.), *Studies in statistical mechanics: Vol. VII. Fluctuation phenomena* (pp. 61–175). Amsterdam: North Holland Publishing.

Morgan, G., Chen, S., & Yen, W. (1986). Transient grating spectroscopy of LaP5O14:Nd3+. *IEEE Journal of Quantum Electronics, 22*(8), 1360–1364.

Mori, H. (1965). Transport, collective motion, and Brownian motion. *Progress of Theoretical Physics, 33*(3), 423–455.

Munn, R. W. (1973). Direct calculation of exciton diffusion coefficient in molecular crystals. *The Journal of Chemical Physics, 58*(8), 3230–3232.

Munn, R. W. (1974). Exciton transport at short times. *Chemical Physics, 6*(3), 469–473.

Muriel, A., & Dresden, M. (1969a). Projection techniques in non-equilibrium statistical mechanics: I. A new hierarchy of equations. *Physica, 43*(3), 424–448.

Muriel, A., & Dresden, M. (1969b). Projection techniques in non-equilibrium statistical mechanics: II. The introduction of outside fields. *Physica, 43*(3), 449–464.

Nakajima, S. (1958). On quantum theory of transport phenomena: Steady diffusion. *Progress of Theoretical Physics, 20*(6), 948–959.

Nasu, K., & Toyozawa, Y. (1981). Tunneling process from free state to self-trapped state of exciton. *Journal of the Physical Society of Japan, 50*(1), 235–245.

Newell, A. C. (1997). The dynamics and analysis of patterns. In H. F. Nijhout, L. Nadel, & D. Stein (Eds.), *Pattern formation in the physical and biological sciences* (Vol. 5, pp. 201–268). Boca Raton, FL: CRC Press.

Nicolis, G. (1995). *Introduction to nonlinear science.* Cambridge: Cambridge University Press.

Nitzan, A., & Jortner, J. (1973). Vibrational relaxation of a molecule in a dense medium. *Molecular Physics, 25*(3), 713–734.

Novikov, S. V., Dunlap, D. H., Kenkre, V. M., Parris, P. E., & Vannikov, A. V. (1998). Essential role of correlations in governing charge transport in disordered organic materials. *Physical Review Letters, 81*(20), 4472.

Odagaki, T., & Lax, M. (1981). Coherent-medium approximation in the stochastic transport theory of random media. *Physical Review B, 24*(9), 5284.

Oxtoby, D. W. (1981). Vibrational relaxation in liquids. *Annual Review of Physical Chemistry, 32*(1), 77–101.

Park, S. H., Peng, H., Parus, S., Taitelbaum, H., & Kopelman, R. (2002). Spatially and temporally resolved studies of convectionless photobleaching kinetics: Line trap. *The Journal of Physical Chemistry A, 106*(33), 7586–7592.

Parris, P. E. (1986). Transport and trapping on a one-dimensional disordered lattice. *Physics Letters A, 114*(5), 250–254.

Parris, P. E. (1987). Site-diagonal T-matrix expansion for anisotropic transport and percolation on bond-disordered lattices. *Physical Review B, 36*(10), 5437.

Parris, P. E. (1989). Exciton diffusion at finite frequency: Luminescence observables for anisotropic percolating solids. *The Journal of Chemical Physics, 90*(4), 2416–2421.

Parris, P. E., & Bookout, B. D. (1996). Effective-medium theory for the electric-field dependence of the hopping conductivity of disordered solids. *Physical Review B, 53*(2), 629.

Parris, P. E., Candia, J., & Kenkre, V. M. (2008). Random-walk access times on partially disordered complex networks: An effective medium theory. *Physical Review E, 77*(6), 061113.

Parris, P. E., & Kenkre, V. M. (1982). Exciton migration with end detectors: Calculation of time-dependent luminescence. *Chemical Physics Letters, 90*(5), 342–345.

Parris, P. E., & Kenkre, V. M. (1984). Non-exponential luminescence intensities in exciton trapping: Quantitative comparison for triplets in 1, 2, 4, 5-tetrachlorobenzene. *Chemical Physics Letters, 107*(4–5), 413–419.

Parris, P. E., & Kenkre, V. M. (1986). Energy transfer for systems possessing long-range rates of capture. *Chemical Physics Letters, 125*(2), 189–193.

Parris, P. E., & Kenkre, V. M. (2004). Variational considerations in the study of carrier transport in organic crystals. *Physical Review B, 70*(6), 064304.

Parris, P. E, & Kenkre, V. M. (2005). Traversal times for random walks on small-world networks. *Physical Review E, 72*(5), 056119.

Parris, P. E., Kenkre, V. M., & Dunlap, D. H. (2001b). Nature of charge carriers in disordered molecular solids: Are polarons compatible with observations? *Physical Review Letters, 87*(12), 126601.

Parris, P. E., Kuś, M., & Kenkre, V. M. (2001a). Fokker–Planck analysis of the nonlinear field dependence of a carrier in a band at arbitrary temperatures. *Physics Letters A, 289*(4–5), 188–192.

Parris, P. E., Phatak, S. M., & Kenkre, V. M. (1984). Motion and capture in the presence of cooperative trap interactions II: Exact calculations for perfect absorbers in one dimension. *Journal of Statistical Physics, 35*(5–6), 749–760.

Parris, P. E., & Silbey, R. (1985). Low temperature tunneling dynamics in condensed media. *The Journal of Chemical Physics, 83*(11), 5619–5626.

Pathria, R. K. (1972). *Statistical mechanics* (Vol. 45). Oxford: Pergamon.

Perrin, F. (1932). Théorie quantique des transferts d'activation entre molécules de même espèce. Cas des solutions fluorescentes. *Annales de Physique, 10*, 283–314. EDP Sciences.

Philipson, K. D, & Sauer, K. (1972). Exciton interaction in a bacteriochlorophyll-protein from Chloropseudomonas ethylicum. Absorption and circular dichroism at 77 °K. *Biochemistry, 11*(10), 1880–1885.

Pollak, M. (1977). On dispersive transport by hopping and by trapping. *Philosophical Magazine, 36*(5), 1157–1169.

Pope, M., & Swenberg, C. E. (1999). *Electronic processes in organic crystals and polymers* (2nd ed.) New York: Oxford University Press.

Powell, R. C., & Soos, Z. G. (1975). Singlet exciton energy transfer in organic solids. *Journal of Luminescence, 11*(1–2), 1–45.

Powles, J. G., Mallett, M. J. D., Rickayzen, G., & Evans, W. A. B. (1992). Exact analytic solutions for diffusion impeded by an infinite array of partially permeable barriers. *Proceedings of the Royal Society of London. Series A: Mathematical and Physical Sciences, 436*(1897), 391–403.

Prigogine, I., & Resibois, P. (1958). On the approach to equilibrium of a quantum gas. *Physica, 24*(6–10), 795–816.

Rabinovich, M. I., Ezersky, A. B., & Weidman, P. D. (2000). *The dynamics of patterns.* Singapore: World Scientific.

Rackovsky, S., & Silbey, R. (1973). Electronic energy transfer in impure solids: I. Two molecules embedded in a lattice. *Molecular Physics, 25*(1), 61–72.

Raghavan, S., Kenkre, V. M., Dunlap, D. H., Bishop, A. R., & Salkola, M. I. (1996). Relation between dynamic localization in crystals and trapping in two-level atoms. *Physical Review A, 54*(3), R1781.

Rahman, T. S., & Knox, R. S. (1973). Theory of singlet-triplet exciton fusion. *Physica Status Solidi (B), 58*(2), 715–720.

Rahman, T. S., Knox, R. S., & Kenkre, V. M. (1979). Theory of depolarization of fluorescence in molecular pairs. *Chemical Physics, 44*(2), 197–211.

Redner, S. (2001). *A guide to first-passage processes.* Cambridge: Cambridge University Press.

Redner, S., & Ben-Avraham, D. (1990). Nearest-neighbour distances of diffusing particles from a single trap. *Journal of Physics A: Mathematical and General, 23*(22), L1169–L1173.

Reichl, L. E. (2009). *A modern course in statistical physics* (3rd ed.) Hoboken, NJ: Wiley.

Reineker, P., Kenkre, V. M., & Kühne, R. (1981). Drift mobility of photo-electrons in organic molecular crystals: Quantitative comparison between theory and experiment. *Physics Letters A, 84*(5), 294–296.

Reineker, P., & Kühne, R. (1980). Exact derivation and solution of the Nakajima-Zwanzig generalized master equation and discussion of approximate treatments for the coupled coherent and incoherent exciton motion. *Physical Review B, 21*(6), 2448.

Rice, S. A. (1985). *Diffusion-limited reactions.* Amsterdam: Elsevier.

Risken, H. (1984). *The Fokker-Planck equation: Methods of solution and applications. Springer series in synergetics* (Vol. 18). Berlin/Heidelberg: Springer-Verlag.

Ritchie, R. H., & Sakakura, A. Y. (1956). Asymptotic expansions of solutions of the heat conduction equation in internally bounded cylindrical geometry. *Journal of Applied Physics, 27*(12), 1453–1459.

Roberts, G. E., & Kaufman, H. (1966). *Table of Laplace transforms.* Philadelphia: Saunders.

Roberts, G. G., Apsley, N., & Munn, R. W. (1980). Temperature dependent electronic conduction in semiconductors. *Physics Reports, 60*(2), 59–150.

Robertson, B. (1966). Spin-echo decay of spins diffusing in a bounded region. *Physical Review, 151*(1), 273.

Robinson, G. W. (1970). Electronic and vibrational excitons in molecular crystals. *Annual Review of Physical Chemistry, 21*(1), 429–474.

Robinson, G. W., & Frosch, R. P. (1962). Theory of electronic energy relaxation in the solid phase. *The Journal of Chemical Physics, 37*(9), 1962–1973.

Robinson, G. W., & Frosch, R. P. (1963). Electronic excitation transfer and relaxation. *The Journal of Chemical Physics, 38*(5), 1187–1203.

Rodriguez, M. A., Abramson, G., Wio, H. S., & Bru, A. (1993). Diffusion-controlled bimolecular reactions: Long-and intermediate-time regimes with imperfect trapping within a Galanin approach. *Physical Review E, 48*(2), 829.

Rose, T. S., Righini, R., & Fayer, M. D. (1984). Picosecond transient grating measurements of singlet exciton transport in anthracene single crystals. *Chemical Physics Letters, 106*(1–2), 13–19.

Rubin, R. J., & Shuler, K. E. (1956). Relaxation of vibrational nonequilibrium distributions. I. Collisional relaxation of a system of harmonic oscillators. *The Journal of Chemical Physics, 25*(1), 59–67.

Rubin, R. J., & Shuler, K. E. (1957). On the relaxation of vibrational nonequilibrium distributions. III. The effect of radiative transitions on the relaxation behavior. *The Journal of Chemical Physics, 26*(1), 137–142.

Rubin, R. J., & Weiss, G. H. (1982). Random walks on lattices. The problem of visits to a set of points revisited. *Journal of Mathematical Physics, 23*(2), 250–253.

Salcedo, J. R., Siegman, A. E., Dlott, D. D., & Fayer, M. D. (1978). Dynamics of energy transport in molecular crystals: The picosecond transient-grating method. *Physical Review Letters, 41*(2), 131.

Sancho, J. M., Romero, A. H., Lacasta, A. M., & Lindenberg, K. (2007). Langevin dynamics of A+ A reactions in one dimension. *Journal of Physics: Condensed Matter, 19*(6), 065108.

Schein, L. B., Duke, C. B., & McGhie, A. R. (1978). Observation of the band-hopping transition for electrons in naphthalene. *Physical Review Letters, 40*(3), 197.

Scher, H., & Lax, M. (1973). Stochastic transport in a disordered solid. I. Theory. *Physical Review B, 7*, 4491–4502.

Scher, H., & Montroll, E. W. (1975). Anomalous transit-time dispersion in amorphous solids. *Physical Review B, 12*, 2455–2477.

Scott, J. E., Kenkre, V. M., & Hurd, A. J. (1998). Nonlocal approach to the analysis of the stress distribution in granular systems. II. Application to experiment. *Physical Review E, 57*(5), 5850–5857.

Seshadri, V., & Kenkre, V. M. (1976). Simultaneous vibrational relaxation and radiative decay of initial Boltzmann distributions. *Physics Letters A, 56*(2), 75–76.

Seshadri, V., & Kenkre, V. M. (1978). Theory of the interplay of luminescence and vibrational relaxation: A master-equation approach. *Physical Review A, 17*(1), 223.

Seshadri, V., & Kenkre, V. M. (1979). Time-dependent effective rates for molecular processes. *Zeitschrift für Physik B Condensed Matter, 33*(3), 289–295.

Sevilla, F. J., & Kenkre, V. M. (2007). Theory of the spin echo signal in NMR microscopy: Analytic solutions of a generalized Torrey–Bloch equation. *Journal of Physics: Condensed Matter, 19*(6), 065113.

Sewell, G. L. (1967). Quantum-statistical theory of irreversible processes. II: The character of closed macroscopic laws. *Physica, 34*(4), 493–514.

Shehu, A., Quiroga, S. D., D'Angelo, P., Albonetti, C., Borgatti, F., Murgia, M., et al. (2010). Layered distribution of charge carriers in organic thin film transistors. *Physical Review Letters, 104*(24), 246602.

Shelby, R. M., Zewail, A. H., & Harris, C. B. (1976). Coherent energy migration in solids: Determination of the average coherence length in one-dimensional systems using tunable dye lasers. *The Journal of Chemical Physics, 64*(8), 3192–3203.

Sheltraw, D., & Kenkre, V. M. (1996). The memory-function technique for the calculation of pulsed-gradient NMR signals in confined geometries. *Journal of Magnetic Resonance, Series A, 122*(2), 126–136.

Sheng, P. (2006). *Introduction to wave scattering, localization and mesoscopic phenomena* (vol. 88). Berlin: Springer Science & Business Media.

Shlesinger, M. F. (1974). Asymptotic solutions of continuous-time random walks. *Journal of Statistical Physics, 10*(5), 421–434.

Shugard, W., & Reiss, H. (1976). Derivation of the continuous-time random walk equation in non-homogeneous lattices. *Journal of Chemical Physics, 65*, 2827.

Shuler, K. E. (1955). On the kinetics of elementary gas phase reactions at high temperatures. In *Symposium (International) on Combustion* (Vol. 5, pp. 56–74). Amsterdam: Elsevier.

Silbey, R. (1976). Electronic energy transfer in molecular crystals. *Annual Review of Physical Chemistry, 27*(1), 203–223.

Silbey, R., & Harris, R. A. (1984). Variational calculation of the dynamics of a two level system interacting with a bath. *The Journal of Chemical Physics, 80*(6), 2615–2617.

Silbey, R., & Munn, R. W. (1980). General theory of electronic transport in molecular crystals. I. Local linear electron–phonon coupling. *The Journal of Chemical Physics, 72*(4), 2763–2773.

E. A. Silinsh & V. Capek (Eds.) (1994). *Organic electronic materials-interaction, localization and transport phenomena.* New York: American Institute of Physics.

Silver, M., Risko, K., & Bässler, H. (1979). A percolation approach to exciton diffusion and carrier drift in disordered media. *Philosophical Magazine B, 40*(3), 247–252.

Simpson, O. (1956). Electronic properties of aromatic hydrocarbons III. Diffusion of excitons. *Proceedings of the Royal Society of London, A238*, 402.

Sköld, K. (1978). Quasielastic neutron scattering studies of metal hydrides. In G. Alefeld & J. Völkl (Eds.), *Hydrogen in metals I. Topics in applied physics* (Vol. 28). Berlin, Heidelberg: Springer. https://doi.org/10.1007/3540087052_49

Sokolov, F. F. (1976). Memory function of Frenkel exciton in the presence of exciton-phonon interaction. *Physica Status Solidi (B), 76*(2), K131–K135.

Soos, Z. G. (1974). Theory of π-molecular charge-transfer crystals. *Annual Review of Physical Chemistry, 25*(1), 121–153.

Soules, T. F., & Duke, C. B. (1971). Resonant energy transfer between localized electronic states in a crystal. *Physical Review B, 3*(2), 262.

Spendier, K. (2012). *Dynamics and distribution of immunoglobolin E receptors : A dialog between experiment and theory.* https://digitalrepository.unm.edu/phyc_etds/65

Spendier, K. (2020). *TIRF Microscopy Image Sequences of Fc? RI-centric Synapse Formation in RBL-2H3 Cells Dataset.* Mendeley Data, **V1**.

Spendier, K., Carroll-Portillo, A., Lidke, K. A., Wilson, B. S., Timlin, J. A., & Thomas, J. L. (2010). Distribution and dynamics of rat basophilic leukemia immunoglobulin E receptors (FcεRI) on planar ligand-presenting surfaces. *Biophysical Journal, 99*(2), 388–397.

Spendier, K., & Kenkre, V. M. (2013). Analytic solutions for some reaction-diffusion scenarios. *Journal of Physical Chemistry B, 117*(49), 15639–15650.

Spendier, K., Sugaya, S., & Kenkre, V. M. (2013). Reaction-diffusion theory in the presence of an attractive harmonic potential. *Physical Review E, 88*, 062142.

Spouge, J. L. (1988). Exact solutions for a diffusion-reaction process in one dimension. *Physical Review Letters, 60*(10), 871.

Springer, T. (1972). Quasielastic neutron scattering for the investigation of diffusive motions in solids and liquids. In *Springer Tracts in Modern Physics* (Vol. 64, pp. 1–100). New York: Springer.

Stehfest, H. (1970). Algorithm 368: Numerical inversion of Laplace transforms [D5]. *Communications of the ACM, 13*(1), 47–49.

Stiles Jr, L. F., & Fitchen, D. B. (1966). F3+ center in NaF. *Physical Review Letters, 17*(13), 689.

Sugaya, S., & Kenkre, V. M. (2018). Analysis of transmission of infection in epidemics: Confined random walkers in dimensions higher than one. *Bulletin of Mathematical Biology, 80*(12), 3106–3126.

Sumi, H. (1979a). Theory of electrical conduction in organic molecular crystals. II. Characteristic effects of electric field and defect scattering on temperature-independent mobilities. *The Journal of Chemical Physics, 71*(8), 3403–3411.

Sumi, H. (1979b). Theory of electrical conduction in organic molecular crystals: Temperature-independent mobilities. *The Journal of Chemical Physics, 70*(8), 3775–3785.

Suna, A. (1970). Kinematics of exciton-exciton annihilation in molecular crystals. *Physical Review B, 1*(4), 1716.

Swenberg, C. E., & Pope, M. (1998). An observation on the mobility and the carrier effective mass in naphthalene. *Chemical Physics Letters, 287*(5–6), 535–536.

Swenberg, Ch. E., & Geacintov, N. E. (1973). Exciton interactions in organic solids. In J. B. Birks (Ed.) *Organic molecular photophysics.* New York: Wiley.

Swenson, R. J. (1963). Generalized master equation and t-Matrix expansion. *Journal of Mathematical Physics, 4*(4), 544–551.

Taitelbaum, H. (1991). Nearest-neighbor distances at an imperfect trap in two dimensions. *Physical Review A, 43*(12), 6592.

Takahashi, Y., & Tomura, M. (1971). Diffusion of singlet excitons in anthracene crystals. *Journal of the Physical Society of Japan, 31*(4), 1100–1108.

Tehver, I., & Hizhnyakov, V. (1975). Radiationless transfer of electronic excitation during vibrational relaxation. *Zhurnal Eksperimental'noi i Teoreticheskoi Fiziki, 69*, 599.

Thiele, E., Stone, J., & Goodman, M. F. (1981). A class of master equations that exhibit a generalized form of canonical invariance and other "displacement" invariances. *The Journal of Chemical Physics, 74*(11), 6394–6406.

Tiwari, M., & Kenkre, V. M. (2014). Approach to equilibrium of a nondegenerate quantum system: Decay of oscillations and detailed balance as separate effects of a reservoir. *The European Physical Journal B, 87*(4), 86.

Toda, M. (2012). *Theory of nonlinear lattices* (Vol. 20). Berlin: Springer Science & Business Media.

Tokmakoff, A., Fayer, M. D., & Dlott, D. D. (1993). Chemical reaction initiation and hot-spot formation in shocked energetic molecular materials. *The Journal of Physical Chemistry, 97*(9), 1901–1913.

Tokmakoff, A., Sauter, B., & Fayer, M. D. (1994). Temperature-dependent vibrational relaxation in polyatomic liquids: Picosecond infrared pump–probe experiments. *The Journal of Chemical Physics, 100*(12), 9035–9043.

Torney, D. C., & McConnell, H. M. (1983). Diffusion-limited reactions in one dimension. *The Journal of Physical Chemistry, 87*(11), 1941–1951.

Touchette, H., Van der Straeten, E., & Just, W. (2010). Brownian motion with dry friction: Fokker–Planck approach. *Journal of Physics A: Mathematical and Theoretical, 43*(44), 445002.

Tsironis, G. P., & Kenkre, V. M. (1988). On the description of the temperature dependence of the transient grating signal in molecular crystals. *Zeitschrift für Physik B Condensed Matter, 73*(2), 231–233.

Tyminski, J. K, Powell, R. C., & Zwicker, W. K. (1984). Investigation of four-wave mixing in NdxLa1− xP5O14. *Physical Review B, 29*(11), 6074.

Uhlenbeck, G. E. (1955). *The statistical mechanics of non-equilibrium phenomena. Lecture notes.* Michigan: University of Ann Arbor.

Van Hove, L. (1954a). Quantum-mechanical perturbations giving rise to a statistical transport equation. *Physica, 21*(1–5), 517–540.

Van Hove, L. (1954b). Time-dependent correlations between spins and neutron scattering in ferromagnetic crystals. *Physical Review, 95*(6), 1374.

Van Hove, L. (1955). Energy corrections and persistent perturbation effects in continuous spectra. *Physica, 21*(6–10), 901–923.

van Kampen, N. G. (1971). Master equations with canonical invariance. *Reports on Mathematical Physics, 2*(3), 199–209.

Van Strien, A. J., Schmidt, J., & Silbey, R. (1982). Exciton dynamics impurity assisted k → k' scattering in a one-dimensional triplet exciton system. *Molecular Physics, 46*(1), 151–160.

Verboven, E. (1960). On the quantum theory of electrical conductivity: The conductivity tensor to zeroth order. *Physica, 26*(12), 1091–1116.

Vidales, A. M., Kenkre, V. M., & Hurd, A. (2001). Simulation of granular compacts in two dimensions. *Granular Matter, 3*(1–2), 141–144.

Vitali, D., & Grigolini, P. (1989). Subdynamics, Fokker-Planck equation, and exponential decay of relaxation processes. *Physical Review A, 39*(3), 1486.

Wade, R. C., Gabdoulline, R. R., Lüdemann, S. K., & Lounnas, V. (1998). Electrostatic steering and ionic tethering in enzyme–ligand binding: Insights from simulations. *Proceedings of the National Academy of Sciences USA, 95*(11), 5942–5949.

Wan, C., Fiebig, T., Schiemann, O., Barton, J. K., & Zewail, A. H. (2000). Femtosecond direct observation of charge transfer between bases in DNA. *Proceedings of the National Academy of Sciences, 97*(26), 14052–14055.

Wang, L. Z., Caprihan, A., & Fukushima, E. (1995). The narrow-pulse criterion for pulsed-gradient spin-echo diffusion measurements. *Journal of Magnetic Resonance, Series A, 117*(2), 209–219.

Wang, M. D., Yin, H., Landick, R., Gelles, J., Block, S. M. (1997). Stretching DNA with optical tweezers. *Biophysical Journal, 72*(3), 1335.

Wang, X., & Dodabalapur, A. (2018). Trapped carrier scattering and charge transport in high-mobility amorphous metal oxide thin-film transistors. *Annalen der Physik, 530*(12), 1800341.

Wang, X., & Dodabalapur, A. (2020). Going beyond polaronic theories in describing charge transport in rubrene single crystals. *Applied Physics Letters, 116*(9), 093301.

Wang, X., Register, L. F., & Dodabalapur, A. (2019). Redefining the mobility edge in thin-film transistors. *Physical Review Applied, 11*(6), 064039.

Wang, Y. K. (1973). Master equation of a mean-field-model ferromagnet. *Physical Review B, 8*(11), 5199.

Wannier, G. H. (1959). *Elements of solid state theory* Cambridge: Cambridge University Press.

Warta, W., & Karl, N. (1985). Hot holes in naphthalene: High, electric-field-dependent mobilities. *Physical Review B, 32*(2), 1172.

West, B., Bologna, M., & Grigolini, P. (2012). *Physics of fractal operators*. Berlin: Springer Science & Business Media.

Wolf, H. C. (1968a). Energy transfer in organic molecular crystals: A survey of experiments. In *Advances in atomic and molecular physics* (Vol. 3, pp. 119–142). New York: Elsevier.

Wolf, H. C. (1968b). Energy transfer in organic molecular crystals: A survey of experiments. In D. R. Bates & I. Eastermann (Eds.), *Advances in atomic and molecular physics* (Vol. 3, pp. 119–142). New York: Elsevier.

Wolf, H. C., & Port, H. (1976). Excitons in aromatic crystals: Trap states, energy transfer and sensitized emission. *Journal of Luminescence, 12*, 33–46.

Wong, Y. M., & Kenkre, V. M. (1979). Momentum-space theory of exciton transport. II. Sensitized luminescence calculations for specific trapping models. *Physical Review B, 20*(6), 2438.

Wong, Y. M., & Kenkre, V. M. (1980). Extension of exciton-transport theory for transient grating experiments into the intermediate coherence domain. *Physical Review B, 22*(6), 3072.

Wong, Y. M., & Kenkre, V. M. (1982). Comments on the effect of disorder on transport with intermediate degree of coherence: Calculation of the mean square displacement. *Zeitschrift für Physik B: Condensed Matter, 46*(2), 185–188.

Wu, H.-L., & Kenkre, V. M. (1989). Generalized master equations from the nonlinear Schrödinger equation and propagation in an infinite chain. *Physical Review B, 39*(4), 2664.

Wu, M. W., & Conwell, E. M. (1997). Transport in α-sexithiophene films. *Chemical Physics Letters, 266*(3–4), 363–367.

Yarkony, D. R., & Silbey, R. (1977). Variational approach to exciton transport in molecular crystals. *The Journal of Chemical Physics, 67*(12), 5818–5827.

Yulmetyev, R. M., Mokshin, A. V., & Hänggi, P. (2003).   Diffusion time-scale invariance, randomization processes, and memory effects in Lennard-Jones liquids. *Physical Review E, 68*(5), 051201.

Yuste, S. B., & Acedo, L. (2001).   Multiparticle trapping problem in the half-line. *Physica A: Statistical Mechanics and Its Applications, 297*(3–4), 321–336.

Zewail, A. H, & Harris, C. B. (1975a). Coherence in electronically excited dimers. II. Theory and its relationship to exciton dynamics. *Physical Review B, 11*(2), 935.

Zewail, A. H., & Harris, C. B. (1975b).   Coherence in electronically excited dimers. III. The observation of coherence in dimers using optically detected electron spin resonance in zero field and its relationship to coherence in one-dimensional excitons. *Physical Review B, 11*(2), 952.

Ziman, J. M. (2001). *Electrons and phonons: The theory of transport phenomena in solids*. Oxford: Oxford University Press.

Zwanzig, R. (1961).   Statistical mechanics of irreversibility.   *Lectures in theoretical physics, 3*, 106–141.

Zwanzig, R. (1964). On the identity of three generalized master equations. *Physica, 30*(6), 1109–1123.

Zwanzig, R. (1982). Non-Markoffian diffusion in a one-dimensional disordered lattice. *Journal of Statistical Physics, 28*(1), 127–133.

Index

A

Anisotropy, 90, 94, 95, 97, 116

Annihilation, 20, 44, 90, 98, 235–239, 243, 317, 322–325, 334, 338, 344

Attractive, 173–175, 226, 228, 245, 259, 261, 264, 265, 267–269, 325, 347

B

Band, 12, 74, 75, 77, 78, 83, 89, 93, 104, 107, 116, 117, 119, 124, 127, 129–132, 160, 321, 349

Bandwidth, 63, 75, 93, 94, 99, 102, 112, 119, 120, 123–127, 129–132, 219

Barrier, 129, 297–301, 303, 306, 308–315

Baths, 15, 42–46, 49, 58, 65–69, 80, 83, 115, 135, 137–139, 142–149, 151, 154–158, 319

Burrows, 325, 330

C

Capture, 16, 117, 213, 215–218, 229, 235–239, 241, 245, 248, 249, 253, 259, 260, 262–264, 266–271, 280, 284, 301, 325, 343–349, 352

Chain, 11, 13, 31, 34, 35, 58, 61–85, 90, 176, 197, 198, 200, 219, 223, 274–277, 285, 286, 289, 291, 292, 298, 301, 303, 304, 309, 310, 314, 318, 322–324, 341, 342, 344, 350

Coalescence, 20, 270–271

Coarse-graining, 8, 39–59, 68, 137, 158, 336, 337, 346

Coherence, 4, 7, 9, 10, 13, 16, 18–21, 23, 35, 47, 53, 55–59, 61, 62, 69, 75, 76, 80–82, 84, 85, 87–113, 127, 137, 141, 142, 152, 155, 156, 201–206, 214, 219, 220, 225, 242, 320–324, 334, 336–340, 346, 349, 352

Coherent, 4–6, 9, 19, 34, 35, 39, 47, 53, 56–59, 61, 62, 64, 68, 69, 80, 82, 88, 92, 93, 95, 96, 99–103, 105–112, 137–139, 141–144, 151, 152, 156, 158, 166, 202–206, 240, 241, 318, 323, 337–340

Compaction, vi, 17, 19, 20, 39, 179–191

Concentration, 19, 102, 219, 221–226, 228, 229, 237, 240, 280, 285, 286, 291, 293, 335, 344

Conductivity, 72, 73, 85, 160–162

Confined, 16, 19, 167, 169, 177, 281, 297, 329

Confinement, 173–177, 259, 281, 297, 301, 329–331, 333, 334

Connection, 12–14, 18, 26, 46, 52, 53, 55, 65, 72–78, 82, 84, 101, 106, 113, 137, 164, 170, 185–189, 195, 200, 211, 217, 314, 335–337, 341

Continuous time, 18, 69–72

Continuous time random walks (CTRW), 18, 69–72

Continuum, 7, 19, 43, 80, 83, 90, 99, 101, 127, 147, 176, 245–271, 304, 305, 310–312, 325

Correlation, 18, 43, 72–78, 84, 85, 116, 146–148, 151, 159–161, 163, 177, 180, 185–189, 203, 224, 226, 227, 274, 280, 281, 288, 289, 320, 321, 336, 342, 349, 350

© The Author(s), under exclusive license to Springer Nature Switzerland AG 2021
V. M. (Nitant) Kenkre, *Memory Functions, Projection Operators, and the Defect Technique*, Lecture Notes in Physics 982,
https://doi.org/10.1007/978-3-030-68667-3

Printed in the United States
by Baker & Taylor Publisher Services